T0183866

Lecture Notes in Computer Science 9493

Commenced Publication in 1973
Founding and Former Series Editors:
Gerhard Goos, Juris Hartmanis, and Jan van Leeuwen

Editorial Board

More information about this series at http://www.springer.com/series/7412

Ana Fred · Maria De Marsico
Mário Figueiredo (Eds.)

Pattern Recognition Applications and Methods

4th International Conference, ICPRAM 2015
Lisbon, Portugal, January 10–12, 2015
Revised Selected Papers

 Springer

Editors
Ana Fred
Instituto de Telecomunicações
Instituto Superior Técnico
University of Lisbon
Lisbon
Portugal

Mário Figueiredo
Instituto de Telecomunicações
Instituto Superior Técnico
University of Lisbon
Lisbon
Portugal

Maria De Marsico
Sapienza Università di Roma
Rome
Italy

ISSN 0302-9743   ISSN 1611-3349   (electronic)
Lecture Notes in Computer Science
ISBN 978-3-319-27676-2   ISBN 978-3-319-27677-9   (eBook)
DOI 10.1007/978-3-319-27677-9

Library of Congress Control Number: 2015957092

LNCS Sublibrary: SL6 – Image Processing, Computer Vision, Pattern Recognition, and Graphics

Printed on acid-free paper

This Springer imprint is published by SpringerNature
The registered company is Springer International Publishing AG Switzerland

Preface

The present book includes extended and revised versions of a set of selected papers from the 4th International Conference on Pattern Recognition Applications and Methods (ICPRAM 2015), held in Lisbon, Portugal, during January 10–12, 2015.

The conference was sponsored by the Institute for Systems and Technologies of Information,Control and Communication (INSTICC) in cooperation with the ACM Special Interest Group on Artificial Intelligence (ACM SIGAI), the ACM Special Interest Group on Applied Computing (SIGAPP), the Association for the Advancement of Artificial Intelligence (AAAI), the European Association for Signal Processing (EURASIP), the Asia Pacific Neural Network Assembly (APNNA), the International Neural Network Society (INNS), and the Italian Association for Artificial Intelligence (AIIA).

The purpose of ICPRAM is to provide a major point of contact between researchers, engineers, and practitioners from several areas of pattern recognition, both from a theoretical and applications perspectives. Associated techniques can achieve relevant results for new research topics that continuously arise as technology advances. These aspects are reflected in the set of papers that make up this book. In particular, the focus is on the application of pattern recognition techniques to emerging real-world problems, interdisciplinary research, experimental and/or theoretical studies yielding new insights that advance pattern recognition methods. The final ambition is to spur new research lines and provide the occasion to start novel collaborations, mostly in interdisciplinary research scenarios.

ICPRAM 2015 received 145 submissions, from 45 countries in all continents, of which 17 % were presented at the conference as full papers, and their authors were invited to submit extended versions of their papers for this book. In order to evaluate each submission, double-blind reviewing was performed by the Program Committee. Finally, only the best 20 papers were included in this book.

We would like to highlight that ICPRAM 2015 also included four plenary keynote lectures, given by internationally distinguished researchers, namely: Hanan Samet (University of Maryland, USA), Nello Cristianini (University of Bristol, UK), Marcello Pelillo (University of Venice, Italy), and Luis Alexandre (University of Beira Interior, Portugal). We must acknowledge the invaluable contribution of all keynote speakers who, as renowned researchers in their areas, presented cutting-edge work, thus contributing toward enriching the scientific content of the conference.

We especially thank the authors, whose research and development efforts are recorded here. The knowledge and diligence of the reviewers were essential to ensure the quality of the papers presented at the conference and published in this book. Finally, a special thanks to all members of the INSTICC team, whose involvement was fundamental for organizing a smooth and successful conference.

September 2015

Ana Fred
Maria De Marsico
Mário Figueiredo

Organization

Conference Chair

Ana Fred — Instituto de Telecomunicações, Instituto Superior Técnico, University of Lisbon, Portugal

Program Co-chairs

Maria De Marsico — Sapienza Università di Roma, Italy
Mário Figueiredo — Instituto de Telecomunicações, Instituto Superior Técnico, University of Lisbon, Portugal

Program Committee

Andrea F. Abate — University of Salerno, Italy
Ashraf AbdelRaouf — Misr International University, MIU, Egypt
Shigeo Abe — Kobe University, Japan
Rahib Abiyev — Near East University, Turkey
Mayer Aladjem — Ben-Gurion University of the Negev, Israel
Rocío Alaiz-Rodríguez — Universidad de Leon, Spain
Andrea Albarelli — Università Ca' Foscari Venezia, Italy
Francisco Martínez Álvarez — Pablo de Olavide University of Seville, Spain
Annalisa Appice — Università degli Studi di Bari Aldo Moro, Italy
Juan Humberto Sossa Azuela — Centro de Investigacion en Computacion-IPN, Mexico
Emili Balaguer-Ballester — Bournemouth University, UK
Vineeth Nallure Balasubramanian — Indian Institute of Technology, India
Subhadip Basu — Jadavpur University, India
Jorge Batista — ISR, Institute of Systems and Robotics, Portugal
Jon Atli Benediktsson — University of Iceland, Iceland
J. Ross Beveridge — Colorado State University, USA
Anastasios Bezerianos — SINAPSE – National University of Singapore, Singapore
Monica Bianchini — Università degli Studi di Siena, Italy
Michael Biehl — University of Groningen, The Netherlands
Isabelle Bloch — Telecom ParisTech - CNRS LTCI, France
Joan Martí Bonmatí — Girona University, Spain
Mohamed-Rafik Bouguelia — LORIA, Lorraine University, France
Nizar Bouguila — Concordia University, Canada
Francesca Bovolo — Fondazione Bruno Kessler, Italy

Paula Brito	Universidade do Porto, Portugal
Hans du Buf	University of the Algarve, Portugal
Tien D. Bui	Concordia University, Canada
Samuel Rota Bulò	Fondazione Bruno Kessler, Italy
Javier Calpe	Universitat de València, Spain
Francesco Camastra	University of Naples Parthenope, Italy
Virginio Cantoni	Università di Pavia, Italy
Ramón A. Mollineda Cárdenas	Universitat Jaume I, Spain
Marco La Cascia	Università Degli Studi di Palermo, Italy
Michelangelo Ceci	University of Bari, Italy
Mehmet Celenk	Ohio University, USA
Amitava Chatterjee	Jadavpur University, India
Snigdhansu Chatterjee	University of Minnesota, USA
Rama Chellappa	University of Maryland, USA
Dmitry Chetverikov	MTA SZTAKI, Hungary
Ioannis Christou	Athens Information Technology, Greece
Miguel Coimbra	Faculdade de Ciências da Universidade do Porto, Portugal
Antoine Cornuejols	AgroParisTech, France
Michel Couprie	LIGM, France
Tom Croonenborghs	KU Leuven, Belgium
Sergio Cruces	University of Seville, Spain
Justin Dauwels	Nanyang Technological University, Singapore
Thorsten Dickhaus	Weierstrass Institute for Applied Analysis and Stochastics, Germany
Gianfranco Doretto	West Virginia University, USA
Gideon Dror	The Academic College of Tel-Aviv-Yaffo, Israel
Andrzej Drygajlo	Swiss Federal Institute of Technology Lausanne (EPFL), Switzerland
Bernard Dubuisson	Université de Technologie de Compiegne, France
Mahmoud El-Sakka	The University of Western Ontario, Canada
Yaokai Feng	Kyushu University, Japan
Francesc J. Ferri	University of Valencia, Spain
Mário Figueiredo	Instituto de Telecomunicações, Instituto Superior Técnico, University of Lisbon, Portugal
Maurizio Filippone	University of Glasgow, UK
Gernot A. Fink	TU Dortmund, Germany
Simone Fiori	Università Politecnica delle Marche, Italy
Damien François	Université Catholique de Louvain, Belgium
Ana Fred	Instituto de Telecomunicações, Instituto Superior Técnico, University of Lisbon, Portugal
Muhammad Marwan Muhammad Fuad	University of Tromsø, Norway

Giorgio Fumera	University of Cagliari, Italy
Langis Gagnon	Centre de Recherche Informatique de Montréal, Canada
Sabrina Gaito	Università degli Studi di Milano, Italy
Vicente Garcia	Universidad Autónoma de Ciudad Juárez, Mexico
Giorgio Giacinto	University of Cagliari, Italy
Fabio Gonzalez	Universidad Nacional de Colombia, Colombia
Bernard Gosselin	University of Mons, Belgium
Sébastien Guérif	University Paris 13 - SPC, France
Amaury Habrard	Laboratoire Hubert Curien, University of St. Etienne, France
Michal Haindl	Institute of Information Theory and Automation, Czech Republic
Barbara Hammer	Bielefeld University, Germany
Robert Harrison	Georgia State University, USA
Makoto Hasegawa	Tokyo Denki University, Japan
Pablo Hennings-Yeomans	Ontario Institute for Cancer Research, Canada
Laurent Heutte	Université de Rouen, France
Kouichi Hirata	Kyushu Institute of Technology, Japan
Sean Holden	University of Cambridge, UK
Geoffrey Holmes	University of Waikato, New Zealand
Qinghua Huang	South China University of Technology, Guangzhou, China
Lazaros S. Iliadis	Democritus University of Thrace, Greece
Jose M. Iñesta	Universidad de Alicante, Spain
Akihiro Inokuchi	Kwansei Gakuin University, Japan
Yuji Iwahori	Chubu University, Japan
Sarangapani Jagannathan	Missouri University of Science and Technology, USA
Yasushi Kanazawa	Toyohashi University of Technology, Japan
Yunho Kim	Ulsan National Institute of Science and Technology, Republic of Korea
Mario Köppen	Kyushu Institute of Technology, Japan
Walter Kosters	Universiteit Leiden, The Netherlands
Constantine Kotropoulos	Aristotle University of Thessaloniki, Greece
Sotiris Kotsiantis	Educational Software Development Laboratory, University of Patras, Greece
Konstantinos Koutroumbas	National Observatory of Athens, Greece
Adam Krzyzak	Concordia University, Canada
Nojun Kwak	Seoul National University, Republic of Korea
Jaerock Kwon	Kettering University, USA
Shang-Hong Lai	National Tsing Hua University, Taiwan
Raffaella Lanzarotti	Università degli Studi di Milano, Italy
Rasmus Larsen	Technical University of Denmark, Denmark
Aristidis Likas	University of Ioannina, Greece
Hantao Liu	University of Hull, UK
Xiaohui Liu	Brunel University, UK

Luca Lombardi	University of Pavia, Italy
Nicolas Loménie	Université Paris Descartes, France
Gaelle Loosli	Clermont Université, France
Alessandra Lumini	Università di Bologna, Italy
Juan Luo	George Mason University, USA
Francesco Marcelloni	University of Pisa, Italy
Elena Marchiori	Radboud University, The Netherlands
Gian Luca Marcialis	Università degli Studi di Cagliari, Italy
Urszula Markowska-Kaczmar	Wroclaw University of Technology, Poland
Maria De Marsico	Sapienza Università di Roma, Italy
J. Francisco Martínez-Trinidad	Instituto Nacional de Astrofísica, Óptica y Electrónica, Puebla, Mexico
Sally Mcclean	University of Ulster, UK
Stephen McKenna	University of Dundee, UK
Hongying Meng	Brunel University London, UK
Erzsébet Merényi	Rice University, USA
Piotr Mirowski	Bell Labs (Alcatel-Lucent), USA
Delia Alexandrina Mitrea	Technical University of Cluj-Napoca, Romania
Giovanni Montana	King's College London, UK
Robert Moskovitch	Columbia University, USA
Marco Muselli	Consiglio Nazionale delle Ricerche, Italy
Laurent Najman	Université Paris-Est, France
Yuichi Nakamura	Kyoto University, Japan
Michele Nappi	Università di Salerno, Italy
Claire Nédellec	MIG, INRA Centre de Jouy-en-Josas, France
Atul Negi	University of Hyderabad, India
Mikael Nilsson	Lund University, Sweden
Il-Seok Oh	Chonbuk National University, Republic of Korea
Simon OKeefe	University of York, UK
Ahmet Okutan	Isik University, Turkey
Yoshito Otake	Johns Hopkins University, USA
Martijn van Otterlo	Radboud Universiteit Nijmegen, The Netherlands
Gonzalo Pajares	Universidad Complutense de Madrid, Spain
Vicente Palazón-González	Universitat Jaume I, Spain
Guenther Palm	University of Ulm, Institute of Neural Information Processing, Germany
Apostolos Papadopoulos	Aristotle University, Greece
Marcello Pelillo	University of Venice, Italy
Luca Piras	University of Cagliari, Italy
Fiora Pirri	University of Rome La Sapienza, Italy
Vincenzo Piuri	Università degli Studi di Milano, Italy
Sylvia Pont	Delft University of Technology, The Netherlands
Philippe Preux	University of Lille 3, France
Lionel Prevost	University of French West Indies and Guiana, France
Hugo Proença	University of Beira Interior, Portugal

Arun K. Pujari	University of Hyderabad, India
Philippe Ravier	University of Orléans, France
Bernardete M. Ribeiro	University of Coimbra, Portugal
Elisa Ricci	University of Perugia, Italy
Daniel Riccio	Univerity of Naples, Federico II, Italy
François Rioult	GREYC CNRS UMR6072 - Université de Caen Basse-Normandie, France
Marcos Rodrigues	Sheffield Hallam University, UK
Juan J. Rodríguez	University of Burgos, Spain
Fernando Rubio	Universidad Complutense de Madrid, Spain
Indrajit Saha	University of Wroclaw, Poland
Lorenza Saitta	Università degli Studi del Piemonte Orientale Amedeo Avogadro, Italy
Antonio-José Sánchez-Salmerón	Universitat Politecnica de Valencia, Spain
Carlo Sansone	University of Naples, Italy
K.C. Santosh	US National Library of Medicine (NLM), National Institutes of Health (NIH), USA
Michele Scarpiniti	Sapienza University of Rome, Italy
Paul Scheunders	University of Antwerp, Belgium
Leizer Schnitman	Universidade Federal da Bahia (UFBA), Salvador, Bahia, Brazil
Bjoern Schuller	Technische Universität München, Germany
Friedhelm Schwenker	University of Ulm, Germany
Katsunari Shibata	Oita University, Japan
Vassilios Stathopolous	University College London, UK
Mu-Chun Su	National Central University, Taiwan
Shiliang Sun	East China Normal University, China
Yajie Sun	Samsung Research America, USA
Zhenan Sun	Institute of Automation, Chinese Academy of Sciences (CASIA), China
Johan Suykens	KU Leuven, Belgium
Alberto Taboada-Crispí	Universidad Central Marta Abreu de Las Villas, Cuba
Andrea Tagarelli	University of Calabria, Italy
Atsuhiro Takasu	National Institute of Informatics, Japan
Ichiro Takeuchi	Nagoya Institute of Technology, Japan
Xiaoyang Tan	Nanjing University of Aeronautics and Astronautics, China
Oriol Ramos Terrades	Centre de Visió per Computador, Universitat Autònoma de Barcelona, Spain
Thomas Tolxdorff	Charité, Germany
Fabien Torre	Lille University, Inria LNE and LIFL, France
Ricardo S. Torres	University of Campinas (UNICAMP), Brazil
Andrea Torsello	Università Ca'Foscari Venezia, Italy
Godfried Toussaint	New York University Abu Dhabi, UAE
Olgierd Unold	Wroclaw University of Technology, Poland

Ernest Valveny	Universitat Autònoma de Barcelona, Spain
Mario Vento	Università Degli Studi di Salerno, Italy
Michel Verleysen	Université Catholique de Louvain, Belgium
Christian Viard-Gaudin	LUNAM Université, Université de Nantes, France
Panayiotis Vlamos	Ionian University, Greece
Asmir Vodencarevic	Reifenhäuser REICOFIL GmbH & Co. KG, Germany
Sviatoslav Voloshynovskiy	University of Geneva, Switzerland
Jian-gang Wang	Institute for Infocomm Research, Singapore
Jonathan Weber	Université de Lorraine, France
Harry Wechsler	George Mason University, USA
Laurent Wendling	LIPADE, France
Slawomir Wierzchon	Polish Academy of Sciences, Poland
Janusz Wnek	Leidos, USA
Xianghua Xie	Swansea University, UK
Xin-Shun Xu	Shandong University, China
Haiqin Yang	Chinese University of Hong Kong, Hong Kong, SAR China
Nicolas Younan	Mississippi State University, USA
Pavel Zemcik	Brno University of Technology, Czech Republic
Albrecht Zimmermann	INSA Lyon, France
Jacek M. Zurada	University of Louisville, USA
Reyer Zwiggelaar	Aberystwyth University, UK

Additional Reviewers

John Arevalo	Universidad Nacional, Colombia
Eugene Borovikov	National Library of Medicine, USA
Bertrand Coüasnon	Irisa/Insa de Rennes, France
Mohamed Dahmane	CRIM, Canada
Ivan Duran-Diaz	University of Seville, Spain
Samuel Foucher	CRIM, Canada
Francesco Gargiulo	University of Naples Federico II, Italy
Rene Grzeszick	TU Dortmund, Germany
Marc Lalonde	CRIM, Canada
Tom Landry	CRIM, Canada
Pedro Martins	ISR, Institute of Systems and Robotics, Portugal
Saeid Motiian	West Virginia University, USA
Jiyong Oh	Graduate School of Convergence Science and Technology, Republic of Korea
Gabriele Piantadosi	Università Federico II di Napoli, Italy
Marco Piccirilli	West Virginia University, USA
Jorge A. Vanegas	MindLab – Universidad Nacional de Colombia, Colombia
Matteo Zignani	Università degli Studi di Milano, Italy

Invited Speakers

Hanan Samet	University of Maryland, USA
Nello Cristianini	University of Bristol, UK
Marcello Pelillo	University of Venice, Italy
Luis Alexandre	University of Beira Interior, Portugal

Contents

Invited Paper

3D Computer Vision: From Points to Concepts

Luís A. Alexandre[✉]

Department of Informatics and Instituto de Telecomunicações,
University of Beira Interior, Covilhã, Portugal
`luis.alexandre@ubi.pt`

Abstract. The emergence of cheap structured light sensors, like the Kinect, opened the door to an increased interest in all matters related to the processing of 3D visual data. Applications for these technologies are abundant, from robot vision to 3D scanning. In this paper we go through the main steps used on a typical 3D vision system, from sensors and point clouds up to understanding the scene contents, including key point detectors, descriptors, set distances, object recognition and tracking and the biological motivation for some of these methods. We present several approaches developed at our lab and some current challenges.

1 Introduction

There are currently many application fields for 3D computer vision (3DCV). One of the recent pushes to the 3D computer vision was the appearance of cheap 3D sensors, such as the Microsoft Kinect. This was not developed for 3D computer vision but for the (console) video gaming industry. 3DCV is used in games as a means to receive user input. Other applications of 3DCV can be found in biometrics, such as for 3D facial and expression recognition, in robotic vision, industrial quality control systems or even in online shopping[1].

We present the current 3D technologies and the most used sensors in Sect. 2. In Sect. 3 the focus will be on keypoint extraction from 3D point clouds. Section 4 discusses 3D descriptors and the following section presents methods used on 3D object recognition. Section 6 presents a 3D tracking method based on keypoint extraction and Sect. 7 indicates some current challenges in this field. The final section contains the conclusion.

2 3D Sensors

There are several possible technologies for obtaining 3D images. These 3D images are in fact sets of points in space called point clouds. These points have, besides

---

L.A. Alexandre—This work was partially financed by FEDER funds through the Programa Operacional Factores de Competitividade - COMPETE and by Portuguese funds through FCT - Fundação para a Ciência e a Tecnologia in the framework of the project PTDC/EIA-EIA/119004/2010.

[1] See http://metail.com/.

A. Fred et al. (Eds.): ICPRAM 2015, LNCS 9493, pp. 3–14, 2015.
DOI: 10.1007/978-3-319-27677-9_1

their 3D coordinates, typically at least a gray scale value or RGB value but can have other measures associated, such has a local curvature. The 3D images can also be represented by two 2D images: one containing the illumination intensity of color values for scene locations and the other the respective depth or distance to the sensor.

A basic approach to obtaining 3D images is by inferring the depth from two different views of a scene (parallax). This can be done by using a single camera and positioning it in different locations (for a static scene) or more commonly, by using two cameras, mimicking the animal's visual sensors (eyes) layout, as in Fig. 1. The major difficulty in this approach is identifying the same scene point in both images to obtain the point disparity. Many approaches have been proposed to achieve this[2].

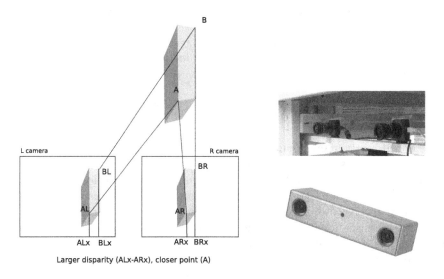

Fig. 1. Parallax-based stereo vision (left) and examples of sensors based on this approach (right-top, a "home-made" stereo vision system and right-bottom, a Bumblebee2 camera).

Another way to obtain 3D visual data is by using active vision and projecting a pattern in the scene that is used to identify the scene points' relative position. This approach is called a structured light approach. Figure 2 presents the idea and shows several sensors based on this approach. The pattern projection is usually made using infrared light such that it doesn't appear in the visible image. A third approach to obtaining 3D images is by inferring the scene points' distance to the sensor by measuring the time light takes to travel from an emitter located near the sensor, to the scene point and returning to the sensor. Since the speed

---

[2] Check for instance the disparity algorithms at the Middlebury Stereo Vision Page http://vision.middlebury.edu/stereo/.

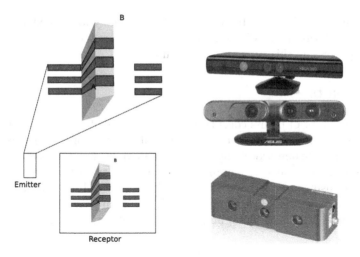

Fig. 2. Projection of a structured light pattern onto the scene (left) and several structured light sensors (from top to bottom): Microsoft Kinect, Asus Xtion Live, IDS Ensenso N20.

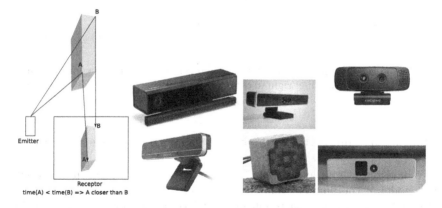

Fig. 3. The principle behind ToF sensors (left) and several sensors (from top left to bottom right): Microsoft Kinect 2, DepthSense, Creative Senz3D, Intel RealSense F200, Fotonic, PMD CamBoard pico[s].

of light in the air is known, the time taken is enough to infer the distance, or depth. Figure 3 illustrates this and presents some commercial available sensors based on this idea.

Size and weight have been falling to the point of currently having a 3D sensor inside a cell phone (see project Tango by Google), something that opens the way to many possible new mobile applications.

These sensors eventually produce a point cloud, typically at 30 fps. For 30 k points with RGB at 30 fps (typical Kinect specification), more than 30 MB/s of data are generated. This can be too much data specially for embedded applications, so some form of sub-sampling must be used to reduce the computational burden of processing this type of data stream.

3   Keypoints

Keypoints are a set of points considered representative of the point cloud. They are extracted from a point cloud when the full data stream is considered too much data for real-time processing. So keypoints are a way to do sub-sampling. Figure 4 presents two different approaches to keypoint extraction: regular spaced sub-sampling using a voxel grid with two different voxel sides (left 1 cm and center 2 cm) and a Harris3D extractor (right). The figure also shows the location of the keypoints (the black dots) and the number of extracted keypoints.

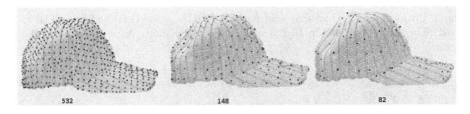

Fig. 4. Examples of keypoint extraction: left and center shows the keypoints obtained using voxel grid sub-sampling; right shows the result of the Harris3D keypoint extractor. Keypoints appear as black dots superimposed on the original point cloud.

Humans don't process every "input pixel", but focus their attention on salient points.

We have recently proposed [6] a 3D keypoint detector based on a computational model of the human visual system (HVS): the Biologically Inspired 3D Keypoint based on Bottom-Up Saliency (BIK-BUS). This approach is inspired on the visual saliency and the method mimics the following HVS mechanisms:

- Center-surround cells: sensitive to the center of their receptive fields and are inhibited by stimuli in its surroundings.
- Color double-opponency: neurons are excited in the center of their receptive field by one color and inhibited by the opponent color (red-green or blue-yellow) while the opposite takes place in the surround.
- Impulse response of orientation-selective neurons is approximated by Gabor filters.
- Lateral inhibition: neighboring cells inhibit each other through lateral connections.

Figure 5 presents a general view of the proposed method. The input point cloud is filtered to obtain color, intensity and normal orientation data. This is then used to build multi-scale representations of these features (Gaussian pyramids) that are combined using a mechanism that simulates center-surround cells and a normalization operator motivated by lateral inhibition to generated feature maps. From these feature maps, new maps, called conspicuity maps, are generated combining information from multiple scales. The three conspicuity maps are combined into a single saliency map. Finally, from the saliency map, and through the use of inhibition mechanisms, the 3D keypoints can be selected.

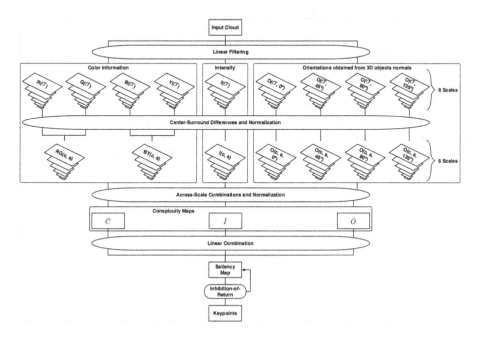

Fig. 5. General view of the BIK-BUS 3D keypoint detection method. Figure from [6].

Table 1. Number of times a keypoint detector came out as the best in the experiments run in [6].

Detector	Category		Object		Total
	AUC	DEC	AUC	DEC	
BIK-BUS	7	9	7	9	32
Curvature	3	2	2	1	8
Harris3D	1	1	0	0	2
ISS3D	2	0	4	2	8
KLT	2	1	1	0	4
Lowe	0	0	1	0	1
Noble	0	1	2	1	4
SIFT3D	0	0	0	1	1
SUSAN	2	1	2	1	6

We evaluated our proposal against 8 state-of-the-art detectors. We performed around 1.6 million comparisons for each pair keypoint detector/descriptor for a total of 135 pairs (9 keypoint detectors × 15 descriptors). The evaluation considered two metrics: area under the ROC curve (AUC) and the decidability (DEC). Table 1 shows the number of times each keypoint detector was the best on the experiments run. BIK-BUS was a clear winner with the second best methods at a considerable distance.

4   Descriptors

4.1   Evaluating Descriptors

A descriptor is a measure extracted from the input data that represents or describes an input data region in a concise manner. They are used to represent input data and allow a system to keep only a condensed representation of the input data (they are the equivalent of features in standard pattern recognition). There is a wide choice of descriptors: which should one use? We made an evaluation of 13 available in PCL [1]. Figure 6 shows the time taken and space used by the evaluated descriptors when they were applied after 3 different keypoint detectors.

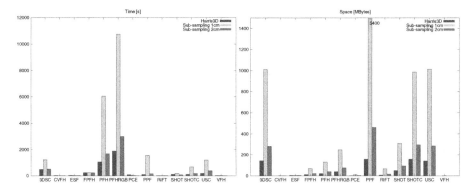

Fig. 6. Evaluation results of 13 descriptors in terms of time and space used in the experiments performed in [1].

Figure 7 shows the Precision-Recall curves for the experiments that used the 1 cm voxel grid sub-sample keypoint detector. Color-based descriptors are better (PFHRGB and SHOTCOLOR). Further details, including the equivalent figures for the remaining 2 keypoint detector approaches can be found in [1].

4.2   Genetic Algorithm-Evolved 3D Point Cloud Descriptor

From the evaluation of the descriptors discussed in the above section, we concluded that accurate descriptors are very computationally intensive and faster

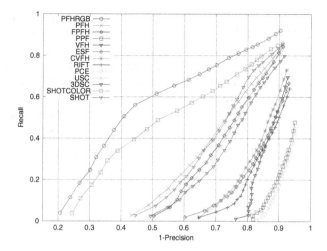

Fig. 7. Precision-Recall curves for the experiments that used the 1 cm voxelgrid sub-sample keypoint detector. Figure from [1].

descriptors use large storage space. For embedded approaches, such as robot-based vision, where computational resources and storage space come at a cost or might not be available in adequate amounts, a simple descriptor is desirable. For this type of application, we developed [8] a genetic algorithm(GA)-based descriptor that is both fast and has a small space footprint, while maintaining an acceptable accuracy.

It works by creating a keypoint cloud by sub-sampling with a voxel grid with leaf size of 2 cm. Two regions around each keypoint are considered: disk (R_1) + ring $(R_2 - R_1)$ (see Fig. 8).

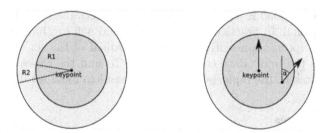

Fig. 8. Left: circular regions considered. Right: angle between the normal at a point in the R_1 region and the normal at the keypoint.

The information stored by the descriptor considers both shape and color information around each keypoint. For the shape, the descriptor records the histogram of angles between normals at keypoint and at each neighbor in region.

For the color information, a (Hue, Saturation) histogram for all points in each region is stored. The used distance between 2 point clouds represented by this descriptor is calculated using: $d = w.d_{shape} + (1 - w).d_{color}$, where the weight w is obtained through the GA optimization procedure. In total, 5 parameters (#shape bins, #color bins, R_1, R_2, w) are searched using the GA on the training data set. The obtained results can be seen in Table 2. This proposal allows for a much faster and lightweight (in terms of space) descriptor, with accuracy comparable to the SHOTCOLOR descriptor, and is thus adequate for use in situations where the computational cost of algorithms is an issue and/ or the available storage space is small.

Table 2. Average error, time and size of the three descriptors evaluated in [8].

Descriptor	Object [%]	Category [%]	Time [s]	Size
PFHRGB	20.25	5.27	2992	250
SHOTCOLOR	26.58	9.28	178	1353
Our	27.43	10.34	72	248

5   3D Object Recognition

The typical 3D object recognition pipeline consists on: obtaining the input data usually in the form of a point cloud; making keypoint detection; finding descriptors at each keypoint that are then grouped into a set that represents the input point cloud. After this, in a test ou deployment phase, incoming point clouds are compared against stored ones in an object database using, for instance, a set distance.

So, each point cloud is represented by a set of descriptors, and each descriptor is n-dimensional. In practice, a given point cloud will can have an arbitrary number of descriptors representing it, so the cardinal of the set of descriptors that represents the input data is not constant. To find the closest object in a database, a match to the input point cloud, we need to use a set distance.

5.1   Set Distances

Set distances are usually built around point distances. Three common point distances are the following: consider $x, y \in \mathbb{R}^n$, then

- City-block:

$$L_1(x, y) = \|x - y\|_1 = \sum_{i=1}^{n} |x(i) - y(i)|$$

– Euclidean:

$$L_2(x,y) = \|x - y\|_2 = \sqrt{\sum_{i=1}^{n}(x(i) - y(i))^2}$$

– Chi-squared:

$$d_{\chi^2}(x,y) = \frac{1}{2}\sum_{i=1}^{n}\frac{(x(i) - y(i))^2}{x(i) + y(i)}$$

Consider that a, b are points and A, B are sets. Let us also consider the following set distances:

– $D_1(A,B) = \max\{\sup\{f(a,B) \mid a \in A\}, \sup\{f(b,A) \mid b \in B\}\}$ with $f(a,B) = \inf\{L_1(a,b),\ b \in B\}$
– $D_2 = $ Pyramid Match Kernel distance [7]
– $D_3(A,B) = L_1(\min_A, \min_B) + L_1(\max_A, \max_B)$ with
  $\min_A(i) = \min_{j=1,\dots,|A|}\{a_j(i)\}, \ i = 1,\dots,n$
  $\max_A(i) = \max_{j=1,\dots,|A|}\{a_j(i)\}, \ i = 1,\dots,n$
  and similarly for $\min_B(i)$ e $\max_B(i)$.
– $D_4(A,B) = L_1(c_A, c_B)$ where c_A, c_B are cloud centroids
– $D_5(A,B) = L_2(c_A, c_B)$
– $D_6(A,B) = D_4(A,B) + L_1(std_A, std_B)$ with
  $std_A(i) = \sqrt{\frac{1}{|A|-1}\sum_{j=1}^{|A|}(a_j(i) - c_A(i))^2}, \ i = 1,\dots,n$ and similarly for std_B.
– $D_7(A,B) = d_{\chi^2}(c_A, c_B) + d_{\chi^2}(std_A, std_B)$
– $D_8(A,B) = \frac{1}{|A||B|}\sum_{i=1}^{|A|}\sum_{j=1}^{|B|}L_1(a_i, b_j)$

We evaluated [2] these 8 distances using 2 descriptors (PFHRBG and SHOT-COLOR). We used a data set with 48 objects from 10 categories and 1421 point clouds. The keypoint detector used was Harris3D. Figure 9 shows the precision-recall curves for the experiments with both descriptors. Table 3 contains the time it took for the evaluation of the test set on a machine running with 12 threads.

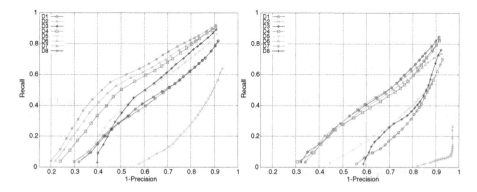

Fig. 9. Results of the distance evaluation using: PFHRBG (left) and SHOTCOLOR (right). Figure from [2].

Table 3. Time in seconds for test set evaluation (12 threads).

Distance	PFHRGB	SHOTCOLOR
D_1	1914	175
D_2	2197	1510
D_3	1889	132
D_4	1876	137
D_5	1886	134
D_6	1885	132
D_7	1883	113
D_8	1914	174

Table 4. Average error and time used on 10 repetitions for the different approaches.

Approach	Error [%]	Time[s]	Approach	Error [%]	Time[s]
RGBD	29.87	714.60	Channel G + TL	37.47	166.60
Channel R	32.15	136.50	Channel B + TL	43.58	95.10
Channel G	44.02	131.60	Channel D + TL	66.32	157.70
Channel B	55.62	110.10	R,G+TL,B+TL,D+TL maj	33.45	555.90
Channel D	65.85	126.30	R,G+TL,B+TL,D+TL mean	**28.80**	555.90
R,G,B,D maj	36.72	**504.50**	R,G+TL,B+TL,D maj	32.63	524.50
R,G,B,D mean	29.58	**504.50**	R,G+TL,B+TL,D mean	29.01	524.50

Simple distances like D_6 and D_7 are a good choice (accurate and fast) better than more common distances such as D_1 and D_2. Additionally, simple distances don't need any parameter search, as is the case with D_2.

5.2 Deep Transfer Learning for 3D Object Recognition

Deep learning is showing great potential in pattern recognition. The idea of transfer learning (TL) is also a very appealing one: learn in one problem and reuse (at least part of) the knowledge in other problems. We used both these ideas in a work where a convolutional neural network learns to recognize objects from 3D data [3]. TL is used from one color channel to the others and also to the depth channel. Decision fusion is used to merge each nets predictions. The results appear in Table 4. As can be seen, the TL approach is successful in obtaining both higher accuracy and shorter time that the baselines considered.

6 3D Object Tracking

The world is dynamic: another step towards understanding it is to follow objects as they move, since movement is a very important visual cue. There are many different approaches to tracking: the most used are particle filter variants [4].

We used a biologically-inspired keypoint extractor to initialize and maintain particles for particle filter-based tracking from 3D [5]. A general overview of the proposed method appears in Fig. 10.

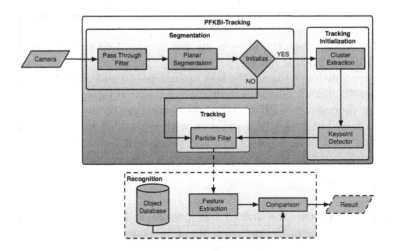

Fig. 10. A general view of PFBIK. Figure from [5].

We compared our tracker against the OpenNI tracker available in PCL. The videos used 10 different moving objects and a total of 3300 point clouds. The results are presented in Table 5. The PFBIK tracker used a much smaller number of particles that enabled it to be much faster during tracking, with the exception of the initialization where it was slower than the OpenNITracker given the necessary keypoint detection (that is only done at the start of the tracking process). The PFBIK was also slightly more accurate as can be seen through the distance error to the tracked object centroid.

Table 5. Results of PFBIK tracking compared to OpenNITracker: number of particles used, initialization and iteration time in seconds and distance error to the tracked centroid in meters.

Method	Particles	Init. time	Track time	Distance error
PFBIK	102 ± 92	0.20 ± 0.16	0.08 ± 0.06	0.045 ± 0.036
OpenNITracker	2132 ± 1988	0.17 ± 0.19	0.19 ± 0.17	0.052 ± 0.022

7   Challenges

Although recent progress in 3DCV has been substantial, there are still many challenges in the field. Some of the current challenges faced by the 3D computer vision community are:

- object representation: in this paper we showed object representations based on sets of descriptors of partial object views (2.5D) but other possibilities might be better (represent an object using a fused view representation, for instance). The best object representation approach may depend on the particular application and is still a important research topic;
- non-rigid object recognition: current keypoint plus descriptor approach is not a good solution when the objects are not rigid. More complex models are needed (3D deformable models);
- activity recognition: what are the best approaches to understand human activities from 3D video? This is currently a hot research topic;
- real-time processing: GPU-based implementations of most algorithms can bring us here but it is still a problem with embedded devices (cloud-based processing requires high bandwidth and permanent connection).

8   Conclusion

This paper summarizes the invited talk present at ICPRAM 2015 where the author reviewed some of the key concepts of 3D computer vision and presented some of the recent work in this field produced by him and his co-authors.

References

1. Alexandre, L.A.: 3D descriptors for object and category recognition: a comparative evaluation. In: Workshop on Color-Depth Camera Fusion in Robotics at the IEEE/RSJ International Conference on Intelligent Robots and Systems (IROS), Vilamoura, Portugal, October 2012
2. Alexandre, L.A.: Set distance functions for 3D object recognition. In: Ruiz-Shulcloper, J., Sanniti di Baja, G. (eds.) CIARP 2013, Part I. LNCS, vol. 8258, pp. 57–64. Springer, Heidelberg (2013)
3. Alexandre, L.A.: 3D object recognition using convolutional neural networks with transfer learning between input channels. In: Menegatti, E., Michael, N., Berns, K., Yamaguchi, H. (eds.) Intelligent Autonomous Systems 13. AISC, pp. 889–898. Springer, Heidelberg (2014)
4. Del Moral, P.: Mean field simulation for monte carlo integration. Chapman and Hall/CRC, Boca Raton (2013)
5. Filipe, S., Alexandre, L.: Pfbik-tracking: Particle filter with bio-inspired keypoints tracking. In: 2014 IEEE Symposium on Computational Intelligence for Multimedia. Signal and Vision Processing (CIMSIVP), pp. 1–8, Florida, USA, December 2014
6. Filipe, S., Itti, L., Alexandre, L.A.: BIK-BUS: biologically motivated 3D keypoint based on bottom-up saliency. IEEE Trans. Image Process. $24(1)$, 163–175 (2015)
7. Grauman, K., Darrell, T.: The pyramid match kernel: efficient learning with sets of features. J. Mach. Learn. Res. 8, 725–760 (2007)
8. Węgrzyn, D., Alexandre, L.A.: A genetic algorithm-evolved 3D point cloud descriptor. In: Ruiz-Shulcloper, J., Sanniti di Baja, G. (eds.) CIARP 2013, Part I. LNCS, vol. 8258, pp. 92–99. Springer, Heidelberg (2013)

Theory and Methods

Density Difference Detection with Application to Exploratory Visualization

Marko Rak[✉], Tim König, Johannes Steffen, Dirk Joachim Lehmann, and Klaus-Dietz Tönnies

Department of Simulation and Graphics, Otto von Guericke University,
Magdeburg, Germany
rak@isg.cs.ovgu.de

Abstract. Identifying differences among the distribution of samples of different observations is an important issue in many research fields. We provide a general framework to detect these difference spots in d-dimensional feature space. Such spots occur not only at various locations, they may also come in various shapes and multiple sizes, even at the same location. We address these challenges by a scale-space representation of the density function difference of the observations in feature space. Using three classification scenarios from UCI Machine Learning Repository we show that interest spots carry valuable information about a data set. To this end, we establish a simple decision rule on top of our framework. Results indicate state-of-the-art performance, underpinning the importance of the information that is carried by the detected spots. Furthermore, we outline that the output of our framework can be used to guide exploratory visualization of high-dimensional feature spaces.

Keywords: Density difference · Kernel density estimation · Scale space · Dendrogram · Blob detection · Affine shape adaption · Exploratory visualization · Orthographic star coordinates

1 Introduction

Sooner or later a large portion of pattern recognition tasks come down to the question *What makes X different from Y?* Some scenarios of that kind are:

Detection of forged money based on image-derived features: *What makes some sort of forgery different from genuine money?*
Comparison of medical data of healthy and non-healthy subjects for disease detection: *What makes the healthy different from the non-healthy?*
Comparison of document data sets for text retrieval purposes: *What makes this set of documents different from another set?*

Apart from this, spotting differences in two or more observations is of interest in fields of computational biology, chemistry or physics. Looking at it from a general perspective, such questions generalize to

© Springer International Publishing Switzerland 2015
A. Fred et al. (Eds.): ICPRAM 2015, LNCS 9493, pp. 17–34, 2015.
DOI: 10.1007/978-3-319-27677-9_2

What makes samples of group X different from the samples of group Y?

This question usually arises when we deal with grouped samples in some feature space. For humans, answering such questions tends to become more challenging with increasing number of groups, samples and feature space dimensions, up to the point where we miss the forest for the trees. This complexity is not an issue to automatic approaches, which, on the other hand, tend to either overfit or underfit patterns in the data. Therefore, semi-automatic approaches are needed to generate a number of interest spots which are to be looked at in more detail.

We address this issue by a scale-space difference detection framework. Our approach relies on the density difference of group samples in feature space. This enables us to identify spots where one group dominates the other. We draw on kernel density estimators to represent arbitrary density functions. Embedding this into a scale-space representation, we are able to detect spots of different sizes and shapes in feature space in an efficient manner. Our framework:

- applies to d-dimensional feature spaces
- is able to reflect arbitrary density functions
- selects optimal spot locations, sizes and shapes
- is robust to outliers and measurement errors
- produces human-interpretable results

Please note that large portions of the subsequent content were already covered in our previous work [16]. Within the current work we go into detail on a second spot detector (complementing the one used previously), provide an extended evaluation and show how the output of our framework can be used to guide the exploratory visualization of high-dimensional feature spaces. The latter may be seen as an intermediate step prior to applying other means of data analysis to the identified interest spots.

Our presentation is structured as follows. We outline the key foundations of our framework in Sect. 2. The specific parts of our framework are detailed in Sect. 3, while Sect. 4 outlines our contribution to exploratory visualization. Section 5 comprises our results on several data sets from UCI Machine Learning Repository. In Sect. 6, we close with a summary of our work, our most important results and an outline of future work.

2   Theoretical Foundations

Searching for differences between the sample distribution of two groups of observations g and h, we, quite naturally, seek for spots where the density function $f^g(\mathbf{x})$ of group g dominates the density function $f^h(\mathbf{x})$ of group h, or vice versa. Hence, we try to find positive-/negative-valued spots of the density difference

$$f^{g-h}(\mathbf{x}) = f^g(\mathbf{x}) - f^h(\mathbf{x}) \tag{1}$$

w.r.t. the underlying feature space $\mathbb{R}^d$ with $\mathbf{x} \in \mathbb{R}^d$. Such spots may come in various shapes and sizes. A difference detection framework should be able to deal with these degrees of freedom. Additionally, it must be robust to various sources of error, e.g. from measurement, quantization and outliers.

We propose to superimpose a scale-space representation to the density difference $f^{g-h}(\mathbf{x})$ to achieve the above-mentioned properties. Scale-space frameworks have been shown to robustly handle a wide range of detection tasks for various types of structures, e.g. text strings [23], persons and animals [8] in natural scenes, neuron membranes in electron microscopy imaging [20] or microaneurysms in digital fundus images [2]. In each of these tasks the function of interest is represented through a grid of values, allowing for an explicit evaluation of the scale-space. However, an explicit grid-based approach becomes intractable for higher-dimensional feature spaces.

In what follows, we show how a scale-space represenation of $f^{g-h}(\mathbf{x})$ can be obtained from kernel density estimates of $f^g(\mathbf{x})$ and $f^h(\mathbf{x})$ in an implicit fashion, expressing the problem by scale-space kernel density estimators. Note that by the usage of kernel density estimates our work is limited to feature spaces with dense filling. We close with a brief discussion on how this can be used to compare observations among more than two groups.

2.1   Scale Space Representation

First, we establish a family $l^{g-h}(\mathbf{x};t)$ of smoothed versions of the densitiy difference $l^{g-h}(\mathbf{x})$. Scale parameter $t \geq 0$ defines the amount of smoothing that is applied to $l^{g-h}(\mathbf{x})$ via convolution with kernel $k_t(\mathbf{x})$ of bandwidth t as stated in

$$l^{g-h}(\mathbf{x};t) = k_t(\mathbf{x}) * f^{g-h}(\mathbf{x}). \tag{2}$$

For a given scale t, spots having a size of about $2\sqrt{t}$ will be highlighted, while smaller ones will be smoothed out. This leads to an efficient spot detection scheme, which will be discussed in Sect. 3. Let

$$l^g(\mathbf{x};t) = k_t(\mathbf{x}) * f^g(\mathbf{x}) \tag{3}$$

$$l^h(\mathbf{x};t) = k_t(\mathbf{x}) * f^h(\mathbf{x}) \tag{4}$$

be the scale-space representations of the group densities $f^g(\mathbf{x})$ and $f^h(\mathbf{x})$. Looking at Eq. 2 more closely, we can rewrite $l^{g-h}(\mathbf{x};t)$ equivalently in terms of $l^g(\mathbf{x};t)$ and $l^h(\mathbf{x};t)$ via Eqs. 3 and 4. This reads

$$l^{g-h}(\mathbf{x};t) = k_t(\mathbf{x}) * f^{g-h}(\mathbf{x}) \tag{5}$$

$$= k_t(\mathbf{x}) * \left[f^g(\mathbf{x}) - f^h(\mathbf{x}) \right] \tag{6}$$

$$= k_t(\mathbf{x}) * f^g(\mathbf{x}) - k_t(\mathbf{x}) * f^h(\mathbf{x}) \tag{7}$$

$$= l^g(\mathbf{x};t) - l^h(\mathbf{x};t). \tag{8}$$

The simple yet powerful relation between the left and the right-hand side of Eq. 8 will allow us to evaluate the scale-space representation $l^{g-h}(\mathbf{x})$ implicitly, i.e. using only kernel functions. Of major importance is the choice of the smoothing kernel $k_t(\mathbf{x})$. According to scale-space axioms, $k_t(\mathbf{x})$ should suffice a number of properties, resulting in the uniform Gaussian kernel of Eq. 9 as the unique choice, cf. [3, 24].

$$\phi_t(\mathbf{x}) = \frac{1}{\sqrt{(2\pi t)^d}} \exp\left(-\frac{1}{2t}\mathbf{x}^\mathrm{T}\mathbf{x}\right) \tag{9}$$

2.2   Kernel Density Estimation

In kernel density estimation, the group density $f^g(\mathbf{x})$ is estimated from its n^g samples by means of a kernel function $K_{\mathbf{B}^g}(\mathbf{x})$. Let $\mathbf{x}_i^g \in \mathbb{R}^{d \times 1}$ with $i = 1, \ldots, n^g$ being the group samples. Then, the group density estimate is given by

$$\hat{f}^g(\mathbf{x}) = \frac{1}{n^g} \sum_{i=1}^{n^g} K_{\mathbf{B}^g}(\mathbf{x} - \mathbf{x}_i^g). \tag{10}$$

Parameter $\mathbf{B}^g \in \mathbb{R}^{d \times d}$ is a symmetric positive-definite matrix, which controls the sample influence to the density estimate. Informally speaking, $K_{\mathbf{B}^g}(\mathbf{x})$ applies a smoothing with bandwidth $\mathbf{B}^g$ to the "spiky sample relief" in feature space.

Plugging kernel density estimator $\hat{f}^g(\mathbf{x})$ into the scale-space representation $l^g(\mathbf{x}; t)$ defines the scale-space kernel density estimator $\hat{l}^g(\mathbf{x}; t)$ to be

$$\hat{l}^g(\mathbf{x}; t) = k_t(\mathbf{x}) * \hat{f}^g(\mathbf{x}). \tag{11}$$

Inserting Eq. 10 into the above, we can trace down the definition of the scale-space density estimator $\hat{l}^g(\mathbf{x}; t)$ to the sample level via transformation

$$\hat{l}^g(\mathbf{x}; t) = k_t(\mathbf{x}) * \hat{f}^g(\mathbf{x}) \tag{12}$$

$$= k_t(\mathbf{x}) * \left[\frac{1}{n^g} \sum_{i=1}^{n^g} K_{\mathbf{B}^g}(\mathbf{x} - \mathbf{x}_i^g)\right] \tag{13}$$

$$= \frac{1}{n^g} \sum_{i=1}^{n^g} (k_t * K_{\mathbf{B}^g})(\mathbf{x} - \mathbf{x}_i^g). \tag{14}$$

Though arbitrary kernels can be used, we choose $K_{\mathbf{B}}(\mathbf{x})$ to be a Gaussian kernel $\Phi_{\mathbf{B}}(\mathbf{x})$ due to its convenient algebraic properties. This (potentially non-uniform) kernel is defined as

$$\Phi_{\mathbf{B}}(\mathbf{x}) = \frac{1}{\sqrt{\det(2\pi\mathbf{B})}} \exp\left(-\frac{1}{2}\mathbf{x}^\mathrm{T}\mathbf{B}^{-1}\mathbf{x}\right). \tag{15}$$

Using the above, the right-hand side of Eq. 14 simplifies further because of the Gaussian's cascade convolution property. Eventually, the scale-space kernel density estimator $\hat{l}^g(\mathbf{x}; t)$ is given by Eq. 16, where $\mathbf{I} \in \mathbb{R}^{d \times d}$ is the identity.

$$\hat{l}^g(\mathbf{x}; t) = \frac{1}{n^g} \sum_{i=1}^{n^g} \Phi_{t\mathbf{I}+\mathbf{B}^g}(\mathbf{x} - \mathbf{x}_i^g) \tag{16}$$

Using this estimator, the scale-space representation $l^g(\mathbf{x}; t)$ of group density $f^g(\mathbf{x})$ and analogously that of group h can be estimated for any $(\mathbf{x}; t)$ in an implicit fashion. Consequently, this allows us to estimate the scale-space representation $l^{g-h}(\mathbf{x}; t)$ of the density difference $f^{g-h}(\mathbf{x})$ via Eq. 7 by means of kernel functions only.

2.3 Bandwidth Selection

When regarding bandwidth selection in such a scale-space representation, we see that the impact of different choices for bandwidth matrix $\mathbf{B}$ vanishes as scale t increases. This can be seen when comparing matrices $t\mathbf{I} + \mathbf{0}$ and $t\mathbf{I} + \mathbf{B}$ where $\mathbf{0}$ represents the zero matrix, i.e. no bandwidth selection at all. We observe that relative differences between them become neglectable once $\|t\mathbf{I}\| \gg \|\mathbf{B}\|$. This is especially true for large sample sizes, because the bandwidth will then tend towards zero for any reasonable bandwidth selector anyway. Hence, we may actually consider setting $\mathbf{B}$ to $\mathbf{0}$ for certain problems, as we typically search for differences that fall above some lower bound for t.

Literature bares extensive work on bandwidth matrix selection, for example, based on plug-in estimators [6, 21] or biased, unbiased and smoothed cross-validation estimators [7, 19]. All of these integrate well with our framework. However, in view of the argument above, we propose to compromise between a full bandwidth optimization and having no bandwidth at all. We define $\mathbf{B}^g = b^g \mathbf{I}$ and use an unbiased least-squares cross-validation to set up the bandwidth estimate for group g. For Gaussian kernels, this leads to the optimization of 17, cf. [7], which we achieved by golden section search over b^g.

$$\underset{\mathbf{B}^g}{\arg\min} \frac{1}{n^g \sqrt{\det(4\pi\mathbf{B}^g)}} + \frac{1}{n^g(n^g - 1)} \sum_{i=1}^{n^g} \sum_{\substack{j=1 \\ j \neq i}}^{n^g} \left(\varPhi_{2\mathbf{B}^g} - 2\varPhi_{\mathbf{B}^g} \right) \left(\mathbf{x}_i^g - \mathbf{x}_j^g \right) \quad (17)$$

2.4 Multiple Groups

If differences among more than two groups shall be detected, we can reduce the comparison to a number of two-group problems. We can consider two typical use cases, namely *one group vs. another* and *one group vs. rest*. Which of the two is more suitable depends on the specific task at hand. Let us illustrate this using two medical scenarios. Assume we have a number of groups which represent patients having different diseases that are hard to discriminate in differential diagnosis. Then we may consider the second use case, to generate clues on markers that make one disease different from the others. In contrast, if these groups represent stages of a disease, potentially including a healthy control group, then we may consider the first use case, comparing only subsequent stages to give clues on markers of the disease's progress.

3   Detection Framework

To identify the positve-/negative-valued spots of a density difference, we apply the concept of blob detection, which is well-known in computer vision, to the scale-space representation derived in Sect. 2. In scale-space blob detection, some blobness criterion is applied to the scale-space representation, seeking for local optima of the function of interest w.r.t. space and scale. This directly leads to an efficient detection scheme that identifies a spot's location and size. The latter corresponds to the detection scale.

In a grid-representable problem we can evaluate blobness densely over the scale-space grid and identify interesting spots directly using the grid neighborhood. This is intractable here, which is why we rely on a more refined three-stage approach. First, we trace the local spatial optima of the density difference through scales of the scale-space representation. Second, we identify the interesting spots by evaluating their blobness along the dendrogram of optima that was obtained during the first stage. Having selected spots and therefore knowing their locations and sizes, we finally calculate an elliptical shape estimate for each spot in a third stage.

Spots obtained in this fashion characterize elliptical regions in feature space as outlined in Fig. 1. The representation of such regions, i.e. location, size and shape, as well as its strength, i.e. its scale-space density difference value, are easily interpretable by humans, which allows to look at them in more detail using some other method. The elliptical nature of the identified regions is also a limitation of our work, because non-elliptical regions may only be approximated by elliptical ones. We now give a detailed description of the three stages.

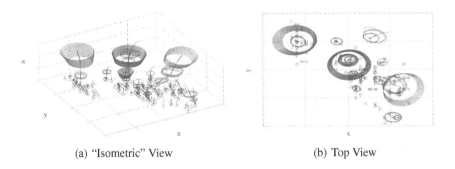

(a) "Isometric" View   (b) Top View

Fig. 1. Detection results for a two-group (red/blue) problem in two-dimensional feature space (xy-plane) with augmented scale dimension s; Red squares and blue circles visualize the samples of each group; Red/blue paths outline the dendrogram of scale-space density difference optima for the red/blue group dominating the other group; Interesting spots of each dendrogram are printed thick; Red/blue ellipses characterize the shape for each of the interest spots (Color figure online).

3.1 Scale Tracing

Assume we are given an equidistant scale sampling, containing non-negative scales $t_1, \ldots, t_n$ in increasing order and we search for spots where group g dominates h. More precisely, we search for the non-negatively valued maxima of $l^{g-h}(\mathbf{x}; t_{i-1})$. The opposite case, i.e. group h dominates g, is equivalent.

Let us further assume that we know the spatial local maxima of the density difference $l^{g-h}(\mathbf{x}; t_{i-1})$ for a certain scale t_{i-1} and we want to estimate those of the current scale t_i. This can be done taking the previous local maxima as initial points and optimizing each w.r.t. $l^{g-h}(\mathbf{x}; t_i)$. In the first scale, we take the samples of group g themselves. As some maxima may be converged to the same location, we merge them together, feeding unique locations as initials into the next scale t_{i+1} only. We also drop any negatively-valued locations as these are not of interest to our task. They will not become of interest for any higher scale either, because local extrema will not enhance as scale increases, cf. [13]. Since derivatives are simple to evaluate for Gaussian kernels, we can use Newton's method for spatial optimization. We can assemble gradient $\frac{\partial}{\partial \mathbf{x}} l^{g-h}(\mathbf{x}; t)$ and Hessian $\frac{\partial^2}{\partial \mathbf{x} \partial \mathbf{x}^{\mathrm{T}}} l^{g-h}(\mathbf{x}; t)$ sample-wise using

$$\frac{\partial}{\partial \mathbf{x}} \Phi_{\mathbf{B}}(\mathbf{x}) = -\Phi_{\mathbf{B}}(\mathbf{x}) \, \mathbf{B}^{-1} \mathbf{x} \qquad \text{and} \tag{18}$$

$$\frac{\partial^2}{\partial \mathbf{x} \partial \mathbf{x}^{\mathrm{T}}} \Phi_{\mathbf{B}}(\mathbf{x}) = \Phi_{\mathbf{B}}(\mathbf{x}) \left(\mathbf{B}^{-1} \mathbf{x} \mathbf{x}^{\mathrm{T}} \mathbf{B}^{-1} - \mathbf{B}^{-1} \right). \tag{19}$$

Iterating this process through all scales, we form a discret dendrogram of the maxima over scales. A dendrogram branching means that a maxima formed from two (or more) maxima from the preceding scale.

3.2 Spot Detection

The maxima of interest are derived from a scale-normalized blobness criterion $c_\gamma(\mathbf{x}; t)$. Two main criteria, namely the determinant of the Hessian [5] given in Eq. 20[1] and the trace of the Hessian [13] given in Eq. 22 have been discussed in literature. In contrast to our previous work [16], we do not focus on a single criterion. Instead, we will later investigate both in comparison.

$$c_\gamma^{\mathrm{det}}(\mathbf{x}; t) = t^{\gamma d} \underbrace{(-1)^d \det \left(\frac{\partial^2}{\partial \mathbf{x} \partial \mathbf{x}^{\mathrm{T}}} l^{g-h}(\mathbf{x}; t) \right)}_{c^{\mathrm{det}}(\mathbf{x}; t)} \tag{20}$$

$$= t^{\gamma d} \qquad\qquad c^{\mathrm{det}}(\mathbf{x}; t) \tag{21}$$

$$c_\gamma^{\mathrm{tr}}(\mathbf{x}; t) = t^{\gamma d} \underbrace{(-1) \mathrm{tr} \left(\frac{\partial^2}{\partial \mathbf{x} \partial \mathbf{x}^{\mathrm{T}}} l^{g-h}(\mathbf{x}; t) \right)}_{c^{\mathrm{tr}}(\mathbf{x}; t)} \tag{22}$$

$$= t^{\gamma d} \qquad\qquad c^{\mathrm{tr}}(\mathbf{x}; t) \tag{23}$$

---

[1] $(-1)^d$ leads to a consistent criterion for even and odd dimensions.

Because the maxima are already spatially optimal, we can search for spots that maximize $c_\gamma(\mathbf{x}; t)$ w.r.t. the dendrogram neighborhood only. Note that we do not require the superscript because the remained is independent of the choice of the blobness criterion. Parameter $\gamma \geq 0$ can be used to introduce a size bias, shifting the detected spot towards smaller or larger scales. The definition of γ highly depends on the type of spot that we are looking for, cf. [12]. This is impractical when we seek for spots of, for example, small and large skewness or extreme kurtosis at the same time.

Addressing the parameter issue, we search for all spots that maximize $c_\gamma(\mathbf{x}; t)$ locally w.r.t. some $\gamma \in [0, \infty)$. Some dendrogram spot s with scale-space coordinates $(\mathbf{x}_s; t_s)$ is locally maximal if there exists a γ-interval such that its blobness $c_\gamma(\mathbf{x}_s; t_s)$ is larger than that of every spot in its dendrogram neighborhood $\mathcal{N}(s)$. This leads to a number of inequalities, which can be written as

$$t_s^{\gamma d} c(\mathbf{x}_s; t_s) \underset{\forall n \in \mathcal{N}(s)}{>} t_n^{\gamma d} c(\mathbf{x}_n; t_n) \qquad \text{or} \qquad (24)$$

$$\gamma \, d \log \frac{t_s}{t_n} \underset{\forall n \in \mathcal{N}(s)}{>} \log \frac{c(\mathbf{x}_n; t_n)}{c(\mathbf{x}_s; t_s)}. \qquad (25)$$

The latter can be solved easily for the γ-interval, if any. We can now identify our interest spots by looking for the maxima along the dendrogram that locally maximize the width of the γ-interval. More precisely, let $w_\gamma(\mathbf{x}_s; t_s)$ be the width of the γ-interval for dendrogram spot s, then s is of interest if the dendrogram Laplacian of $w_\gamma(\mathbf{x}; t)$ is negative at $(\mathbf{x}_s; t_s)$, or equivalently, if

$$w_\gamma(\mathbf{x}_s; t_s) > \frac{1}{|\mathcal{N}(s)|} \sum_{n \in \mathcal{N}(s)} w_\gamma(\mathbf{x}_n; t_n). \qquad (26)$$

Intuitively, a spot is of interest if its γ-interval width is above neighborhood average. This is the only assumption we can make without imposing limitations on the results. Interest spots indentified in this way will be dendrogram segments, each ranging over a number of consecutive scales.

3.3   Shape Adaption

Shape estimation can be done in an iterative manner for each interest spot. The iteration alternatingly updates the current shape estimate based on a measure of anisotropy around the spot and then corrects the bandwidth of the scale-space smoothing kernel according to this estimate, eventually reaching a fixed point. The second moment matrix of the function of interest is typically used as an anisotropy measure, e.g. in [14] and [15]. Since it requires spatial integration of the scale-space representation around the interest spot, this measure is not feasible here.

We adapted the Hessian-based approach of [10] to d-dimensional problems. The aim is to make the scale-space representation isotropic around the interest spot, iteratively moving any anisotropy into the symmetric positive-definite shape matrix $\mathbf{S} \in \mathbb{R}^{d \times d}$ of the smoothing kernel's bandwidth $t\mathbf{S}$. Thus, we lift the problem into a generalized representation $lg^{-h}(\mathbf{x}; t\mathbf{S})$ of anisotropic scale-space kernels, which requires us to replace the definition of $\phi_t(\mathbf{x})$ by that of $\Phi_{\mathbf{B}}(\mathbf{x})$.

Starting with the isotropic $\mathbf{S}_1 = \mathbf{I}$, we decompose the current Hessian via

$$\frac{\partial^2}{\partial \mathbf{x} \partial \mathbf{x}^{\mathrm{T}}} lg^{-h}(\,\cdot\,; t\mathbf{S}_i) = \mathbf{V}\mathbf{D}^2\mathbf{V}^{\mathrm{T}} \tag{27}$$

into its eigenvectors in columns of $\mathbf{V}$ and eigenvalues on the diagonal of $\mathbf{D}^2$. We then normalize the latter to unit determinant via

$$\mathbf{D} = \sqrt[d]{\det(\mathbf{D})}\mathbf{D} \tag{28}$$

to get a relative measure of anisotropy for each of the eigenvector directions. Finally, we move the anisotropy into the shape estimate via

$$\mathbf{S}_{i+1} = \left(\mathbf{V}^{\mathrm{T}}\mathbf{D}^{-\frac{1}{2}}\mathbf{V}\right)\mathbf{S}_i\left(\mathbf{V}\mathbf{D}^{-\frac{1}{2}}\mathbf{V}^{\mathrm{T}}\right) \tag{29}$$

and start all over again. Iteration terminates when isotropy is reached. More precisely: when the ratio of minimal and maximal eigenvalue of the Hessian approaches one, which usually happens within a few iterations.

4   Exploratory Visualization

As mentioned introductory, exploratory visualization may be a reasonable intermediate step prior to directly applying other means of data analysis to the interest spots. There are plenty of visualization techniques that aim at identification of interesting patterns in the distribution of samples in high-dimensional feature spaces. For this work, we focus on a recent in-house development namely orthographic star coordinates [11]. We next give a short introduction to the topic and discuss how outputs of our framework can be used to guide the visual exploration process.

4.1   Star Coordinate Visualization

Star coordinate visualizations make use of projections from d-dimensional feature spaces to a two-dimensional projection plane which is then visualized. Such projections are characterized by a projection matrix $\mathbf{P} \in \mathbb{R}^{2 \times d}$ the columns of which can be interpreted as d points in two-dimensional space. Modifying these so-called anchor points is equivalent to manipulation of the projection plane itself, which the star coordinate visualization exploits by an interactive interface like that shown in Fig. 2.

In general, star coordinates allow for arbitrary projections thus potentially introducing arbitrary distortions to the visualization of the high-dimensional content. This is not desirable for various reasons, therefore [11] proposed to restrict the interaction to orthographic projections. Orthography is directly

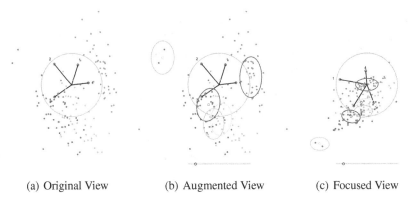

(a) Original View   (b) Augmented View   (c) Focused View

Fig. 2. Exploratory visualization of a three-group (red/green/blue) problem in 4-dimensional feature space by orthographic star coordinates; Original orthographic star coordinates (left) augmented with output of our framework (middle) and focused on a particular interest spot (right); Moveable anchor points are connected to the origin by thick black line segments; A slider for scale selection is located at the bottom of the interface; The remaining visual content is discussed in the text (Color figure online).

related to d-dimensional rotation, enforcing this property thus provides an intuitive way to "rotate" the high-dimensional content in front of a user's viewpoint. This directly targets the human's ability to interpret spatial relations from a steerable sequence of projections which is pretty much what we do with two-dimensional visualizations of three-dimensional content on a daily basis.

4.2   Preserving Orthography

Regarding orthography, we have to address two main issues. First, how to recover an orthographic projection when starting from an arbitrary projection. Second, how to reinforce orthography during interactive anchor movement. A sufficient condition for orthography of some anchor point constellation $\mathbf{P}_o \in \mathbb{R}^{2 \times d}$ is that

$$\mathbf{P}_o\mathbf{P}_o{}^{\mathrm{T}} = \mathbf{I}, \tag{30}$$

whereby $\mathbf{I} \in \mathbb{R}^{2 \times 2}$ is the identity matrix, cf. [11]. Therefore, given an arbitrary non-orthographic $\mathbf{P}$ we may seek to make $\mathbf{PP}^{\mathrm{T}} \in \mathbb{R}^{2 \times 2}$ identity. Since the latter Gramian matrix is almost certainly positive-definite in practice,[2] we can obtain it's Cholesky factor $\mathbf{L} \in \mathbb{R}^{2 \times 2}$ and manipulate the decomposition as follows

$$\mathbf{LL}^{\mathrm{T}} = \mathbf{PP}^{\mathrm{T}} \tag{31}$$

$$\mathbf{I} = \underbrace{\mathbf{L}^{-1}\mathbf{P}}\,\underbrace{\mathbf{P}^{\mathrm{T}}\mathbf{L}^{-\mathrm{T}}} \tag{32}$$

$$\mathbf{I} = \quad \mathbf{P}_o \quad \mathbf{P}_o{}^{\mathrm{T}} \tag{33}$$

---

[2] Rare semi-definite cases are avoided by regularization $\mathbf{PP}^{\mathrm{T}} + \epsilon\mathbf{I}$ for some small ϵ.

with $\mathbf{P}_o$ being the recovered orthographic projection.[3] Regarding the second issue, we can simply take the steps just outlined, continuously reinforcing orthography during interactive movement of particular anchors. Note how the anchor points of the given non-orthographic $\mathbf{P}$ are all transformed in the same manner by the (inverse of the) Cholesky factor $\mathbf{L}$ to obtain the orthographic anchor points $\mathbf{P}_o$. This avoids any experience of "arbitrariness" during user interaction.

4.3   Guiding Explorations

As already discussed in [11], there are certain open questions associated to star coordinate visualizations. This includes suitable anchor point constellations, centers of "rotation", i.e. the choice of the origin in d-dimensional feature space prior to projection, as well as a reasonable zoom into the data after projection. Otherwise put, we need to know where to look at and how. The interest spots detected by our framework can be used to address these issues, thereby also providing an interactive mechanism to switch among potentially interesting structures.

As show in Fig. 2, we have augmented the star coordinate visualization by a scale selection slider, letting the user choose the size (scale) of structures he/she is interested in. Based on his/her selection, the visualization is overlayed with the output of our framework that corresponds to the selected scale. Specifically, we transparently visualize the locations of maxima that were found during scale tracing (see Sect. 3.1) and their respective shapes, which were estimated during shape adaption (see Sect. 3.3). In case a maximum was found interesting (see Sect. 3.2), it's location and shape is highlighted opaquely instead.

When interactively selecting a maxima, the visualization is changed to put focus on the selection. Specifically, the origin of the d-dimensional feature space is shifted to the maxima's location thereby making it the center of "rotation". The user can then change the zoom to a multiple of the maxima's scale by keyboard bindings if desired. By another binding, he/she may also align the projection plane with the two most significant axes of the shape estimate to get a reasonable initial constellation of anchor points. To this end, the unit eigenvectors that correspond to the two largest eigenvalues of the shape estimate are used to fill the rows of the projection matrix.

We combined the above with a binding that resets the visualization to just before focusing a selection which allows to rapidly explore several potentially interesting spots before the user eventually moves on to differently sized structures. Changing the scale selection slider steadily, the course of locations and shapes of the maxima gives an impression on how the data is structured from coarse to fine without missing any highlighted interest spot.

5   Experiments

We next demonstrate that interest spots carry valuable information about a data set. Due to the lack of data sets that match our particular detection task a ground

---

[3] This formulation is another view on the Gram-Schmidt process used in [11].

truth comparison is impossible. Certainly, artificially constructed problems are an exception. However, the generalizability of results is at least questionable for such problems. Therefore, we chose to benchmark our approach indirectly via a number of classification tasks. The rational is that results that are comparable to those of well-established classifiers should underpin the importance of the identified interest spots.

Next we show how to use these interest spots for classification using a simple decision rule and detail the data sets that were used. We then investigate parameters of our approach and discuss the results of the classification tasks in comparison to decision trees, Fisher's linear discriminant analysis, k-nearest neighbors with optimized k and support vector machines with linear and cubic kernels. All experiments were performed via leave-one-out cross-validation.

5.1 Decision Rule

To perform classification we establish a simple decision rule based on interest spots that were detected using the *one group vs. rest* use case. Therefore, we define a group likelihood criterion as follows. For each group g, having the set of interest spots $\mathcal{I}^g$, we define

$$p^g(\mathbf{x}) = \max_{s \in \mathcal{I}^g} l^{g-h}(\mathbf{x}_s; t_s \mathbf{S}_s) \cdot \exp\left(-\frac{1}{2}(\mathbf{x} - \mathbf{x}_s)^{\mathrm{T}}(t_s \mathbf{S}_s)^{-1}(\mathbf{x} - \mathbf{x}_s)\right). \quad (34)$$

This is a quite natural trade-off, where the first factor favors spots s with high density difference, while the latter factor favors spots with small Mahalanobis distance to the location $\mathbf{x}$ that is investigated. We may also think of $p_g(\mathbf{x})$ as an exponential approximation of the scale-space density difference using interesting

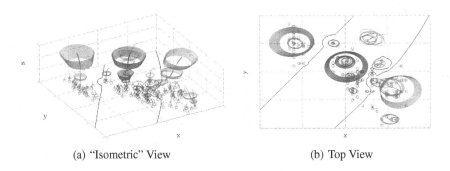

(a) "Isometric" View   (b) Top View

Fig. 3. Feature space decision boundaries (black plane curves) obtained from group likelihood criterion for the two-dimensional two-group problem of Fig. 1 using c_γ^{det} for spot detection; Red squares and blue circles visualize the samples of each group; Red/blue paths outline the dendrogram of scale-space density difference optima for the red/blue group dominating the other group; Interesting spots of each dendrogram are printed thick; Red/blue ellipses characterize the shape for each of the interest spots (Color figure online).

spots only. Given this, our decision rule simply takes the group that maximizes the group likelihood for the location of interest $\mathbf{x}$. Figure 3 illustrate the decision boundary obtained from this rule.

5.2   Data Sets

We carried out our experiments on three classification data sets taken from UCI Machine Learning Repository. A brief summary of them is given in Table 1. In the first task, we distinguish between benign and malign breast cancer based on manually graded cytological charateristics, cf. [22]. In the second task, we distinguish between genuine and forged money based on wavelet-transform-derived features from photographs of banknote-like specimen, cf. [9]. In the third task, we differentiate among normal, spondylolisthetic and disc-herniated vertebral columns based on biomechanical attributes derived from shape and orientation of the pelvis and the lumbar vertebral column, cf. [4].

Table 1. Data sets from UCI Machine Learning Repository.

	Breast cancer	Banknote authentication	Vertebral column
Groups	benign/malign	genuine/forged	normal/spondylo./herniated
Samples	444/239	762/610	100/150/60
Dimensions	10	4	6

5.3   Parameter Investigation

Before detailing classification results, we investigate two aspects of our approach. Firstly, we inspect the importance of bandwidth selection, benchmarking no kernel density bandwidth against the least-squares cross-validation technique that we use. Secondly, we determine the influence of the scale sampling rate. For the latter we space $n + 1$ scales for various n equidistantly from zero to

$$t_n = F_{\chi^2}^{-1}(1 - \epsilon|d)\max_g \left(\sqrt[d]{\det\left(\Sigma_g\right)} \right), \tag{35}$$

where $F_{\chi^2}^{-1}(\,\cdot\,|d)$ is the cumulative inverse-χ^2 distribution with d degrees of freedom and Σ_g is the covariance matrix of group g. Intuitively, t_n captures the extent of the group with largest variance up to a small ϵ, i.e. here $1.5 \cdot 10^{-8}$.

To investigate the two aspects, we compare classification accuracies with and without bandwidth selection as well as sampling rates ranging from $n = 100$ to $n = 300$ in steps of 25. From the results, which are given in Table 2, we observe that bandwidth selection is almost neglectable for the Breast Cancer (BC) and the Banknote Authentication (BA) data set no matter which criterion is used for spot detection. However, the impact is substantial throughout all scale sampling rates for the Vertebral Column (VC) data set for both criteria. This may be due to the comparably small number of samples per group for this data set.

Table 2. Classification accuracy of our decision rule in $\lfloor\%\rfloor$ for data sets of Table 1 for both detectors with/without bandwidth selection.

c_γ^{det}-based decision rule	Scale sampling rate n								
	100	125	150	175	200	225	250	270	300
Breast Cancer	65/65	97/97	97/97	95/95	97/97	95/95	97/97	96/96	97/97
Banknote Authentication	96/94	96/96	96/96	98/98	98/98	98/98	98/98	98/98	99/99
Vertebral Column	87/82	88/83	88/84	88/83	88/85	88/85	88/86	88/86	88/87
c_γ^{tr}-based decision rule	Scale sampling rate n								
	100	125	150	175	200	225	250	270	300
Breast Cancer	96/96	97/97	96/96	96/96	96/96	96/96	96/96	96/96	96/96
Banknote Authentication	95/95	97/97	98/98	97/97	98/98	98/98	98/98	98/99	99/99
Vertebral Column	89/81	89/80	89/80	89/83	89/83	89/84	89/83	89/84	89/83

Regarding the second aspect, we observe that for both criteria the BA and VC data set classification accuracy increases only slightly when the scale sampling rate rises. On the BC data set, accuracy remains stable, except for the lower rates when c_γ^{det} is used for spot detection. There is no such drop for the c_γ^{tr}-derived results, indicating a higher sensitivity of the latter for sparser samplings. Apart from that, the differences between the results of both criteria are minor for all data sets and sampling rates. From the results we conclude that bandwidth selection is a necessary part for interest spot detection. We further recommend $n \geq 200$, because accuracy is saturated at this point for all data sets independently of the choice of the spot detection criterion. For the remaining experiments we used bandwidth selection and a sampling rate of $n = 200$.

5.4 Classification Results

A comparison of classification accuracies of our decision rule against the afore-mentioned classifiers is given in Table 3. For the BC data set we observe that except for the support vector machine (SVM) with cubic kernel all approaches were highly accurate, scoring between 94 % and 97 % with our c_γ^{det}-based decision rule being topmost and the c_γ^{tr}-derived results being only slightly worse. Even more similar to each other are results for the BA data set, where all approaches score between 97 % and 99 %, with ours lying in the middle of this range. Results are most diverse for the VC data sets. Here, the SVM with cubic kernel again performs significantly worse than the rest, which all score between 80 % and 85 %, while our c_γ^{det}/c_γ^{tr}-based decision rules peak at 88 % and 89 % respectively. Other research showed similar scores on the given data sets. For example the artificial neural networks based on pareto-differential evolution in [1] obtained 98 % accuracy for the BC data set, while [18] achieved 83 % to 85 % accuracy on the VC data set with SVMs with different kernels. These results suggest that our interest points carry information about a data set that are similarly important than the information carried by the well-established classifiers.

Table 3. Classification accuracies of different classifiers in $\lfloor\%\rfloor$ for data sets of Table 1.

	Breast cancer	Banknote authen.	Verteral column
decision tree	94	98	82
k-nearest neighbors	97	99	80
Fisher's discriminant	96	97	80
linear/cubic kernel SVM	96/90	99/98	85/74
$c_\gamma^{\mathrm{det}}/c_\gamma^{\mathrm{tr}}$-based decision rule	97/96	98/98	88/89

Confusion tables for our approach are given in Table 4 for all data sets. As can be seen, our $c_\gamma^{\mathrm{det}}/c_\gamma^{\mathrm{tr}}$-based decision rules gave balanced inter-group results for the BC and the BA data set. We obtained only small inaccuracies for the recall of the benign (96 %/96 %) and genuine (97 %/96 %) groups as well as for the precision of the malign (94 %/93 %) and forged (96 %/95 %) groups. Results for the VC data set were more diverse. Here, a number of samples with disc herniation were mistaken for being normal, lowering the recall of the herniated group (86 %/86 %) noticeably. However, more severe inter-group imbalances were caused by the normal samples, which were relatively often mistaken for being spondylolisthetic or herniated discs. Thus, recall for the normal group (76 %/80 %) and precision for the herniated group (74 %/76 %) decreased significantly. The latter is to some degree caused by a handful of strong outliers from the normal group that fall into either of the other groups, which can already be seen from the group likelihood plots in Fig. 4. This finding was made by others as well, cf. [17].

The other classifiers performed similarly balanced on the BA and BC data set. Major differences occured on the VC data set only. A precision/recall comparison of all classifiers on the VC data set is given in Table 5. We observe that the

Table 4. Confusion table for predicted/actual groups of our $c_\gamma^{\mathrm{det}}/c_\gamma^{\mathrm{tr}}$-based decision rule for data sets of Table 1

(a) Breast Cancer

pred. \ act.	benign	malign	precision
benign	429/429	4/6	99/98
malign	15/15	235/233	94/93
recall	96/96	98/97	$\lfloor\%\rfloor$

(b) Banknote Authentication

pred. \ act.	genuine	forged	precision
genuine	742/736	0/0	100/100
forged	20/26	610/610	96/95
recall	97/96	100/100	$\lfloor\%\rfloor$

(c) Vertebral Column

pred. \ act.	normal	spondylo.	herniated	precision
normal	76/80	1/1	6/7	91/90
spondylo.	10/8	145/145	2/1	92/94
herniated	14/12	4/4	52/52	74/76
recall	76/80	96/96	86/86	$\lfloor\%\rfloor$

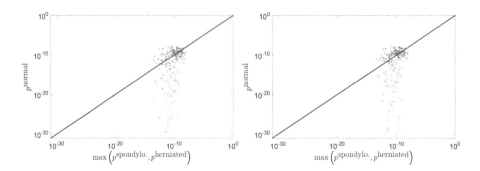

Fig. 4. Sample group likelihoods and decision boundary (black diagonal line) for the Vertebral Column data set of Table 1 using c_γ^{det} (left) and c_γ^{tr} (right) for spot detection; Normal, spondylolisthetic and herniated discs in blue, magenta and red, respectively (Color figure online).

precision of the normal and the herniated group are significantly lower (gap > 12 %) than that of the spondylolisthetic group for all classifiers except for our decision rules, for which at least the normal group is predicted with a similar precision by both rules. Regarding the recall we note an even more unbalanced behavior. Here, a strict ordering from spondylolisthetic over normal to herniated disks occurs. The differences of the recall of spondylolisthetic and normal are significant (gap > 16 %) and those between normal and herniated are even larger (gap > 18 %) among all classifiers that we compared against. The recalls for our decision rules are distributed differently, ordering the herniated before the normal group. Also the magnitude of differences is less significant (gaps ≈ 10 %/6 %) for both decision rules. Results of the comparison indicate that the information that is carried by our interest points tends to be more balanced among groups than the information carried by the well-established classifiers that we compared against. The final question which interest spot detection criterion (c_γ^{det} or c_γ^{tr}) should be recommended cannot be answered satisfactorily based solely on our evaluation, because result differ only insignificantly. Yet, we advocate c_γ^{det} since it has been shown to provide better scale selection properties under affine transformation of the feature space, cf. [13].

Table 5. Classification precision and recall of different classifiers in ⌊%⌋ for the Vertebral Column data set of Table 1.

	Normal group		Spondylo. group		Herniated group	
	Precision	Recall	Precision	Recall	Precision	Recall
decision tree	69	83	97	95	68	50
k-nearest neighbors	70	74	96	96	58	55
Fisher's discriminant	70	80	87	92	74	48
linear/cubic kernel SVM	76/59	85/82	97/90	96/91	72/52	61/18
$c_\gamma^{\mathrm{det}}/c_\gamma^{\mathrm{tr}}$-based decision rule	91/90	76/80	92/94	96/96	74/76	86/86

6   Conclusion

We proposed a detection framework that is able to identify differences among the sample distributions of different observations. Potential applications are manifold, touching fields such as medicine, biology, chemistry and physics. Our approach bases on the density function difference of the observations in feature space, seeking to identify spots where one observation dominates the other. Superimposing a scale-space framework to the density difference, we are able to detect interest spots of various locations, size and shapes in an efficient manner.

Our framework is intended for semi-automatic processing, providing human-interpretable interest spots for further investigation of some kind. We outlined how the output of our framework can be used to guide exploratory visualization of high-dimensional feature spaces as an intermediate step prior to other means of data analysis. Furthermore, we showed that the detected interest spots carry valuable information about a data set on a number of classification tasks from the UCI Machine Learning Repository. To this end, we established a simple decision rule on top of our framework. Results indicate state-of-the-art performance of our approach, which underpins the importance of the information that is carried by the detected interest spots.

In the future, we plan to extend our work to support repetitive features such as angles, which currently is a limitation of our approach. Modifying our notion of distance, we would then be able to cope with problems defined on, e.g. a sphere or torus. Future work may also include the migration of other types of scale-space detectors to density difference problems. This includes the notion of ridges, valleys and zero-crossings, leading to richer sources of information.

Acknowledgements. This research was partially funded by the project "Visual Analytics in Public Health" (TO 166/13-2) of the German Research Foundation.

References

1. Abbass, H.A.: An evolutionary artificial neural networks approach for breast cancer diagnosis. Artif. Intell. Med. **25**, 265–281 (2002)
2. Adal, K.M., Sidibe, D., Ali, S., Chaum, E., Karnowski, T.P., Meriaudeau, F.: Automated detection of microaneurysms using scale-adapted blob analysis and semi-supervised learning. Comput. Methods Programs Biomed. **114**, 1–10 (2014)
3. Babaud, J., Witkin, A.P., Baudin, M., Duda, R.O.: Uniqueness of the gaussian kernel for scale-space filtering. IEEE Trans. Pattern Anal. Mach. Intell. **8**, 26–33 (1986)
4. Berthonnaud, E., Dimnet, J., Roussouly, P., Labelle, H.: Analysis of the sagittal balance of the spine and pelvis using shape and orientation parameters. J. Spinal Disord. Tech. **18**, 40–47 (2005)
5. Bretzner, L., Lindeberg, T.: Feature tracking with automatic selection of spatial scales. Comput. Vis. Image Underst. **71**, 385–392 (1998)
6. Duong, T., Hazelton, M.L.: Plug-in bandwidth matrices for bivariate kernel density estimation. J. Nonparametric Stat. **15**, 17–30 (2003)

7. Duong, T., Hazelton, M.L.: Cross-validation bandwidth matrices for multivariate kernel density estimation. Scand. J. Stat. **32**, 485–506 (2005)
8. Felzenszwalb, P.F., Girshick, R.B., McAllester, D., Ramanan, D.: Object detection with discriminatively trained part-based models. IEEE Trans. Pattern Anal. Mach. Intell. **32**, 1627–1645 (2010)
9. Glock, S., Gillich, E., Schaede, J., Lohweg, V.: Feature extraction algorithm for banknote textures based on incomplete shift invariant wavelet packet transform. In: Denzler, J., Notni, G., Süße, H. (eds.) Pattern Recognition. LNCS, vol. 5748, pp. 422–431. Springer, Heidelberg (2009)
10. Lakemond, R., Sridharan, S., Fookes, C.: Hessian-based affine adaptation of salient local image features. J. Math. Imaging Vis. **44**, 150–167 (2012)
11. Lehmann, D.J., Theisel, H.: Orthographic star coordinates. IEEE Trans. Vis. Comput. Graph. **19**, 2615–2624 (2013)
12. Lindeberg, T.: Edge detection and ridge detection with automatic scale selection. In: IEEE Computer Society Conference on Computer Vision and Pattern Recognition, pp. 465–470 (1996)
13. Lindeberg, T.: Feature detection with automatic scale selection. Int. J. Comput. Vis. **30**, 79–116 (1998)
14. Lindeberg, T., Garding, J.: Shape-adapted smoothing in estimation of 3-d depth cues from affine distortions of local 2-d brightness structure. Eur. Conf. Comput. Vis. **800**, 389–400 (1994)
15. Mikolajczyk, K., Schmid, C.: Scale & affine invariant interest point detectors. Int. J. Comput. Vis. **60**, 63–86 (2004)
16. Rak, M., König, T., Tönnies, K.D.: Spotting differences among observations. In: International Conference on Pattern Recognition Applications and Methods, pp. 5–13 (2015)
17. da Rocha Neto, A.R.: A.R., Barreto, G.A.: On the application of ensembles of classifiers to the diagnosis of pathologies of the vertebral column: A comparative analysis. IEEE Lat. Am. Trans. **7**, 487–496 (2009)
18. da Rocha Neto, A.R., Sousa, R., de A. Barreto, G., Cardoso, J.S.: Diagnostic of pathology on the vertebral column with embedded reject option. In: Vitrià, J., Sanches, J.M., Hernández, M. (eds.) IbPRIA 2011. LNCS, vol. 6669, pp. 588–595. Springer, Heidelberg (2011)
19. Sain, S.R., Baggerly, K.A., Scott, D.W.: Cross-validation of multivariate densities. J. Am. Stat. Assoc. **89**, 807–817 (1992)
20. Seyedhosseini, M., Kumar, R., Jurrus, E., Giuly, R., Ellisman, M., Pfister, H., Tasdizen, T.: Detection of neuron membranes in electron microscopy images using multi-scale context and Radon-like features. Int. Conf. Med. Image Comput. Comput. Assist. Intervention **6891**, 670–677 (2011)
21. Wand, M.P., Jones, M.C.: Multivariate plug-in bandwidth selection. Comput. Stat. **9**, 97–116 (1994)
22. Wolberg, W., Mangasarian, O.: Multisurface method of pattern separation for medical diagnosis applied to breast cytology. In: National Academy of Sciences, pp. 9193–9196 (1990)
23. Yi, C., Tian, Y.: Text string detection from natural scenes by structure-based partition and grouping. IEEE Trans. Image Process. **20**, 2594–2605 (2011)
24. Yuille, A.L., Poggio, T.A.: Scaling theorems for zero crossings. IEEE Trans. Pattern Anal. Mach. Intell. **8**, 15–25 (1986)

Identifying and Mitigating Labelling Errors in Active Learning

Mohamed-Rafik Bouguelia$^{(\boxtimes)}$, Yolande Belaïd, and Abdel Belaïd

Université de Lorraine - LORIA, UMR 7503, 54506 Vandoeuvre-les-Nancy, France
{mohamed.bouguelia,yolande.belaid,abdel.belaid}@loria.fr

Abstract. Most existing active learning methods for classification, assume that the observed labels (i.e. given by a human labeller) are perfectly correct. However, in real world applications, the labeller is usually subject to labelling errors that reduce the classification accuracy of the learned model. In this paper, we address this issue for active learning in the streaming setting and we try to answer the following questions: (1) which labelled instances are most likely to be mislabelled? (2) is it always good to abstain from learning when data is suspected to be mislabelled? (3) which mislabelled instances require relabelling? We propose a hybrid active learning strategy based on two measures. The first measure allows to filter the potentially mislabelled instances, based on the degree of disagreement among the manually given label and the predicted class label. The second measure allows to select (for relabelling) only the most informative instances that deserve to be corrected. An instance is worth relabelling if it shows highly conflicting information among the predicted and the queried labels. Experiments on several real world data show that filtering mislabelled instances according to the first measure and relabelling few instances selected according to the second measure, greatly improves the classification accuracy of the stream-based active learning.

Keywords: Label noise · Active learning · Classification · Data stream

1 Introduction

In usual supervised learning methods, a classification model is built by performing several passes over a static dataset with sufficiently many labelled data. Firstly, this is not possible in the case of data streams where data is massively and continuously arriving from an infinite-length stream. Secondly, manual labelling is costly and time consuming. Active learning reduces the manual labelling cost, by querying from a human labeller only the class labels of data which are informative (usually uncertain instances). Active learning methods [5] are convenient for data stream classification. Several active learning methods [2–5] and stream-based active learning methods [1,6,7] have been proposed. Most of these methods assume that the queried label is perfect. However, in real world scenarios this

© Springer International Publishing Switzerland 2015
A. Fred et al. (Eds.): ICPRAM 2015, LNCS 9493, pp. 35–51, 2015.
DOI: 10.1007/978-3-319-27677-9_3

assumption is often not satisfied. Indeed, it is difficult to obtain completely reliable labels, because the labeller is prone to mislabelling errors. Mislabelling may occur for several reasons: inattention or accidental labelling errors, uncertain labelling knowledge, subjectivity of classes, etc.

Usually, the active learner queries labels of instances that are uncertain. These instances are likely to improve the classification model if we assume that their queried class label is correct. Under such assumption, the active learner aims to search for instances that reduce its uncertainty. However, when the labeller is noisy, mislabelling errors cause the learner to incorrectly focus the search on poor regions of the feature space. This represents an additional difficulty for active learning to reduce the label complexity [8]. If the potential labelling errors are not detected and mitigated, the active learner can easily fail to converge to a good model. Therefore, label noise is harmful for active learning and dealing with it is an important issue.

Detecting label noise is not trivial in stream-based active learning, mainly for two reasons. Firstly, in a data stream setting, the decision to filter or not a potentially mislabelled instance should be taken immediately. Secondly, because the learning is active, the mislabelled instances are necessarily among those that the classifier is uncertain about their class.

Usual methods to deal with label noise like those surveyed in [9], assume that a static dataset is available beforehand and try to clean it before training occurs by repeatedly removing the most likely mislabelled instances among all instances of the dataset. A method proposed in [10] is designed for cleansing noisy data streams by removing the potentially mislabelled instances from a stream of labelled data. However, the method does not consider an active learning setting where the mislabelling errors concern uncertain instances. Moreover, they divide the data stream into large chunks and try to detect mislabelled instances in each chunk, which makes the method partially online and reduces the importance of its streaming nature. The method in [11] considers an active label correction, but the learning itself is not active. Rather, the method iteratively selects the top k likely mislabelled instances from a labelled dataset and presents them to an expert for correction rather than discarding them. Some other methods like [12,13] are designed for active learning with label noise and are intended only for label noise whose source is the uncertain labelling knowledge of the labeller. Generally speaking, they try to model the knowledge of the labeller and avoid asking for the label of an instance if it belongs to the uncertain knowledge set of the labeller. However, this may lead to discarding many informative data. Moreover, the method implicitly assumes that the labeller is always the same (since his knowledge is modelled). Methods like [14,15] can be applied to active learning but they try to mitigate the effect of label noise differently: rather than trying to detect the possibly mislabelled instances, they repeatedly ask for the label of an instance from noisy labellers using crowd-sourcing techniques [16]. However, all these methods require multiple labellers that can provide redundant labels for each queried instance and are not intended to be used with a single alternative labeller.

The method we propose is different. We consider a stream-based active learning with label noise. The main question we address is whether some possibly mislabelled instances among the queried ones are worth relabelling more than others. A potentially mislabelled instance is filtered as soon as it is received according to a mislabelling likelihood. An alternative expert labeller can be used to correct the filtered instance. The method is able to select (for relabelling) only those instances that deserve correction according to an informativeness measure.

This paper is organized as follows. In Sect. 2 we give background on stream-based active learning with uncertainty and its sensibility to label noise. In Sect. 3 we firstly propose two measures to characterize the mislabelled instances. Then, we derives an informativeness measure that determines to which extent a possibly mislabelled instance would be useful if corrected. In Sect. 4 we present different strategies to mitigate label noise using the proposed measures. In Sect. 5 we present the experiments. Finally, we conclude and present some future work in Sect. 6.

2   Background

2.1   Active Learning with Uncertainty

Let X be the input space of instances and Y the output space. We consider a stream-based active learning where at each time step t, the learner receives a new unlabelled instance $x_t \in X$ from an infinite-length data stream and has to make the decision (at time t) of whether or not to query the corresponding class label $y_t \in Y$ from a labeller. Each $x \in X$ is presented in a p-dimensional space as a feature vector $x \overset{\text{def}}{=} (x_{f_1}, x_{f_2}, ..., x_{f_p})$, where $x_{f_i} \in \mathbb{R}$ and f_i is the i'th feature.

If the label y_t of x_t is queried, the labelled instance (x_t, y_t) is used to update a classification model h. Otherwise, the classifier outputs the predicted label $y_t = h(x_t)$. In this way, the goal is to learn an efficient classification model $h : X \to Y$ using a minimal number of queried labels. In order to decide whether or not to query the label of an instance, many active learning strategies have been studied [5]. The most common ones are the uncertainty sampling based strategies. The instances that are selected for manual labelling are typically those for which the model h is uncertain about their class. If an uncertain instance x is labelled manually and correctly, two objectives are met: (i) the classifier avoids to output a probable prediction error, and (ii) knowing the true class label of

---

Algorithm 1. Stream-based active learning (δ).

1: **Input:** uncertainty threshold δ, underlying classification model h
2: **for** each new data point x from the stream **do**
3:     **if** the uncertainty $\Delta_x^h > \delta$ **then**
4:         query y the label of x from a labeller
5:         train h using (x, y)
6:     **end if**
7:     **else** output the predicted label $y = h(x)$
8: **end for**

---

x would be useful to improve h and reduce its overall uncertainty (x is said to be informative). A simple uncertainty measure that selects instances with a low prediction probability can be defined as $\Delta_x^h = 1 - \max_{y \in Y} P_h(y|x)$, where $P_h(y|x)$ is the probability that x belongs to class y. A general stream-based active learning process is described by Algorithm 1. Any base classifier can be used to learn the model h. The algorithm queries labels of instances with an uncertainty beyond a given threshold δ.

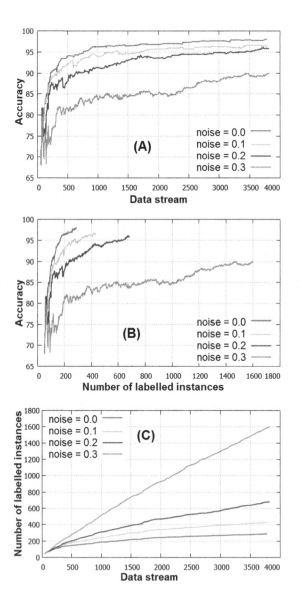

Fig. 1. A stream-based active learning with different levels of label noise (σ). The data set used is optdigits (UCI repository). SVM is used as a base classifier.

2.2   Impact of Label Noise

As mentioned previously, we are not always guaranteed to obtain a perfectly reliable label when querying it from a human labeller. We consider a random label noise process where the noisy labeller has a probability σ for giving a wrong answer and $1 - \sigma$ for giving the correct answer, each time a label is queried.

Figure 1 shows the results obtained using Algorithm 1 in the presence of label noise with different intensities σ and compared to the noise-free setting $\sigma = 0$. Figure 1(A) shows the accuracy of the model h on a test set, according to the number of instances from the stream. As for the usual supervised learning, it is not surprising to see that in active learning, label noise also reduces the overall classification accuracy. Figure 1(B) shows the accuracy according to the number of queried labels (manually labelled instances). We can see that in addition to achieving a lower accuracy, more label noise also causes the active learner to make more queries. This is confirmed in Fig. 1(C) that shows the number of instances whose label is queried, according to the number of instances seen from the stream. This is explained by the fact that the most uncertain instance can be informative if we obtain its true class label, but may easily become the most misleading one if it is mislabelled. Therefore, mislabelled instances causes the active learner to incorrectly focus the query on poor regions of the feature space, and deviates it from querying the truly informative instances.

In summary, stream-based active learning is very sensitive to label noise since it not only impacts the predictive capabilities of the learned model but also leads it to query labels of instances which are not necessarily informative. This results in more queried instances and represents a bottleneck for minimizing the label complexity of active learning [8].

3   Characterizing Mislabelled Instances

In this section we propose measures for characterizing mislabelled instances and their importance. First, we present in Sect. 3.1 a disagreement coefficient that reflects the mislabelling likelihood of instances. Then, we derive in Sect. 3.2 an informativeness measure that reflects the importance of the instance, which is later used to decide if the instance merits to be relabelled.

Let x be a data point whose label is queried. The class label given by the labeller is noted y_g. Let $y_p = \underset{y \in Y}{argmax}\, P_h(y|x)$ be the class label of x which is predicted by the classifier. If $y_p = y_g$ then the label given by the labeller is trusted and we consider that it is not a mislabelling error. Otherwise, a mislabelling error may have occurred.

3.1   Mislabelling Likelihood

Assume that $y_p \neq y_g$. We express how likely x is mislabelled by estimating the degree of disagreement among the predicted class y_p and the observed class y_g, which is proportional to the difference in probabilities of y_p and y_g.

Let $p_p = P(y_p|x)$ and $p_g = P(y_g|x)$. As in the silhouette coefficient [17] and given that $p_p \geq p_g$, we define the degree of disagreement $D_1(x) \in [0,1]$ as:

$$D_1(x) = \frac{p_p - p_g}{\max(p_p, p_g)} = \frac{p_p - p_g}{p_p} = 1 - \frac{P(y_g|x)}{P(y_p|x)}$$

The higher the value of $D_1(x)$, the more likely that x has been incorrectly labelled with y_g, because the probability that x belongs to y_g would be small relatively to y_p.

Inspired by multi-view learning [18], we present a second measure to estimate the degree of disagreement. In multi-view learning, classifiers are learned on different views of data using different feature subsets. Data points of one view (using a feature f_1) are scattered differently in a second view (using a different feature f_2). Therefore, it is possible for some instance x that we are uncertain if its label is y_p or y_g in one view, to be less uncertain in another view.

Let us take features separately. Each feature value f_i of an instance x has a contribution $q_y^{f_i}$ for classifying x into a class y. As an analogy, a textual document contains terms (features) that attracts towards a given class more strongly than another one. $q_y^{f_i}$ can be considered as any score that shows how much the feature value f_i attracts x towards class y. For example, let x_{f_i} be the instance x restricted to the feature f_i. Let $d_y^{f_i}$ be the mean distance from x_{f_i} to its k nearest neighbours belonging to class y, restricted to feature f_i. Then, $q_y^{f_i}$ can be defined as inversely proportional to the distance $d_y^{f_i}$, e.g. $q_y^{f_i} = \frac{1}{d_y^{f_i}}$.

Considering the predicted class y_p and the given class y_g, $q_{y_p}^{f_i}$ and $q_{y_g}^{f_i}$ represent how much the feature f_i is likely to contribute at classifying x into y_p and y_g respectively. Let F_p be the set of features that contributes at classifying x in the predicted class more than the given class, and inversely for F_g:

$$F_p = \{f_i | q_{y_p}^{f_i} > q_{y_g}^{f_i}\} \qquad F_g = \{f_i | q_{y_p}^{f_i} \leq q_{y_g}^{f_i}\}$$

The amount of information reflecting the membership of x to y_p (resp. y_g) is:

$$q_p = \sum_{f_i \in F_p} (q_{y_p}^{f_i} - q_{y_g}^{f_i}) \qquad q_g = \sum_{f_i \in F_g} (q_{y_g}^{f_i} - q_{y_p}^{f_i})$$

Note that $q_p \in [0, +\infty)$ and $q_g \in [0, +\infty)$. Again, by applying the silhouette coefficient, a degree of disagreement $D_2'(x) \in [-1, 1]$ among y_p and y_g can be expressed as:

$$D_2'(x) = \frac{q_p - q_g}{\max(q_p, q_g)}$$

Note that D_2' can be normalized to be in $[0,1]$ rather than $[-1,1]$ simply as $D_2 = \frac{D_2'+1}{2}$.

Instances distributed according to D_1 and D_2 are shown on Fig. 2. A final mislabelling score D can be expressed either by D_1 or D_2 or by using possible combinations of both including: the average $\frac{D_1+D_2}{2}$, $\max(D_1, D_2)$, or $\min(D_1, D_2)$. In order to decide whether an instance x is potentially mislabelled, a usual way is to define a threshold (that we denote t_D) on D.

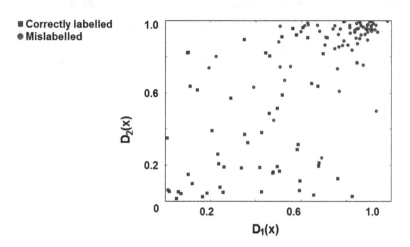

Fig. 2. Mislabelled and correctly labelled instances distributed according to D_1 and D_2.

In summary, the presented disagreement measure only expresses how likely x has been incorrectly labelled. Strong information reflecting the predicted class label and low information reflecting the given class label, indicates a mislabelling error. For example, most terms in a textual document strongly attract towards a class, but the other words weakly attracts towards the class given by the labeller. Nonetheless, the mislabelling score does not give information about the importance of an instance or how much its queried label deserves to be reviewed. This is discussed in the next section.

3.2   Informativeness of Possibly Mislabelled Instances

In active learning with uncertainty sampling, instances for which the model is uncertain on how to label, are designated as informative and their label is queried. In this section we are not referring to informativeness in terms of uncertainty (the considered instances are already uncertain). Rather, we are trying to determine to which extent a possibly mislabelled instance would be useful if corrected.

It is possible for the mislabelling likelihood using D_1 and/or D_2 to be uncertain if an instance x is mislabelled or not. This appears on Fig. 2 as the overlapped region of the mislabelled and the correctly labelled instances. This happens essentially in the presence of either strong or weak conflicting informations in x with respect to y_p and y_g, which leads to $P(y_p|x) \simeq P(y_g|x)$ and $q_p \simeq q_g$. Let us again consider the example of a textual document:

- *Strongly conflicting information*: some terms strongly attract the document towards y_p and other terms attract it with the same strength towards y_g. In this case q_p and q_g are both high and close to each other.
- *Weakly conflicting information*: terms equally but weakly attract the document towards y_p and y_g, that is, there is no persuasive information for y_p or y_g. In this case q_p and q_g are both low and close to each other.

In both the above cases $q_p - q_g$ would be low. However, instances showing strongly conflicting information are more informative if their true class label is available, and deserve to be reviewed and corrected more than the other instances. Therefore, in addition to the mislabelling likelihood (Sect. 3.1), we define the informativeness measure $I \in [0, +\infty)$ as

$$I = \sqrt{q_p \times q_g}$$

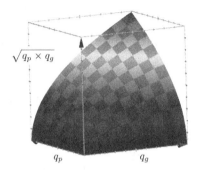

Fig. 3. $\sqrt{q_p \times q_g}$ is high when both q_{y_p} and q_{y_g} are high and close to each other.

The higher I, the more its queried label deserves to be reviewed (and eventually corrected). Figure 3 justifies the choice of $I = \sqrt{q_p \times q_g}$ since it gives a high value when both q_{y_p} and q_{y_g} are high and close to each other. The measure I is unbounded but can be normalized, for example by dividing it on a maximum value of I which can be computed on a validation set. This way, a threshold (that we denote t_I) can be defined on the informativeness measure I.

4   Mitigating Label Noise

When an instance x is detected as potentially mislabelled, the next step is to decide what to do with this instance. In usual label noise cleansing methods, the dataset is available beforehand and the methods just discard or remove the instances that are likely to be mislabelled or predict their labels. In order to mitigate the impact of label noise on the stream-based active learning, we study three strategies including discarding, weighting and relabelling by an expert labeller, and we show a hybrid approach.

4.1   Discard, Weight and Relabel

What not to do. In a stream-based active learning, correcting a mislabelled instance by predicting its labels is not the right way to go. Indeed, updating the model h using the predicted label of x (i.e. $y = h(x)$) rather than the queried label, is usually more harmful to active learning than the label noise itself. This is due to the fact that the mislabelled instances are those instances for which the model h was primarily uncertain how to classify. Therefore, predicting their labels will more likely result in an error that the model would be unable to detect (otherwise it would have avoided that error).

Discarding. When $D(x) > t_D$, x is considered as mislabelled, otherwise it is considered as correctly labelled. This way, if x is identified as being mislabelled, it can be just discarded, that is, we do not update the classification model h with x.

Weighting. Depending on the base classifier, the instances can be weighted so that the classifier learns more from instances with a higher weight. Therefore, a possible alternative to mitigate label noise without defining a threshold on D, is to update the model h using every instance x with its queried label y_g weighted by $w = 1 - D(x)$ which is inversely proportional to its mislabelling likelihood D. Indeed, instances with a high mislabelling likelihood will have a weight closer to 0 and will not affect the classification model h too much, unlike an instance with a low mislabelling likelihood.

Relabelling. If an alternative reliable labeller is available, the label of the potentially mislabelled instance (having a mislabelling likelihood $D > t_D$) can be verified and eventually corrected. Then, the model h is updated using the instance and its corrected label.

   Figure 4 shows the results obtained using the different strategies. It is obvious that the best alternative is to relabel correctly the instances that are identified as mislabelled. However, this is done under a cost by an expert labeller which is assumed to be reliable. Discarding the possible mislabelled instances may improve the classification accuracy. However, informative instances that were correctly labelled may be lost, especially if many instances are wrongly discarded (depending on threshold t_D). Rather than discarding possible mislabelled instances, weighting all the instances with an importance which is inversely proportional to their mislabelling likelihood may improve the performances of the active learner without loosing informative instances (at the risk of underweighting some informative instances). Finally, Fig. 4 confirms that it is very harmful to predict the label of a filtered instance x and updating the model using the instance x with its predicted label. Indeed, some dataset cleansing methods for supervised learning propose to predict the label of the potentially mislabelled instances. However, in an active learning configuration, this becomes very harmful, because the queried labels are precisely those of uncertain instances (about which the classifier is uncertain about its prediction).

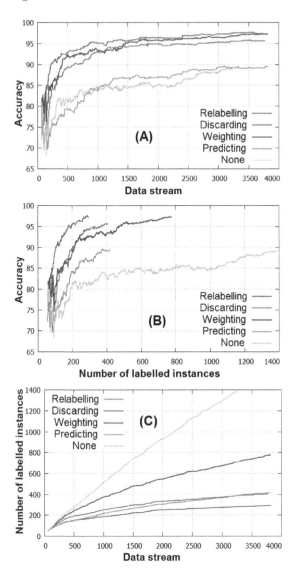

Fig. 4. Active learning from a data stream with label noise intensity $\sigma = 0.3$ and $t_D = 0.6$. The dataset used is optdigits (UCI repository).

Hybrid Strategy. Correcting mislabelled instances using their true class label gives the best results. However, this requires an expert labeller which implies a high cost since it is assumed to be a reliable labeller. We present a hybrid approach that minimizes the cost required by using the alternative expert labeller. Since relabelling is costly, we assume that we have a limited budget B for relabelling, that is, the expert can review and relabel no more than B instances. Given the budget B, the problem can be stated as which instances are worth to be relabelled. Actually, relabelling instances that are informative according to the measure I (see Sect. 3.2) is more likely to improve the classification accuracy. Therefore, if an instance x is identified as being mislabelled and has a high informativeness $I(x) > t_I$, then it is relabelled. Otherwise, either the discarding or the weighting strategy can be used.

5   Experiments

We use for our experiments different public datasets obtained from the UCI machine learning repository[1]. We also consider a real administrative documents dataset provided by the ITESOFT[2] company. Each document was processed by an OCR and represented as a bag-of-words which is a sparse feature-vector containing the occurrence counts of words in the document. Without loss of generality, the label noise intensity is set to $\sigma = 0.3$. An SVM is used as a base classifier (we use the python implementation available on scikit-learn [21]). Threshold t_D is fixed to 0.6 for all datasets to allow reproducibility of the experiments, although a better threshold can be found for each dataset.

5.1   Mislabelling Likelihood

This experiment figures-out the ability of the proposed mislabelling likelihood to correctly characterize mislabelled instances as proposed in Sect. 3.1, without considering a stream-based active learning. We corrupt the labels of each dataset such that instances with a low prediction probability are more likely to be mislabelled. Then, instances of each dataset are ranked according to degrees of disagreement defined in terms of D_1 and D_2. Results are compared with an entropy measure E which is commonly used in active learning. Instances with a low entropy implies a confident classification, thus, instances for which the classifier disagrees with their queried label are more likely to be mislabelled when they have a low entropy $E = - \sum_{y \in Y} P_h(y|x) \times \log P_h(y|x)$

We select the top n ranked instances of each dataset and we compute 2 types of errors: e_1 represents the correctly labelled instances that are erroneously selected, and e_2 the mislabelled instances that are not selected:

$$e_1 = \frac{\text{correct\_selected}}{\text{correct}} \qquad e_2 = \frac{\text{mislabelled\_unselected}}{\text{mislabelled}}$$

where "correct_selected" is the number of correctly labelled instances that are selected as being potentially mislabelled. "mislabelled_unselected" is the number of mislabelled instances that are not selected. "correct" and "mislabelled" are the total number of correctly labelled and mislabelled instances respectively.

---

[1] http://archive.ics.uci.edu/ml/.
[2] http://www.itesoft.com/.

We also compute the percentage of selected instances that are actually mis-labelled *prec*, which represents the noise detection precision:

$$prec = \frac{\text{mislabelled\_selected}}{\text{selected}}$$

where "mislabelled_selected" is the number of mislabelled instances that are selected, and "selected" is the number of selected instances.

Table 1 shows the obtained results. We can see that for *documents* and *pendigits* datasets, D_2 achieves better results than D_1 and inversely for *opt-digits* and *letter-recognition* datasets. However combining D_1 and D_2 using $D = \max(D_1, D_2)$ may yield better results than both D_1 and D_2, and always achieves better results than the entropy measure E. This is confirmed by the average results obtained over all the datasets. For the reminder of the experiments we use the D which represents a convenient disagreement measure for almost all datasets.

Table 1. Mislabelling likelihood measures.

	D_1	D_2	D	E
Optdigits dataset				
e_1	**0.67**%	1.49%	1.12%	1.27%
e_2	**1.22**%	3.14%	2.26%	2.61%
prec	**98.44**%	96.53%	97.4%	97.05%
Documents dataset				
e_1	3.51	3.29%	**2.96**%	4.50%
e_2	3.07	2.56%	**1.79**%	5.38%
prec	92.2	92.69%	**93.42**%	90.0%
Pendigits dataset				
e_1	2.34%	2.19%	**2.07**%	2.61%
e_2	1.37%	1.02%	**0.75**%	2.00%
prec	94.75%	95.10%	**95.35**%	94.15%
Letter-recognition dataset				
e_1	**1.82**%	3.2%	1.93%	2.29%
e_2	**21.1**%	24.3%	21.35%	22.18%
prec	**94.87**%	91.03%	94.6%	93.57%
Average results over all datasets				
e_1	2.08%	2.54%	**2.02**%	2.66%
e_2	6.69%	7.75%	**6.53**%	8.04%
prec	95.07%	93.84%	**95.20**%	93.70%

5.2 Label Noise Mitigation

In this experiment we consider a stream-based active learning where the label noise is mitigated according to different strategies: relabelling, discarding and weighting. A hybrid strategy is also considered where only a small number of instances are manually relabelled according to a relabelling budget B. For the hybrid strategy, without any loss of generality, we used in our experiments a budget of $B = 20$ instances allowed to be relabelled. Results with others values of B lead to similar conclusions but are not reported due to the space limitation. The considered strategies are listed below:

- Full relabelling: relabelling every instance x that is identified as mislabelled (i.e. if $D(x) > t_D$, then x is relabelled)
- Full discarding: discarding every instance x that is identified as mislabelled
- Full weighting: using every instance and its queried label (x, y_g) weighted by $w = 1 - D(x)$ to update the classification model.
- Hybrid discarding and relabelling: consists in relabelling an instance that is identified as mislabelled only if it shows a high informativeness $(I(x) > t_I)$ and the budget B is not yet exhausted. Otherwise, the instance is discarded.
- Hybrid weighting and relabelling: same as the above hybrid strategy but using weighting instead of discarding.

The results obtained for each strategy are illustrated in Fig. 5 and Table 2. The classification accuracy obtained on a test set is shown on Fig. 5 according to the number of labelled instances. Table 2 shows the final classification accuracy, the final number of instances N_1 whose label was queried from the first (unreliable) labeller, and the number of instances N_2 that are relabelled by the alternative expert labeller (fixed to $N_2 = B = 20$ for the hybrid strategies). Let c_1 and c_2 respectively be the cost required by the first labeller and the expert labeller to label a single instance. It is assumed that $c_2 > c_1$ since the expert labeller is supposed to be reliable (or much more reliable than the first labeller). Then, the labelling cost is $c_1 \times N_1$, the relabelling cost is $c_2 \times N_2$, and the overall cost is $C = c_1 \times N_1 + c_2 \times N_2$.

Firstly, the results on Fig. 5 confirm that the "full relabelling" is obviously the most effective strategy in terms of classification accuracy, since all instances that are identified as being mislabelled are relabelled by the expert. Secondly, the results obtained on all datasets show that discarding the possibly mislabelled instances is not better than the weighting strategy. Actually, it has been observed in many works on label noise cleansing [9,19,20] that learning with mislabelled instances harms more than removing too many correctly labelled instances, but this is not true in the active learning setting, as it is confirmed by the obtained results. This is due to the fact that in the active learning setting, the discarded instances are more likely to improve the classification model if they are correctly labelled, thus, discarding them may negatively impact the performances of the active learning. Fore the same reason, we can see that the "hybrid weighting and relabelling" strategy performs better than the "hybrid discarding and relabelling" strategy. Also, Fig. 5 shows for the "hybrid weighting and relabelling"

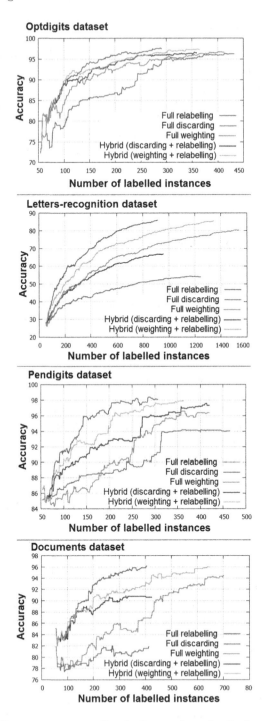

Fig. 5. Classification accuracy according to the number of actively queried labels using different strategies.

strategy that relabelling only $B = 20$ instances with a high value of I, greatly improves the accuracy compared to the "discarding" or the "weighting" strategy. We can see on Table 2 that the "hybrid weighting and relabelling" strategy achieves a final classification accuracy which is pretty close to that of the "full relabelling" strategy. For example, the final accuracy achieved by the "hybrid weighting and relabelling" strategy for the optdigits dataset is 97.21 %, whereas the one achieved by the "full relabelling" strategy is 97.49 %. As explained in Sect. 2.2, mislabelling errors causes the active learner to ask for more labelled instances. This explains why N_1 in Table 2 is smaller in the "full relabelling" strategy. However, by taking into consideration the cost induced by relabelling the mislabelled instances, the "full relabelling" strategy will have a higher overall cost than the other strategies, since all the instances that are identified as being mislabelled are relabelled.

Finally, we can conclude that although the "hybrid weighting and relabelling" strategy has a low relabelling cost, it achieves a final classification accuracy which is close to the one achieved by the "full relabelling" strategy. Therefore, if a limited relabelling budget is available, then this budget should be devoted to relabelling instances with a high informativeness I.

Table 2. Final accuracy and cost according to different strategies.

	Relabel	Discard	Discard and relabel	Weight	Weight and relabel
Optdigits dataset					
Accuracy	97.49 %	96.21 %	96.43 %	96.66 %	97.21 %
N_1	290	432	359	412	364
N_2	92	0	20	0	20
Pendigits dataset					
Accuracy	98.11 %	94.02 %	97.31 %	96.36 %	97.88 %
N_1	307	464	420	420	363
N_2	105	0	20	0	20
Letters-recognition dataset					
Accuracy	85.98 %	53.95 %	66.79 %	80.26 %	85.52 %
N_1	914	1242	956	1537	1344
N_2	211	0	20	0	20
Documents dataset					
Accuracy	96.1 %5	81.84 %	90.61 %	94.46 %	96.0 %
N_1	406	413	424	701	646
N_2	104	0	20	0	20

6   Conclusion and Future Work

In this paper we addressed the label noise detection and mitigation problem in stream-based active learning for classification. In order to identify the potentially mislabelled instances, we proposed a mislabelling likelihood based on the disagreement among the probabilities and the quantity of information that the instance carries for the predicted and the queried class labels. Then, we derived an informativeness measure that reflects how much a queried label would be useful if it is corrected. Our experiments on real datasets show that the proposed mislabelling likelihood is more efficient in characterizing label noise compared to the commonly used entropy measure. The experimental evaluation also shows that the potentially mislabelled instances with high conflicting information are worth relabelling.

Nonetheless, one limitation of the current hybrid label noise mitigation strategy is that it requires a threshold on the informativeness measure I which depends on the data and its automatic adaptation constitute one of our perspectives. As future work, we want to minimize the correction cost by defining and optimizing a multi-objective function that combines together (i) the mislabelling likelihood, (ii) the informativeness, and (iii) the cost of relabelling instances. Also, in the current work we observed that manually relabelling few instances chosen according to their informativeness I can improve results, but figuring out the number of labelled instances that are required to achieve closer accuracy to the case where all instances are relabelled still constitute one of our future work.

References

1. Zliobaite, I., Bifet, A., Pfahringer, B., Holmes, G.: Active learning with drifting streaming data. IEEE Trans. Neural Netw. Learn. Syst. **25**(1), 27–39 (2014)
2. Kremer, J., Steenstrup Pedersen, K., Igel, C.: Active learning with support vector machines. In: Wiley Interdisciplinary Reviews: Data Mining and Knowledge Discovery, pp. 313–326 (2014)
3. Huang, L., Liu, Y., Liu, X., Wang, X., Lang, B.: Graph-based active semi-supervised learning: a new perspective for relieving multi-class annotation labor. In: IEEE International Conference on Multimedia and Expo, pp. 1–6 (2014)
4. Kushnir, D.: Active-transductive learning with label-adapted kernels. In: ACM SIGKDD International Conference on Knowledge Discovery and Data Mining, pp. 462–471 (2014)
5. Settles, B.: Active learning. In: Synthesis Lectures on Artificial Intelligence and Machine Learning, pp. 1–114 (2012)
6. Bouguelia, M-R., Belaïd, Y., Belaïd, A.: A stream-based semi-supervised active learning approach for document classification. In: IEEE International Conference on Document Analysis and Recognition, pp. 611–615 (2013)
7. Goldberg, A., Zhu, X., Furger, A., Xu, J.M.: OASIS: online active semi-supervised learning. In: AAAI Conference on Artificial Intelligence, pp. 1–6 (2011)
8. Dasgupta, S.: Coarse sample complexity bounds for active learning. In: Neural Information Processing Systems (NIPS), pp. 235–242 (2005)
9. Frénay, B., Verleysen, M.: Classification in the presence of label noise: a survey. IEEE Trans. Neural Netw. Learn. Syst. **25**(5), 845–869 (2013)
10. Zhu, X., Zhang, P., Wu, X., He, D., Zhang, C., Shi, Y.: Cleansing noisy data streams. In: IEEE International Conference on Data Mining, pp. 1139–1144 (2008)

11. Rebbapragada, U., Brodley, C.E., Sulla-Menashe, D., Friedl, M.A.: Active label correction. In: IEEE International Conference on Data Mining, pp. 1080–1085 (2012)
12. Fang, M., Zhu, X.: Active learning with uncertain labeling knowledge. Pattern Recogn. Lett. **43**, 98–108 (2013)
13. Tuia, D., Munoz-Mari, J.: Learning user's confidence for active learning. IEEE Trans. Geosci. Remote Sens. **51**(2), 872–880 (2013)
14. Sheng, V.S., Provost, F., Ipeirotis, P.G.: Get another label? improving data quality and data mining using multiple noisy labelers. In: ACM Conference on Knowledge Discovery and Data Mining, pp. 614–622 (2008)
15. Ipeirotis, P.G., Provost, F., Sheng, V.S., Wang, J.: Repeated labeling using multiple noisy labelers. In: ACM Conference on Knowledge Discovery and Data Mining, pp. 402–441 (2014)
16. Yan, Y., Fung, G.M., Rosales, R., Dy, J.G.: Active learning from crowds. In: International Conference on Machine Learning, pp. 1161–1168 (2011)
17. Rousseeuw, P.J.: Silhouettes: a graphical aid to the interpretation and validation of cluster analysis. Comput. Appl. Math. **20**, 53–65 (1987)
18. Sun, S.: A survey of multi-view machine learning. Neural Comput. Appl. **23**(7–8), 2031–2038 (2013)
19. Gamberger, D., Lavrac, N., Dzeroski, S.: Noise elimination in inductive concept learning: a case study in medical diagnosis. In: Arikawa, Setsuo, Sharma, A.K. (eds.) ALT 1996. LNCS, vol. 1160, pp. 199–212. Springer, Heidelberg (1996)
20. Brodley, C.E., Friedl, M.A.: Identifying mislabeled training data. J. Artif. Intell. Res. **11**, 131–167 (1999)
21. Pedregosa, F., et al.: Scikit-learn: machine learning in python. J. Mach. Learn. Res. **12**, 2825–2830 (2011)

A Holistic Classification Optimization Framework with Feature Selection, Preprocessing, Manifold Learning and Classifiers

Fabian Bürger[(✉)] and Josef Pauli

Lehrstuhl für Intelligente Systeme, Universität Duisburg-Essen, Bismarckstraße 90,
47057 Duisburg, Germany
{fabian.buerger,josef.pauli}@uni-due.de
http://www.is.uni-due.de/

Abstract. All real-world classification problems require a carefully designed system to achieve the desired generalization performance. Developers need to select a useful feature subset and a classifier with suitable hyperparameters. Furthermore, a feature preprocessing method (e.g. scaling or pre-whitening) and a dimension reduction method (e.g. Principal Component Analysis (PCA), Autoencoders or other manifold learning algorithms) may improve the performance. The interplay of all these components is complex and a manual selection is time-consuming. This paper presents an automatic optimization framework that incorporates feature selection, several feature preprocessing methods, multiple feature transforms learned by manifold learning and multiple classifiers including all hyperparameters. The highly combinatorial optimization problem is solved with an evolutionary algorithm. Additionally, a multi-classifier based on the optimization trajectory is presented which improves the generalization. The evaluation on several datasets shows the effectiveness of the proposed framework.

Keywords: Feature selection · Model selection · Evolutionary optimization · Representation learning

1 Introduction

A classifier system that learns the connections from feature data to discrete class labels is useful for many applications such as medical diagnose systems or image-based object recognition. Several powerful methods have been established, like Support Vector Machines (SVM), that perform well on a large amount of tasks. However, the no-free-lunch theorem [1] states that there will never be a single best machine learning concept for all tasks. In practice, a lot of expertise is required for the development of a classification system to meet the generalization requirements. Numerous challenges occur in real-world applications, like high-dimensional and noisy feature data, too few training samples or suboptimal hyperparameters[1].

---

[1] Hyperparameters control the learning algorithm itself – e.g. the number of hidden layers in a neural network.

© Springer International Publishing Switzerland 2015
A. Fred et al. (Eds.): ICPRAM 2015, LNCS 9493, pp. 52–68, 2015.
DOI: 10.1007/978-3-319-27677-9_4

Furthermore, the feature data itself has a huge impact on the classification performance. There are three aspects that need to be considered: First, the selection of a reasonable subset of features is needed. A large amount of irrelevant, redundant or too noisy features tend to disturb classifiers due to the curse of dimensionality [2]. Secondly, the preprocessing of features usually improves the performance [3], especially when the distribution and value ranges differ greatly. Popular methods are e.g. feature scaling or pre-whitening. And third, representation learning with the goal of automatic feature construction out of low-level data is helpful [4]. This approach has gained more importance in the field of deep learning. Manifold learning is one variant of learning a simpler, low-dimensional representation from high-dimensional data. A great variety of such algorithms has been introduced, but their individual performance is highly dependent on the learning task.

Automatic optimization frameworks are designed to help the developer of machine learning systems to find an optimized combination of features, classifiers and hyperparameters. This paper presents an extended version of the classification pipeline framework presented in [5] and contains the fully automatic selection of features, feature preprocessing methods, manifold learning methods and classifiers. Furthermore, all hyperparameters – of the classifiers and the manifold learning methods – are optimized as well. As the interplay of all these components is complex an evolutionary optimization algorithm with an adapted variant of cross-validation is used to find good pipeline configurations.

Additionally, the optimization trajectory is exploited for a multi-pipeline classifier as well as graphical statistics to get deep insights into the classification problem itself. We compare our framework performance with the state-of-the-art optimization framework Auto-WEKA [6] with respect to classification accuracy and optimization speed.

2   Automatic Optimization Frameworks

When the supervised classification task is considered, a training dataset $T = \{(\mathbf{x}_i, y_i)\}$ with $1 \leq i \leq m$ training samples is provided. It contains feature vectors $\mathbf{x} \in \mathbb{R}^{d_{in}}$ and class labels $y_i \in \{\omega_1, \omega_2, \ldots, \omega_c\}$. The goal is to find a classifier model that predicts the correct class labels of previously unseen instances.

Automatic machine learning optimization frameworks try to find a suitable classifier model or a complete classification processing pipeline that fits to a given dataset T. A standard approach is the tuning of hyperparameters of a single classifier, e.g. the SVM. This problem is well discussed in many papers,e.g. in [7–9], to name a few. Usually, search-based approaches are used that evaluate different system configurations and hyperparameters with the goal to optimize the classification accuracy. Methods like cross-validation are used to estimate the generalization of a chosen algorithm [2].

Feature selection is one approach to dimension reduction with the strategy to remove irrelevant dimensions to overcome disturbing effects due to the peaking phenomenon [2]. Some frameworks, like [10–12], involve feature selection and hyperparameter optimization using evolutionary algorithms (see Sect. 4.1).

When more machine learning components should be optimized, the combinatorial complexity basically explodes. Furthermore, especially categorical variables and multiple algorithm portfolios introduce variable dependencies. The combinatorial optimization community developed sophisticated general purpose heuristics to handle problems with numerous and heterogeneous parameters which is also known as the *algorithm configuration problem*. With this respect, three approaches have to be mentioned. The ParamILS framework [13] is a local-search-based algorithm that limits the time spent for evaluating single configurations. Sequential model-based optimization (SMBO) [14] is a form of Bayesian optimization that keeps track of all knowledge of the objective function and its uncertainties to evaluate the next most promising configurations. SMBO is used in the Auto-WEKA framework [6] that optimizes features, classifiers and hyperparameters. A gender-based evolutionary approach is presented in [15] that handles variable dependencies in a tree structure.

However, there is no framework that targets holistic machine learning optimization covering feature selection, preprocessing, manifold learning and dimension reduction, classifier selection and hyperparameter tuning.

3 Classification Pipeline

In order to include all the aforementioned machine learning components into a holistic framework, a classification pipeline structure with 4 elements is proposed which is depicted in Fig. 1. Generally, the processing works like the pipes and filters pattern [16] while the pipeline has two modes: the *training mode* in which the training dataset T is needed and the *classification mode* in which new samples can be classified. A key design principle is a consecutive dimensionality reduction of the feature vectors while they pass through the pipeline. All pipeline elements contain important processing steps and multiple degrees of freedom that are summarized in the pipeline's configuration θ. It describes a set of important hyperparameters which have to be optimized for each learning task (see Sect. 4). The pipeline elements and their functionality are described in the following.

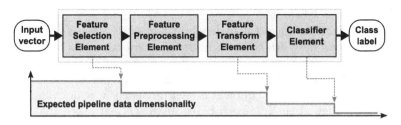

Fig. 1. Classification pipeline structure with expected data dimensionality.

3.1 Feature Selection Element

The first element is the feature selection element which removes irrelevant and noisy feature dimensions that could disturb any following algorithm. In training and classification mode, it selects a subset $S_{FeatSet} \in \mathcal{P}(\{1, 2, ..., d_{in}\}) \setminus \emptyset$ of features. Feature selection is a difficult problem as $\mathcal{O}(2^{d_{in}})$ possible combinations exist and it has a great impact on the classification performance. The feature subset $S_{FeatSet}$ is included in θ.

3.2 Feature Preprocessing Element

The second element of the pipeline handles the preprocessing of the feature vectors. Almost all machine learning algorithms perform better when the numeric properties and value ranges of the features are stable. The preprocessing element uses a portfolio set $S_{PreProc}$ of commonly used feature preprocessing methods:

- *Rescaling* scales each feature dimension to a specific range, such that all values of all dimensions lie in the range of $[0, 1]$.
- *Standardization* is similar to rescaling, but the data is scaled to a normal distribution with zero mean and a standard deviation of 1.
- *L2-Normalization* scales each feature vector independently to unit length which is equivalent to an L_2-norm of 1.
- *Pre-Whitening* is a more complex preprocessing step that performs a decorrelation transformation resulting to a feature matrix with zero mean and having a covariance matrix equal to the identity matrix [17].
- The *identity* function does not change any feature data in this pipeline element. This leads to the best results for some datasets.

The preprocessing method $f_{PreProc} \in S_{PreProc}$ is a part of the pipeline configuration θ. In training mode, the selected method $f_{PreProc}$ is used to extract model variables from the training dataset T, such as minimum and maximum values in case of the rescaling method. In classification mode, incoming feature vectors are processed with $f_{PreProc}$.

3.3 Feature Transform Element

The third element is the feature transform element which uses manifold learning algorithms to obtain a transform for a better suitable feature representation. Manifold learning describes a family of linear and nonlinear dimensionality reduction algorithms that analyze the topological properties of the feature data distribution to build a transformation function which embeds feature data into a low-dimensional space. In order to use manifold learning for real-world applications the following definition from [18] is used that requires just a set of D-dimensional data vectors. The assumption is that the datapoints lie on a lower-dimensional manifold with an intrinsic dimensionality $d < D$ which is embedded in the D-dimensional space. In practice, the target dimensionality d is not known and must be estimated. The manifold maybe non-Riemannian or disconnected which is likely for noisy real-world data. The goal is to find a feature transform function that embeds sample vectors into the lower dimensional vector space using $\tilde{\mathbf{x}}_i = f_{trans}(\mathbf{x}_i) \in \mathbb{R}^d$, while $\mathbf{x}_i \in \mathbb{R}^D$, without losing important information about the geometrical structure and distribution.

The pipeline element contains a portfolio set of possible transformations $S_{FeatTrans}$. Currently we use a set of 30 – mostly unsupervised – transforms provided by [19] which are listed in the appendix. Examples of linear transforms are e.g. Principal Component Analysis (PCA) or Linear Discriminant Analysis (LDA). Nonlinear techniques are e.g. Isomap, Kernel-PCA, Local Linear Embedding (LLE) or Autoencoders. References to these methods can be found e.g. in [18] or [20].

Furthermore, many of these manifold learning algorithms have – like classifiers – hyperparameters (see Sect. 3.5 for definition) that influence the performance. An example is the number of neighbors for neighborhood-based graph algorithms like Isomap [21]. The set of hyperparameters of a specific algorithm $f_{FeatTrans} \in S_{FeatTrans}$ is denoted as $S_{\mathbb{H}}(f_{FeatTrans})$. The choice of a method $f_{FeatTrans}$, the corresponding target dimensionality d and the hyperparameters $S_{\mathbb{H}}(f_{FeatTrans})$ are included into the pipeline configuration θ.

In training mode, the incoming feature vectors (and labels for the supervised methods like LDA) are used to learn the parameters of the manifold learning algorithm. In classification mode, previously unseen vectors need to be transformed to the new feature space. Unfortunately, not all methods directly support the so-called out-of-sample embedding. A direct extension is available only for parametric methods [18], e.g. PCA and Autoencoders. For spectral methods, like LLE, Isomap or Laplacian Eigenmaps, the Nyström theorem [22] can be used for an extension. For all other methods a rather naive non-parametric out-of-sample extension can be used (see [18,19]). It requires the storage of the complete set of base vectors and their corresponding transformed vectors. For each new feature vector, the nearest vector in the training dataset T and its corresponding transformed vector are determined which are used for an estimated linear projection.

3.4 Classifier Element

The last element is the classifier element which uses a classifier $f_{Classifier} \in S_{Classifiers}$. The framework currently contains 6 "popular" classifier concepts: the naive Bayes classifier, k-nearest neighbors (kNN), Support Vector Machine (SVM) with different kernels (linear, polynomial and Gaussian), random forest, extreme learning machine (ELM) and multilayer perceptron (MLP). References to these concepts can be found e.g. in [17,23]. Each classifier can have an arbitrary number of hyperparameters (see Sect. 3.5 for definition) which are tuned during the optimization phase. The selection of the classifier $f_{Classifier}$ and its hyperparameters $S_{\mathbb{H}}(f_{Classifier})$ are included into θ.

In training mode, the chosen classifier is trained using the data processed by all previous pipeline elements while the labels stay the same as in the training set T. In classification mode, the classifier classifies the incoming vectors.

3.5   Hyperparameters

Most machine learning algorithms have hyperparameters that need to be carefully adapted to the current learning task. This is also true for manifold learning algorithms. Almost all hyperparameters can be categorized into two basic types:

- A *numerical hyperparameter* h_{num} can be either real or integer values $h_{num} \in \{\mathbb{R}, \mathbb{Z}\}$ that are usually bounded within a reasonable minimum and maximum value $h_{min} \leq h_{num} \leq h_{max}$.
- For *categorical hyperparameters* an item $h_{cat} \in H_{cat}$ has to be selected out of a set H_{cat}.

In order to facilitate a simpler notation, $S_{\mathbb{H}}(f) = \{h_1, h_2, \ldots, h_N\}$ with $h_j \in \{h_{num}, h_{cat}\}$ denotes the set of hyperparameters of an algorithm f. Note that the sets of hyperparameters of different algorithms are independent.

4   Optimization of the Pipeline Configuration

The pipeline configuration finally contains all important hyperparameters

$$\theta = (S_{FeatSet}, f_{PreProc}, f_{FeatTrans}, S_{\mathbb{H}}(f_{FeatTrans}), d_{FeatTrans},$$
$$f_{Classifier}, S_{\mathbb{H}}(f_{Classifier})) \tag{1}$$

which have to be optimized for each learning task. First, a suitable evaluation metric has to be involved to estimate the predictive performance of a pipeline configuration. Secondly, the highly combinatorial search problem to find the best configuration has to be solved within a reasonable time. Furthermore, the objective function is expected to be non-smooth with numerous local optima.

4.1   Extended Evolution Strategies

We choose evolutionary optimization because it is well-suited to solve complex optimization problems and can easily be parallelized. These algorithms are inspired by Darwin's Theory of Evolution [24] in which the fitness in terms of adaptation to the environment has a great impact on the survival and reproduction of individuals in a species. A key part of these algorithms is randomization which is helpful for objective functions with many local optima. Especially evolution strategies (ES) with extensions are suitable for the optimization of heterogeneous hyperparameters [25,26]. ES uses sets of solution candidates which are called population of individuals. Evolutionary operators for random initial population of individuals, selection, recombination and mutation perform the actual optimization. The variables of an optimization problem can conveniently be coded directly as real or integer number search space $V_{\mathbb{R}}$ and $V_{\mathbb{Z}}$. Extensions allow the use of a binary search space $V_{\mathbb{B}}$ to form bitstrings as well as a discrete set search space $V_{\mathbb{S}}$ to model categorical parameters [27]. Furthermore, numeric hyperparameters with an exponential value range occur (e.g. Gaussian kernel parameters). In this case, the exponent $log_{10}(h_{num})$ is used for optimization in combination with $V_{\mathbb{R}}$.

The parameters for the ES strategies can be coded in the $(\mu, \kappa, \lambda, \rho)$ notation. The number of individuals that survive in each generation is denoted as μ. In each generation λ children from ρ parents are derived. The selection operator selects individuals for mating depending on their fitness which is directly connected to the objective function (see Sect. 4.2). Finally, individuals have a limited lifetime of κ generations.

Pipeline Configuration Coding. The classification pipeline configuration θ is transformed into a suitable ES variable representation in the following way (see Fig. 2). The feature subset is coded as binary mask consisting of d_{in} binary variables of type $V_{\mathbb{B}}$ which is the same idea than in [10]. All algorithm selection problems, namely feature preprocessing, feature transform and classifier, are handled as categorical variables $V_{\mathbb{S}}$. For the target dimensionality a factor $\alpha \in [0, 1]$ is coded as $V_{\mathbb{R}}$ genotype. It determines the fraction of the number of dimensions delivered by the feature selection that should be used as target dimensionality $d = \lfloor \alpha \cdot |S_{FeatSet}| \rfloor$ and $d \geq 1$.

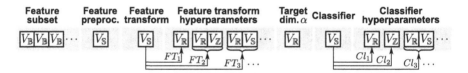

Fig. 2. Coding schema of a pipeline configuration θ for ES with four variable types $V_{\mathbb{R}}$, $V_{\mathbb{Z}}$, $V_{\mathbb{B}}$ and $V_{\mathbb{S}}$. Note that FT_i refers to the ith feature transform and Cl_j to the jth classifier.

The handling of hyperparameters is more difficult as they depend on the selection of the corresponding algorithm. In [5], the pipeline configuration only contained *one* set of hyperparameters, namely the classifier's. Two variants have been proposed: a combination of ES with grid search and a "complete" evolutionary optimization. Grid search is likely not feasible to optimize *both* hyperparameter sets from classifier and feature transform. Therefore, a similar approach as the proposed complete evolutionary strategy (denoted as *CES*) is used that concatenates all hyperparameters of all algorithms in a linear way with their corresponding type (typically $V_{\mathbb{R}}, V_{\mathbb{Z}}$ or $V_{\mathbb{S}}$). Another benefit is that numeric values can be adapted much finer compared to the grid search approach. All hyperparameters are evolved in parallel, but the selection of a specific algorithm acts as a switch that activates only the corresponding values when the configuration is used for a classification pipeline.

Evolutionary Parameters. First, an initial population of 200 random individuals is generated to get a reasonably large sample of the huge search space.

In each generation $\lambda = 100$ individuals from $\rho = 3$ parents are generated. The selection operator selects $\mu = 25$ individuals out of the newly generated offspring and the parent generation. In order to increase the diversity of solutions to overcome local optima, all individuals have a limited lifespan of $\kappa = 5$ generations.

The algorithm terminates when the improvement of the best fitness is less than $\epsilon = 10^{-3}$ (equal to 0.1% of accuracy) after at least three consecutive generations. However, to prevent a premature end of the optimization, at least 10 generations are evaluated.

Mutation. Standard ES algorithms just define a mutation operator for real vectors. However, the proposed system needs to be extended to mutate the heterogeneous variable types. For numerical variables $V_{\mathbb{R}}$ and $V_{\mathbb{Z}}$, an additive, normally distributed random variable is used whose standard deviation is initialized depending on the expected value boundaries of the corresponding hyperparameter $\sigma = 0.2 \cdot (h_{max} - h_{min})$. Each of the categorical and binary variables $V_{\mathbb{S}}, V_{\mathbb{B}}$ have a probability variable p_{mut} which is initialized with $p_{mut} = 0.1$. This probability defines a mutation to select either a random item or a bit flip, respectively. All mutation parameters are adapted during the optimization process as well. However, the originally proposed correlation between mutation variables [25] is not considered due to high number of additional variables that would be needed.

Initial Population Improvement of the Feature Subset. Due to the large search space – especially due to the feature selection – an improvement of the initial subsets compared to pure random subsets is expected to improve both optimization runtime and accuracy. Before every main optimization with the full set of framework components, a fast pre-optimization is performed that usually is done within less than a minute. This pre-optimization just contains feature selection and preprocessing methods in combination with the hyperparameter-free naive Bayes classifier. The initial population size contains 200 individuals, while 5 generations are performed using $\lambda_{pre} = 50$, $\rho_{pre} = 3$ and $\mu_{pre} = 20$. The feature subsets of the last surviving individuals of the pre-optimization are used as initial pool of feature subsets for the main optimization. One of the "good" subsets from the pool is chosen randomly for each individual of the initial generation of the main optimization.

4.2   Optimization Target Function

The evaluation metric of a configuration θ plays a central role to evaluate the generalization of the whole pipeline and is, of course, needed for the fitness of individuals. A common way to minimize the risk of overfitting of classifiers is k-fold cross-validation [2]. In the proposed pipeline multiple processing steps influence the generalization. Especially, the feature transform element with its out-of-sample function has a special role as the "intelligence" is potentially moved from the classifier to the feature transform: A highly nonlinear feature transform might work best with a simple, e.g. linear classifier. Furthermore, the preprocessing element also extracts model parameters from the training data and processes unseen feature vectors with these parameters. Problems may occur even with the simple rescaling method to $[0, 1]$: single outlier values with a very high value compared to the others make this method unstable.

Adapted Cross-Validation. In order estimate the generalization of a pipeline configuration θ, all components of the pipeline have to be involved into an adapted cross-validation process which is depicted in Fig. 3. The training set T is separated into $K = 5$ cross-validation tuples with disjoint training and validation datasets $\{(T_{train,l}, T_{valid,l})\}$. In each round, all models and parameters are estimated using only the training data $T_{train,l}$. The validation dataset is processed separately and the predictions of the classifier are used to calculate the accuracy acc_l. Finally, the average cross validation accuracy $acc_{avg} = \frac{1}{K} \sum_{l=1}^{K} acc_l$ and the standard deviation $acc_{sd} = \sqrt{\frac{1}{K} \sum_{l=1}^{K} (acc_l - acc_{avg})^2}$ are computed.

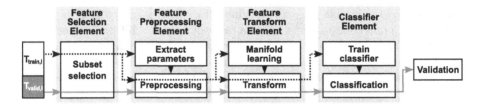

Fig. 3. Processing of the lth cross-validation round with training and validation data $T_{train,l}$ and $T_{valid,l}$ that incorporates feature selection, preprocessing, manifold learning and classifier into the generalization estimation. Note that the validation set is never used to estimate parameters or train any algorithm.

Early Rejection. The random character of ES leads to a relatively high fraction of suboptimal solutions that need to be evaluated. We propose an early rejection strategy that discards inferior configurations as soon as possible during cross-validation to save computation time for potentially better solutions. Two criteria lead to an early rejection: At first, if any configuration performs worse than guessing, thus $acc_l < 1/number\ classes$, the cross-validation is stopped. The second criterion uses a statistical method to test if a configuration will likely be equal or better than the currently best one. The cross-validation performance value $acc_{avg,best}$ and $acc_{sd,best}$ of the overall best configuration are stored. In each cross-validation round, a one-sided confidence interval for the accuracy mean is calculated that determines a minimum average accuracy of the lth round $acc_{min,l} = acc_{avg,best} - z \cdot acc_{sd,best}/\sqrt{l}$. The parameter z is the confidence level; we use $z = 1.96$ which is equal to an error probability of 2.5 %. The cross-validation is stopped if any $acc_l < acc_{min,l}$.

4.3 Multi-pipeline Classifier

The presented optimization method leads to a result list of N_{Res} configurations $R = \{(\theta_j, q_j)\}, 1 \le j \le N_{Res}$, with a corresponding fitness q_j. The configurations can be sorted by their fitness q_j and, at first glance, the configuration with the highest fitness is the most interesting result. However, this solution could be randomly picked and therefore quite "unusual" and also potentially overfitted to the training dataset, even though cross-validation is used.

The distribution of the top-n configurations can be used to generate a multi-pipeline classifier. Multi-classifier systems have the potential to improve the generalization capabilities compared to a single classifier when the diversity of the different models is large enough [28]. A multi-pipeline classifier is defined such that the top-n configurations are used to set up n pipelines with the corresponding configuration θ_j. In classification mode, all pipelines are classifying the input vector parallelly and finally, a majority voting is performed to select the most frequent label.

5   Experiments

For the evaluation of the presented framework 11 classification problems from the UCI database [29] have been used with different dimensionalities and classes (see Table 1). In order to test the generalization capabilities the instances of all datasets have been divided randomly into 50 % train and 50 % test sets. The proposed optimization algorithm, denoted as *CES (complete evolutionary strategy)*, is evaluated and compared to a *baseline* classifier. For the baseline we choose a popular standard approach with an SVM with a Gaussian kernel, in combination with the full feature set (no feature selection), no preprocessing, no feature transform and grid-based tuned hyperparameters.

The proposed evolutionary algorithm uses random components which may lead to non-reproducible results and local optima. In order to overcome this problem in the evaluations, all experiments have been repeated 3 times. Currently, the framework is implemented in Matlab 2014 using the Parallel Computing Toolbox and is run on an Intel Xeon workstation with 6 × 2.5 Ghz and 32 GB of RAM[2].

Table 1. Dataset information from the UCI database [29].

Index	Tataset	Dimensions	Classes	Index	Dataset	Dimensions	Classes
1	iris	4	3	7	australian	14	2
2	diabetes	8	2	8	vehicle	18	4
3	breast-cancer	9	2	9	ionosphere	34	2
4	contraceptive	9	3	10	sonar	60	2
5	glass	9	6	11	semeion-digits	256	10
6	statlogheart	13	2				

---

[2] The average memory consumption of the proposed system is below 8 GB.

5.1 Evaluation of the Optimization Process

First, the optimization process on the training datasets itself is evaluated. Table 2 shows the average cross-validation accuracies compared to the baseline SVM. The accuracy values for CES are significantly higher compared to the SVM for all datasets – the average accuracy improvement is $6.36 \pm 6.02\,\%$. These results show that the proposed classification pipeline is able to adapt very well to any learning task due do its large repertoire of algorithms and hyperparameters. The results are mostly stable, however the standard deviation is slightly higher for three datasets, namely *glass*, *vehicle* and *sonar*. This indicates that the optimization algorithm was stuck in local optima.

The optimization runtimes for each dataset can be found in Table 3. The average runtime for the 11 datasets is $131.4 \pm 96.0\,\text{min}$. There is no general link between the runtime and the dimensionality of the dataset. The main runtime heavily depends on the complexity of the selected algorithms in the pipeline configuration. Especially some feature transforms need much more time than others. Note that the runtime of the algorithms is not considered in the optimization process yet.

Table 2. Average training cross-validation accuracies compared to the baseline SVM.

Dataset	Baseline	CES	Dataset	Baseline	CES
1	96.00	**98.67 ± 0.00**	7	68.76	**88.35 ± 0.43**
2	76.81	**81.17 ± 0.60**	8	75.94	**82.55 ± 2.05**
3	97.07	**98.15 ± 0.45**	9	92.65	**95.48 ± 0.56**
4	52.70	**55.68 ± 0.27**	10	85.71	**89.21 ± 1.98**
5	68.96	**80.43 ± 2.24**	11	92.12	**93.21 ± 0.69**
6	74.07	**87.90 ± 1.54**			

Table 3. Average optimization runtimes in minutes.

Dataset	CES	Dataset	CES	Dataset	CES
1	29.1 ± 5.0	5	59.8 ± 36.0	9	131.4 ± 18.6
2	132.0 ± 30.3	6	39.4 ± 22.1	10	47.6 ± 1.1
3	182.0 ± 73.8	7	124.8 ± 48.7	11	287.5 ± 118.5
4	313.3 ± 55.6	8	99.0 ± 26.9		

5.2 Analysis of the Optimization Trajectory

In the following the optimization processes of two datasets, namely *statlogheart*[3] and *glass*[4], are discussed in detail. Figures 4 (a) and 5 (a) show the fitness

---

[3] The *statlogheart* dataset is a medical application in which diagnoses are correlated with absence of presence of serious heart diseases.

[4] The *glass* dataset is a forensic application in which glass is classified by oxide contents with the goal to identify the origins of the glass.

developments during the optimization. In both cases, a fast and steep gain of the mean fitness of the populations can be observed. The best fitness values per generation start at a relatively high level and increase much slower than the mean fitness. This indicates that the initial population already contains very well performing individuals which are fine-tuned in the main optimization.

Additionally, the effectiveness of the early rejection strategy during cross-validation can be seen in Figs. 4 (b) and 5 (b). These graphs depict the percentage of saved cross-validation rounds depending on the generation. The generation number zero is the initial population of the main optimization and during these evaluations, only the "worse-than-guessing" criterion is applied. In generation one, the fitness values of the best individuals from the initial generation are available to apply the confidence interval criterion to save more cross-validation rounds. As a large fraction of individuals perform inferiorly in the first generation, the ratio of saved evaluations is maximal there – for both datasets. Totally, 34.6 % of cross-validation rounds have been saved for the *statlogheart* dataset and and 41.5 % for the *glass* dataset, respectively.

5.3   Top Configuration Graph

The trajectory of the optimization can be exploited to get insight into the classification problem and its solutions. After the optimization terminates, the sorted result list R (see Sect. 4.3) is available. However, it is hard to analyze the configurations in text or table form. One way of visually analyze the solutions in R is the top configuration graph which is depicted in Fig. 6. The graph shows the distribution of frequencies of features, feature preprocessing methods, feature transforms and classifiers with a different shading. Additionally, components are considered as connected if they appear in the same configuration. These connections are shown as edges which are also shaded according to their frequency. The idea behind this graph is that features and algorithms which have been selected more often play a more important role for the classification problem.

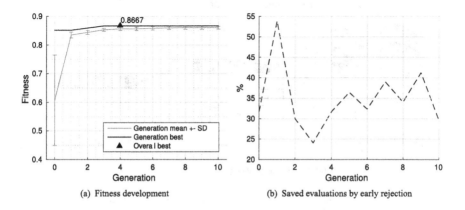

(a) Fitness development   (b) Saved evaluations by early rejection

Fig. 4. Exemplary optimization statistics for the *statlogheart* dataset.

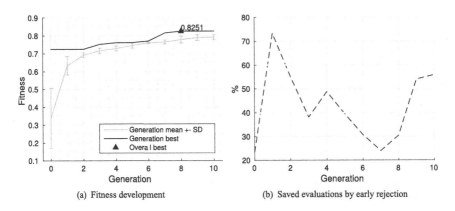

(a) Fitness development

(b) Saved evaluations by early rejection

Fig. 5. Exemplary optimization statistics for the *glass* dataset.

The two examples found in Fig. 6 show the top-50 configuration graphs for the *statlogheart* dataset in (a) and the *glass* dataset in (b).

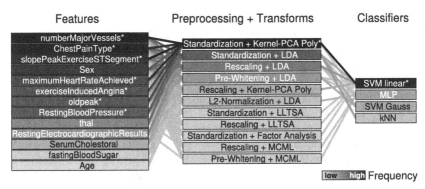

(a) Top-50 configuration graph for the *statlogheart* dataset.

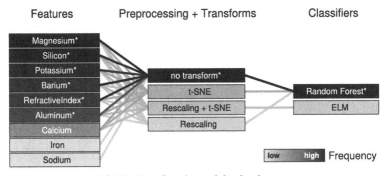

(b) Top-50 configuration graph for *glass* dataset.

Fig. 6. Examples of graphical analyses of the distribution of the top-50 configurations. The asterisk (*) denotes elements that occur in the overall best configuration.

For the *statlogheart* dataset, many different feature preprocessing and transforms perform well. Standardization, Kernel-PCA with a polynomial kernel and the LDA methods perform best. The feature preprocessing and transforms seem to make the problem linearly separable, as a linear SVM performs best in most cases. The shading of the features indicate their importance and, e.g. the feature *numberMajorVessels* and *ChestPainType* are very relevant, while the features *fastingBloodSugar* and *Age* seem to be the most irrelevant for predicting heart diseases.

For the *glass* dataset, no feature preprocessing method and transform have been successful. The distribution of features seems to be highly non-smooth as only the random forest performed well in most cases. The feature analysis reveals that the features *Magnesium, Silicon, Potassium, Barium, RefractiveIndex* and *Aluminium* are much more important than *Calcium, Iron* and *Sodium* to classify glass samples.

5.4   Evaluation of the Generalization

The huge accuracy improvements during cross-validation are promising, but the risk of overfitting is evident. Table 4 shows the accuracies of the proposed classification pipelines on the test datasets which have not been used during cross-validation. The generalization of a single classification pipeline using the best configuration in terms of fitness (denoted as top-1) is in many cases better than the generalization of the baseline classifier, but the average improvement is marginal (0.73 % with a high standard deviation). This would usually not justify the optimization time of several hours. The multi-pipeline classifier improves the accuracy greatly for many datasets; the average accuracy improvement

Table 4. Average generalization accuracies on test datasets compared to baseline and Auto-WEKA (24h of time budget). The number of pipelines for the best top-n multi-pipeline classifier is denoted in parentheses (n).

Dataset	Baseline	Top-1	Top-10	Best Top-n	AutoWEKA
1	97.33	97.33 ± 0.00	98.22 ± 1.54	**99.11** ± 1.54 (5)	92.27
2	73.96	74.91 ± 0.15	74.57 ± 0.60	74.91 ± 0.15 (1)	**75.83**
3	**96.77**	86.12 ± 17.96	96.29 ± 0.45	**96.77** ± 0.78 (15)	96.72
4	54.29	56.19 ± 0.59	57.23 ± 1.19	**57.55** ± 1.30 (11)	57.17
5	63.81	65.71 ± 7.19	68.57 ± 7.44	69.52 ± 8.30 (4)	**74.86**
6	72.59	81.73 ± 1.86	82.22 ± 2.22	82.96 ± 2.96 (23)	**83.70**
7	71.51	85.47 ± 0.77	85.95 ± 0.17	**86.24** ± 0.61 (6)	85.29
8	77.49	80.25 ± 4.12	81.99 ± 0.24	**83.10** ± 1.68 (30)	81.18
9	96.57	92.57 ± 1.98	95.24 ± 0.87	**96.76** ± 0.87 (50)	96.11
10	**87.38**	79.94 ± 1.48	82.52 ± 3.88	83.17 ± 2.24 (21)	85.63
11	95.21	94.75 ± 0.65	95.71 ± 0.44	**96.47** ± 0.13 (38)	94.13
	Delta Baseline:	0.73 ± 6.88	2.87 ± 5.38	3.60 ± 5.29	3.27 ± 6.11

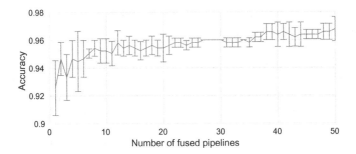

Fig. 7. Exemplary accuracy on test dataset of a multi-pipeline classifier for the *ionosphere* dataset depending on the number of pipelines n. The plot shows means and standard deviations of the repetitions.

compared to the baseline is 2.87 % if the 10 best pipelines are used. However, the optimal number of pipelines with the best generalization depends on the dataset. It can be observed for many datasets that a fusion of a higher number of pipelines leads to a better classification performance. Figure 7 shows the increasing performance depending on the number of fused pipelines for the *ionosphere* dataset.

The classifiers optimized by the Auto-WEKA framework with a time budget of 24 h also show a good performance on many datasets that is in the range of the best multi-pipeline classifiers. However, it performed rather poorly on the very simple *iris*[5] dataset. This shows that overfitting is also an issue for Auto-WEKA.

6   Conclusions

In this work, a holistic classification pipeline framework with feature selection, portfolios of feature preprocessing methods, feature transforms and classifiers is presented. An evolutionary algorithm is used that optimizes the configuration of the pipeline consisting of feature subset, algorithm selection and hyperparameters relatively efficiently. An adapted variant of cross-validation is proposed that incorporates the generalization performance of feature preprocessing, feature transforms and the classifier. A multi-pipeline classifier is used to improve the generalization of the classifier system. The framework does not require expert knowledge to reach state-of-the-art performance within a few hours. Additionally, graphical analyses of the best configurations help to reveal information about latent properties of the learning task.

The evaluation of the framework shows that overfitting is still a problem even though cross-validation is used. Sometimes, the standard SVM without special preprocessing generalizes best. However, the fusion of multiple pipelines shows a much better generalization performance, but introduces a higher computational cost. On the other hand, the datasets were rather low-dimensional and the performance improvement is expected to be larger in higher dimensional feature

---

[5] The popular *iris* dataset correlates variants of the iris plant with leaf dimensions.

spaces. In future work, the generalization estimation needs to be improved with advanced methods like e.g. bootstrapping. The runtime of the selected algo-rithms should also be considered during the optimization process to make it faster.

Acknowledgements. This work was funded by the European Commission within the Ziel2.NRW programme "NanoMikro+Werkstoffe.NRW".

Appendix

List of feature transforms and manifold learning methods that are used in the portfolio $S_{FeatTrans}$ of the feature transform element:
Autoencoder, CFA (Coordinated Factor Analysis), Diffusion Maps, Factor Analysis, FastICA (Independent Component Analysis), GPLVM (Gaussian Process Latent Variable Models), Hessian LLE, Identity (no transform), Isomap, Kernel-LDA (extension to LDA with e.g. Gaussian or polynomial kernels), Kernel-PCA (extension to PCA with e.g. Gaussian or polynomial kernels), Landmark Isomap, Laplacian Eigenmaps, LDA (Linear Discriminant Analysis / Fisher Discriminant Analysis / FDA), LLC (Locally Linear Coordination), LLE (Locally Linear Embedding), LLTSA (Linear LTSA), LMNN (Large-Margin Nearest Neighbor), LPP (Locality Preserving Projection), LTSA (Local Tangent Space Analysis), Manifold Charting, MCML (Maximally Collapsing Metric Learning), NCA (Neighborhood Components Analysis), NPE (Neighborhood Preserving Embedding), PCA (Principal Component Analysis), Sammon Mapping, SNE (Stochastic Neighbor Embedding), SPE (Structure Preserving Embedding), Symmetric SNE, t-SNE (parametric).

References

1. Wolpert, D.H.: The lack of a priori distinctions between learning algorithms. Neural Comput. **8**, 1341–1390 (1996)
2. Jain, A.K., Duin, R.P.W., Mao, J.: Statistical pattern recognition: a review. IEEE Trans. Pattern Anal. Mach. Intell. **22**, 4–37 (2000)
3. Juszczak, P., Tax, D., Duin, R.: Feature scaling in support vector data description. In: Proceedings ASCI, pp. 95–102. Citeseer (2002)
4. Bengio, Y., Courville, A., Vincent, P.: Representation learning: a review and new perspectives. IEEE Trans. Pattern Anal. Mach. Intelli. **35**, 1798–1828 (2013)
5. Bürger, F., Pauli, J.: Representation optimization with feature selection and manifold learning in a holistic classification framework. In: De Marsico, M., Fred, A., eds.: Proceedings of the International Conference on Pattern Recognition Applications and Methods ICPRAM 2015, vol. 1, pp. 35–44. INSTICC, SCITEPRESS, Lisbon (2015)
6. Thornton, C., Hutter, F., Hoos, H.H., Leyton-Brown, K.: Auto-WEKA: combined selection and hyperparameter optimization of classification algorithms. In: Proceedings of KDD-2013, pp. 847–855 (2013)
7. Bengio, Y.: Gradient-based optimization of hyperparameters. Neural Comput. **12**, 1889–1900 (2000)

8. Bergstra, J., Bardenet, R., Bengio, Y., Kégl, B., et al.: Algorithms for hyper-parameter optimization. In: 25th Annual Conference on Neural Information Processing Systems (NIPS 2011) (2011)
9. Bergstra, J., Bengio, Y.: Random search for hyper-parameter optimization. J. Mach. Learn. Res. **13**, 281–305 (2012)
10. Huang, C.L., Wang, C.J.: A GA-based feature selection and parameters optimization for support vector machines. Expert Syst. Appl. **31**, 231–240 (2006)
11. Huang, H.L., Chang, F.L.: ESVM: evolutionary support vector machine for automatic feature selection and classification of microarray data. Biosyst. **90**, 516–528 (2007)
12. Åberg, M., Wessberg, J.: Evolutionary optimization of classifiers and features for single trial EEG discrimination. Biomed. Eng. Online **6**, 32 (2007)
13. Hutter, F., Hoos, H.H., Leyton-Brown, K., Stützle, T.: ParamILS: an automatic algorithm configuration framework. J. Artif. Intell. Res. **36**, 267–306 (2009)
14. Hutter, F., Hoos, H.H., Leyton-Brown, K.: Sequential model-based optimization for general algorithm configuration. In: Coello, C.A.C. (ed.) LION 2011. LNCS, vol. 6683, pp. 507–523. Springer, Heidelberg (2011)
15. Ansótegui, C., Sellmann, M., Tierney, K.: A gender-based genetic algorithm for the automatic configuration of algorithms. In: Gent, I.P. (ed.) CP 2009. LNCS, vol. 5732, pp. 142–157. Springer, Heidelberg (2009)
16. Buschmann, F., Meunier, R., Rohnert, H., Sommerlad, P., Stal, M., Stal, M.: Pattern-Oriented Software Architecture: A System of Patterns, vol. 1, Wiley, New York (1996)
17. Bishop, C.M., Nasrabadi, N.M.: Pattern recognition and machine learning. vol. 1, Springer, New York (2006)
18. Van der Maaten, L., Postma, E., Van Den Herik, H.: Dimensionality reduction: a comparative review. J. Mach. Learn. Res. **10**, 1–41 (2009)
19. Van der Maaten, L.: Matlab Toolbox for Dimensionality Reduction (2014). http://homepage.tudelft.nl/19j49/Matlab_Toolbox_for_Dimensionality_Reduction.html
20. Ma, Y., Fu, Y.: Manifold Learning Theory and Applications. CRC Press, Boca Raton (2011)
21. Tenenbaum, J.B., De Silva, V., Langford, J.C.: A global geometric framework for nonlinear dimensionality reduction. Science **290**, 2319–2323 (2000)
22. Bengio, Y., Paiement, J.f., Vincent, P., Delalleau, O., Roux, N.L., Ouimet, M.: Out-of-sample extensions for LLE, Isomap, MDS, eigenmaps, and spectral clustering. In: Advances in Neural Information Processing Systems (2003)
23. Huang, G.B., Zhu, Q.Y., Siew, C.K.: Extreme learning machine: theory and applications. Neurocomputing **70**, 489–501 (2006)
24. Darwin, C.: On the Origins of Species by Means of Natural Selection. Murray, London (1859)
25. Bäck, T.: Evolutionary algorithms in theory and practice. Oxford UniversityPress (1996)
26. Beyer, H.G., Schwefel, H.P.: Evolution strategies - a comprehensive introduction. Nat. Comput. **1**, 3–52 (2002)
27. Müller, M.: Ein Entwurfsmuster für die multikriterielle Parameteradaption mit Evolutionsstrategien in der Bildverarbeitung. VDI-Verlag (2012)
28. Ranawana, R., Palade, V.: Multi-classifier systems: review and a roadmap for developers. Int. J. Hybrid Intell. Syst. **3**, 35–61 (2006)
29. Bache, K., Lichman, M.: UCI machine learning repository (2013). http://archive.ics.uci.edu/ml/

Feature Extraction and Learning
Using Context Cue and Rényi Entropy
Based Mutual Information

Hong Pan[1,2(✉)], Søren Ingvor Olsen[1], and Yaping Zhu[1]

[1] Department of Computer Science, University of Copenhagen,
2100 København Ø, Denmark
{hong.pan,ingvor,y.zhu}@di.ku.dk
[2] School of Automation, Southeast University, Nanjing 210096, China

Abstract. Feature extraction and learning play a critical role for visual perception tasks. We focus on improving the robustness of the kernel descriptors (KDES) by embedding context cues and further learning a compact and discriminative feature codebook for feature reduction using Rényi entropy based mutual information. In particular, for feature extraction, we develop a new set of kernel descriptors—Context Kernel Descriptors (CKD), which enhance the original KDES by embedding the spatial context into the descriptors. Context cues contained in the context kernel enforce some degree of spatial consistency, thus improving the robustness of CKD. For feature learning and reduction, we propose a novel codebook learning method, based on a Rényi quadratic entropy based mutual information measure called Cauchy-Schwarz Quadratic Mutual Information (CSQMI), to learn a compact and discriminative CKD codebook. Projecting the original full-dimensional CKD onto the codebook, we reduce the dimensionality of CKD while preserving its discriminability. Moreover, the latent connection between Rényi quadratic entropy and the mapping data in kernel feature space further facilitates us to capture the geometric structure as well as the information about the underlying labels of the CKD using CSQMI. Thus the resulting codebook and reduced CKD are discriminative. We verify the effectiveness of our method on several public image benchmark datasets such as YaleB, Caltech-101 and CIFAR-10, as well as a challenging chicken feet dataset of our own. Experimental results show that our method has promising potential for visual object recognition and detection applications.

Keywords: Context Kernel Descriptors · Cauchy-Schwarz Quadratic Mutual Information · Feature extraction and learning · Object classification and detection

1 Introduction

Recognition and detection of real-world objects are challenging, because it is difficult to model objects with significant variations in color, shape and texture. In addition, the backgrounds in which the objects exist are often complex and cluttered, and we have to account for changes of illumination, pose, size, and number of objects in the most

© Springer International Publishing Switzerland 2015
A. Fred et al. (Eds.): ICPRAM 2015, LNCS 9493, pp. 69–88, 2015.
DOI: 10.1007/978-3-319-27677-9_5

contrived situations. Currently, local based image representations [1–13] prevail in the state-of-the-art object recognition and detection algorithms. These local based image representations follow the bag-of-features framework [5, 6]. It first extracts low-level patch descriptors over a dense grid or salient points, then encodes them into mid-level features in a unsupervised way using mix of Gaussian, K-means or sparse coding, and finally derives the image-level representation using spatial pooling schemes [5–7]. Usually, carefully designed descriptors such as SIFT [8], SURF [9], LBP [10] and HOG [11] are used as low-level descriptors to gather statistics of pixel attributes within local patches. However, design of hand-crafted descriptors is non-trivial as sufficient prior knowledge is required and well-tuned parameters are necessary to achieve a good performance. Besides, we still lack a deep understanding on the design rules behind them. Recently, Bo et al. [1, 2] tried to answer how SIFT and HOG measure the similarity between image patches and interpret the design philosophy behind them from a kernel's view. They showed that the inner product of orientation histogram applied in SIFT and HOG is a particular match kernel over image patches. This insight provides a general way to turn pixel-level attributes into patch-level features with match kernels comparing similarities between image patches. Based on that, they designed a set of low-level descriptors called kernel descriptors (KDES) and kernel principal component analysis (KPCA) [14, 15] was used to reduce the dimensionality of KDES. However, KPCA only captures second-order statistics of KDES and cannot preserve its high-order statistics. It inevitably degrades the distinctiveness of KDES for nonlinear clustering and recognition where high-order statistics are needed. Wang et al. [4] merged the image label into the design of patch-level KDES and derived a variant KDES called supervised kernel descriptors (SKDES). Guiding KDES under a supervised framework with the large margin nearest neighbor criterion and low-rank regularization, SKDES reported an improved performance on object recognition.

In this work, we focus on improving the original KDES by embedding context cues into the descriptors and further learning a compact and discriminative Context Kernel Descriptors (CKD) codebook for object recognition and detection using information theoretic learning techniques. In particular, for feature extraction, we develop a set of CKD that enhance the KDES with embedded spatial context. Context cues enforce some degree of spatial consistency which improves the robustness of the resulting descriptors. For feature learning, we adopt the Rényi entropy based Cauchy-Schwarz Quadratic Mutual Information (CSQMI) [28], as an information theoretic measure, to learn a compact and discriminative CKD codebook from a rich and redundant CKD dictionary. In our method, codebook learning involves two steps including the codebook selection and refinement. In the first step, a group of compact and discriminative basis vectors are selected out of all available basis vectors to construct the codebook. By maximizing the CSQMI between the selected basis vectors in the codebook and the remaining basis vectors in the dictionary, we obtain a compact CKD codebook. By maximizing the CSQMI between the low-dimensional CKD generated from the codebook and their class labels, we also boost the discriminability of the learned CKD codebook. In the second step, we further refine the codebook for improved discriminability and low approximation error with a gradient ascent method that maximizes the CSQMI between the low-dimensional CKD and their class labels, given the constraint on a sufficient approximation accuracy. Projecting the full-dimensional CKD onto the learned CKD codebook, we derive the final low-dimensional discriminative CKD for feature representation. Evaluation results on standard recognition benchmarks, and a challenging chicken feet dataset show that our proposed CKD model outperforms the original KDES as well as carefully tuned descriptors like SIFT and some sophisticated deep learning methods.

The low-level patch features used in our work is built upon the KDES. Conceptually, it is related to [1], but our work departs from it in two distinct ways that improve the robustness and discriminability of our feature representation. First, we propose an enhanced match kernel called context match kernel (CMK). CMK strengthens the spatial consistency of the original match kernel by embedding the extra neighboring information into it. Spatial occurrence constraints implicit in the CMK significantly improve the robustness of similarity matching between feature sets, even for ambiguous or impaired features generated from partially occluded objects. Second, rather than using KPCA for reduction of the feature dimensionality, we perform the feature dimensionality reduction by projecting the original high-dimensional CKD onto a compact and discriminative CKD codebook. The CKD codebook is learned from a novel information theoretic feature selection algorithm based on the CSQMI. Because CSQMI is derived from the Rényi quadratic entropy, we can efficiently approximate it using a Parzen window [28]. In addition, considering the geometric interpretation of the CSQMI [28], it allows us to learn a discriminative CKD codebook that captures the cluster structure of input samples as well as the information about their underlying labels. Hence, the low-dimensional CKD derived from our model is more discriminative than the original KDES derived from KPCA.

2 Feature Extraction Using CKD

We enhance the original match kernel in [1] by embedding neighborhood constraints into it. As neighborhood defines an adjacent set of pixels surrounding the center pixel, neighborhood information can be regarded as the spatial context of the center pixel. So we refer to this enhanced match kernel as Context Match Kernel and the resulting descriptors as Context Kernel Descriptors. Intuition behind CMK is that pixels with similar attributes from two patches should have a high probability to have neighboring pixels whose attributes are also similar. Considering the spatial co-occurrence constraint, our CMK significantly improve the matching accuracy. CMK can be easily applied to develop a set of local descriptors using any pixel attributes, such as gradient, color, texture, and shape, etc. Next we derive the CMK, then we introduce several specific CMKs used in this work.

2.1 Formulation of CMK

An image patch can be modelled as a set of pixels $X = \{x_i\}_{i=1}^{n}$, where x_i is the coordinate of the ith pixel. Let a_i be attribute vector of the ith pixel x_i. The k-neighborhood N_k^i of pixel x_i in X is defined as a group of pixels (including itself) that are closest to it. Mathematically, $N_k^i = \{x_j \in X | \; \|x_i - x_j\| \leq k; \; k \geq 1\}$. To eliminate the image noise, we smooth the image using a Haar wavelet filter and compute the local gradient in the k-neighborhood. For the k-neighborhood centered at x_p, we first normalize the neighborhood's attribute by voting the pixel's attribute in N_k^p with its gradient magnitude weighted by a Gaussian function centered at x_p. The width of Gaussian function, which normalizes the attributes contributed from off-center pixels, is controlled by the neighborhood size k. Similarly, we also normalize the attribute in the k-neighborhood centered at x_q. With the normalized attributes in N_k^p and N_k^q, we then define the context kernel of attributes a between x_p and x_q as

$$\kappa_{con}[(x_p, a_p), (x_q, a_q)] = \kappa_a(\bar{a}_p, \bar{a}_q)$$

$$\bar{a}_p = \frac{1}{|N_k^p|} \sum_{x_u \in N_k^p} a_u m_u \exp\left(-\frac{8\|x_u - x_p\|^2}{k^2}\right), \bar{a}_q = \frac{1}{|N_k^q|} \sum_{x_v \in N_k^q} a_v m_v \exp\left(-\frac{8\|x_v - x_q\|^2}{k^2}\right)$$

$$(1)$$

where m_u and m_v are the gradient magnitudes at pixels x_u and x_v, respectively; $\bar{a}_p$ and $\bar{a}_q$ are the normalized image attributes in k-neighborhood centered at x_p and x_q, respectively; $\kappa_a(\bar{a}_p, \bar{a}_q) = exp(-\gamma_a\|\bar{a}_p - \bar{a}_q\|^2) = \varphi_a(\bar{a}_p)^T \varphi_a(\bar{a}_q)$ is the Gaussian kernel measuring the similarity of normalized attributes $\bar{a}_p$ and $\bar{a}_q$. The context kernel κ_{con} provides a normalized measure of the attribute similarity between two k-neighborhoods centered at pixels x_p and x_q. Merging κ_{con} into match kernels [1] and replacing the attribute a in Eq. (1) with specific attributes, we can derive a set of ad hoc attribute based CMKs.

For example, let θ'_p and m'_p be normalized orientation and normalized magnitude of the image gradient at pixel x_p, such that $\theta'_p = (\sin\theta_p, \cos\theta_p)$ and $m'_p = m_p / \sqrt{\sum_{p \in P} m_p^2 + \tau}$, with τ being a small positive number. To compare the similarity of gradients between patches P and Q from two different images, the gradient CMK K_{gck} can be defined as

$$K_{gck}(P, Q) = \sum_{p \in P} \sum_{q \in Q} m'_p m'_q \kappa_o(\theta'_p, \theta'_q) \kappa_s(x_p, x_q) \kappa_{con}[(x_p, \theta'_p), (x_q, \theta'_q)] \qquad (2)$$

where $\kappa_o(\theta'_p, \theta'_q) = exp(-\gamma_o\|\theta'_p - \theta'_q\|^2) = \varphi_o(\theta'_p)^T \varphi_o(\theta'_q)$ is the orientation kernel measuring the similarity of normalized orientations at two pixels x_p and x_q; $\kappa_s(x_p, x_q) = \exp(-\gamma_s\|x_p - x_q\|^2) = \varphi_s(x_p)^T \varphi_s(x_q)$ is the spatial kernel measuring how close two pixels are spatially; and $\kappa_{con}[(x_p, \theta'_p), (x_q, \theta'_q)]$ is given by Eq. (1).

Similarly, to measure the similarity of color attributes between P and Q, the color CMK K_{cck} can be defined as

$$K_{cck}(P, Q) = \sum_{p \in P} \sum_{q \in Q} \kappa_c(c_p, c_q) \kappa_s(x_p, x_q) \kappa_{con}[(x_p, c_p), (x_q, c_q)] \qquad (3)$$

where $\kappa_c(c_p, c_q) = \exp(-\gamma_c \| c_p - c_q \|^2) = \varphi_c(c_p)^T \varphi_c(c_q)$ is the color kernel measuring the similarity of color values c_p and c_q. For color images, we use normalized *rgb* vector as color value, whereas intensity value is used for grayscale images.

For the texture attribute, we derive the texture CMK, K_{lbpck}, based on Local Binary Patterns (*lbp*) [10]

$$K_{lbpck}(P, Q) = \sum_{p \in P} \sum_{q \in Q} \sigma'_p \sigma'_q \kappa_{lbp}(lbp_p, lbp_q) \kappa_s(x_p, x_q) \kappa_{con}[(x_p, lbp_p), (x_q, lbp_q)] \quad (4)$$

where $\sigma'_p = \sigma_p / \sqrt{\sum_{p \in N_3} \sigma_p^2 + \tau}$ is the normalized standard deviation of pixel values within a 3×3 window around x_p; $\kappa_{lbp}(lbp_p, lbp_q) = \exp(-\gamma_{lbp}\|lbp_p - lbp_q\|^2)$ is a Gaussian match kernel for *lbp* operator.

As shown in Eqs. (2)-(4), each attribute based CMK consists of four terms: (1) normalized linear kernel, e.g. $m'_p m'_q$ for K_{gck}; 1 for K_{cck} and $\sigma'_p \sigma'_q$ for K_{lbpck}, weighting the contribution of each pixel to the final attribute based CMK; (2) attribute kernel evaluating the similarity of pixel attributes; (3) spatial kernel κ_s measuring the relative distance between two pixels; (4) context kernel κ_{con} comparing the spatial co-occurrence of pixel attributes. In this sense, we formulate these attribute CMKs, defined in Eqs. (2)-(4), in a unified way as

$$K(P, Q) = \sum_{p \in P} \sum_{q \in Q} w_p w_q \kappa_a(a_p, a_q) \kappa_s(x_p, x_q) \kappa_{con}[(x_p, a_p), (x_q, a_q)] \quad (5)$$

where $w_p w_q$ and κ_a correspond to normalized linear weighting kernel and attribute kernel, respectively.

2.2   Approximation of CMK

Using the inner product representation, we rewrite the match kernel matrix K as

$$K(P, Q) = \langle \psi(Q), \psi(P) \rangle = \psi(P)^T \psi(Q)$$
$$\psi(P) = \sum_{p \in P} w_p \varphi_a(a_p) \otimes \varphi_s(x_p) \otimes \varphi_{con}(x_p, a_p), \ \psi(Q) = \sum_{q \in Q} w_q \varphi_a(a_q) \otimes \varphi_s(x_q) \otimes \varphi_{con}(x_q, a_q)$$

$$(6)$$

where $\otimes$ is the tensor product and $\psi(\cdot)$ gives the mapping features in kernel space, namely the CKD. Note that the dimensions of φ_a, φ_s and φ_{con} are all infinite, since Gaussian kernel is used. To obtain an accurate approximation of K, we have to uniformly sample φ_a, φ_s and φ_{con} using a dense grid along sufficient basis vectors. In particular, for φ_a and φ_{con}, we discretize a into G bins and approximate them with their projections onto the subspaces spanned by G basis vectors $\{\varphi_a(a^g)\}$ ($g = 1...G$). Similarly, for space vector x, we discretize spatial basis vectors into L bins and sample along L basis vectors spatially. Finally, we can approximate $\psi(\cdot)$ by its projections onto the $G \times L \times G$ joint basis vectors: $\{\phi_l\}_{l=1}^{G \times L \times G} = \{\varphi_a(a^1) \otimes \varphi_s(x^1) \otimes \varphi_{con}(a^1), \ldots, \varphi_a(a^G) \otimes \varphi_s(x^L) \otimes \varphi_{con}(a^G)\}$.

$$\psi(\cdot) \simeq \sum_{l=1}^{G \times L \times G} f_l \phi_l \tag{7}$$

where f_l is the projection coefficient onto the lth joint basis vector ϕ_l. Thus, dimensionality of the resulting CKD ψ is $G \times L \times G$. Uniform sampling provides a set of representative joint basis vectors, but does not guarantee their compactness. Projecting onto these basis vectors usually yield a group of redundant CKD. Next, we show how to learn a CKD codebook by selecting and refining a subset of compact and discriminative joint basis vectors using a CSQMI based information theoretic feature learning scheme. Projecting the original CKD ψ onto the codebook reduces the redundancy of ψ and gives a low-dimensional discriminative CKD representation.

3 Feature Learning Using CSQMI

Shannon entropy and its related measures, such as mutual information and Kullback-Leibler divergence (KLD) are widely used in feature learning [16–26]. However, Shannon entropy based feature learning methods share the common weakness of high evaluation complexity involved in the estimation of probability density function (*pdf*) in Shannon entropy [16]. Recently, Rényi entropy [27, 28] has attracted more attentions in information theoretic learning. The most impressive advantage of Rényi entropy is its moderate computational complexity because the estimate of Rényi entropy can be efficiently implemented by the kernel density estimation [29] (e.g. the Parzen windowing). Several novel information theoretic metrics derived from Rényi entropy are introduced in feature learning [30–33].

3.1 Rényi Entropy and CSQMI

Let $S \in \mathcal{R}^d$ be a discrete random variable which has a *pdf* of $p(s)$, then its Rényi entropy is defined as [27]

$$H_\alpha(S) = \frac{1}{1-\alpha} \log_2 \sum_{s \in S} p^\alpha(s) \tag{8}$$

Rényi entropy defines a family of functions that quantify the diversity in a data distribution. Standard Shannon entropy can be treated as a special case of Rényi entropy as $\alpha \rightarrow 1$. Rényi entropy of order $\alpha = 2$, given in Eq. (9), is called Rényi quadratic entropy $H_2(S)$.

$$H_2(S) = -\log_2 \sum_{s \in S} p^2(s) \tag{9}$$

Similar to KLD defined using Shannon entropy, Cauchy-Schwarz divergence (CSD) based on Rényi quadratic entropy also defines a measure of divergence between different *pdfs*. Given two discrete random variables S_1 and S_2, with S_1 having a *pdf* of $p_1(s_1)$ and S_2 having a *pdf* of $p_2(s_2)$, the CSD [28, 31] of p_1 and p_2 is given by

$$CSD(p_1; p_2) = -\log_2 \frac{\left(\sum\limits_{s_1 \in S_1, s_2 \in S_2} p_1(s_1)p_2(s_2) \right)^2}{\sum\limits_{s_1 \in S_1} p_1^2(s_1) \sum\limits_{s_2 \in S_2} p_2^2(s_2)} = 2H_2(S_1, S_2) - H_2(S_1) - H_2(S_2)$$

(10)

where $H_2(S_1, S_2) = -\log_2 \sum\limits_{s_1 \in S_1, s_2 \in S_2} p_1(s_1)p_2(s_2)$ measures the similarity (distance) between the two *pdfs* and can be considered as the Rényi quadratic cross entropy. We can interpret $H_2(S_1, S_2)$ as the information gain from observing p_2 with respect to the "true" density p_1, and vice versa. Hence, the CSD derived from Rényi quadratic entropy is semantically similar to Shannon's mutual information. Note that $CSD (p_1; p_2) \geq 0$ is a symmetric measure that equals zero if and only if $p_1(s) = p_2(s)$, and increases towards positive infinity as the two *pdfs* are apart further and further. Based on $CSD (p_1; p_2)$, the Cauchy-Schwarz Quadratic Mutual Information between two discrete random variables S_1 and S_2 is defined as [28].

$$I_{CSD}(S_1; S_2) = CSD(p_{12}(s_1, s_2); p_1(s_1)p_2(s_2))$$
$$= \log_2 \sum\limits_{s_1 \in S_1 s_2 \in S_2} p_{12}^2(s_1, s_2) + \log_2 \sum\limits_{s_1 \in S_1 s_2 \in S_2} p_1^2(s_1)p_2^2(s_2) - 2\log_2 \sum\limits_{s_1 \in S_1 s_2 \in S_2} p_{12}(s_1, s_2)p_1(s_1)p_2(s_2)$$

(11)

where $p_{12}(s_1, s_2)$ is the joint *pdf* of (S_1, S_2), and $p_1(s_1)$ and $p_2(s_2)$ are marginal *pdf* of S_1 and S_2, respectively. $I_{CSD}(S_1; S_2) \geq 0$ meets the equality if and only if S_1 and S_2 are independent. So $I_{CSD}(S_1; S_2)$ is a measure of independence that reflects the information shared between S_1 and S_2. In other words, it measures how much knowing S_1 reduces the uncertainty about S_2, and vice versa.

To calculate CSD and I_{CSD}, we have to estimate marginal *pdf* $p(\cdot)$ and joint *pdf* $p_{12}(\cdot, \cdot)$. Fortunately, Principe [28] showed that, for Rényi quadratic entropy and its induced measures such as CSD and I_{CSD}, these marginal and joint *pdfs* can be efficiently estimated with a Parzen window density estimator [29], even in a high-dimensional feature space like CDK. Whereas, it is not possible for Shannon entropy [28]. This explains why we choose the Rényi quadratic entropy based I_{CSD}, instead of the Shannon entropy based mutual information, as information theoretic measure in our codebook learning algorithm.

In addition, recent findings from Jenssen et al. [30, 31] uncovered the latent connections between Rényi quadratic entropy and mapping features in the kernel space. It shows that, when applying a Gaussian Parzen window estimator, Rényi quadratic entropy estimator is equivalent to $\|m\|^2$, where $m = \frac{1}{M} \sum\limits_{s_t \in S} \varphi(s_t)$ is the mean vector of mapping data samples $\varphi(s_t)$ $(t = 1, \cdots, M)$ in the kernel feature space. Meanwhile, the

CSD estimator is directly associated with the angle between the mean vectors m_1 and m_2 of the clusters of mapping data samples in the kernel feature space. These clusters correspond to the mapping data samples yielded from $p_1(s)$ and $p_2(s)$, respectively. Consequently, CSQMI, measuring the CSD between a joint *pdf* and the product of two marginal *pdf*s, also relates to the cluster structure in the kernel feature space. The relationships between Rényi quadratic entropy, CSD/CSQMI and the mean vector of mapping features in the kernel space provide us the geometric interpretation behind $H_2(S)$ and CSD/CSQMI. It means that the Rényi quadratic entropy based measures are very suitable for the analysis of nonlinear data (even in high-dimensional spaces) because they are able to capture the geometric structure of the data. In contrast, the Shannon entropy and the KLD do not have such good properties.

3.2 Codebook Selection and Refinement Using CSQMI

As mentioned in Sect. 2.2, we approximate the original CKD ψ with a group of redundant joint basis vectors $\{\phi_l\}_{l=1}^{G \times L \times G}$. We define these joint basis vectors as dictionary, and represent it as Φ (Φ has a cardinality of $G \times L \times G$). Assuming that we are given CKD, $\psi^1, \cdots, \psi^M$, of M samples from C classes, for each class c ($c = 1, \cdots, C$), it has M_c samples and the corresponding CKD are denoted as $\Psi_c = [\psi_c^1, \cdots, \psi_c^{Mc}]$. Then we formulate the CKD of all samples as $\Psi = \{\Psi_c\}_{c=1}^C$. Similarly, we denote $F = \{F_c\}_{c=1}^C$, where $F_c = [F_c^1, \cdots, F_c^{Mc}] = \left[(f_{c1}^1, \cdots, f_{cG \times L \times G}^1)^T, \cdots, (f_{c1}^{Mc}, \cdots, f_{cG \times L \times G}^{Mc})^T \right]$. Then, Eq. (7) can be represented as $\Psi = \Phi F$, where $\Phi = [\phi_1, \cdots, \phi_{G \times L \times G}]$ and $F = \begin{bmatrix} f_{11}^1 & \cdots & f_{C1}^{Mc} \\ \vdots & & \vdots \\ f_{1G \times L \times G}^1 & \cdots & f_{CG \times L \times G}^{Mc} \end{bmatrix}$ is the projection coefficients matrix. Given a CKD ψ from a random sample, the uncertainty of its class label L in terms of the class prior probabilities can be measured by $H_2(L)$, given in Eq. (9). Whereas, the CSQMI $I_{CSD}(\psi; L)$ defined in Eq. (11) measures the decrease in uncertainty of the pattern ψ due to the knowledge of the underlying class label L.

Given Ψ and an initial dictionary Φ, we aim to learn a compact and discriminative subset of joint basis vectors Φ^* out of Φ, such that *cardinality* $(\Phi^*) <$ *cardinality* (Φ). We refer to Φ^* as codebook. Projecting the original CKD Ψ onto the codebook Φ^* gives a low-dimensional CKD, $\Psi^* = \Phi^* F^*$. We expect Ψ^* should be compact and discriminative. To learn a compact codebook, we maximize the CSQMI between Φ^* and the unselected basis vectors $\Phi - \Phi^*$ in Φ, i.e. $I_{CSD}(\Phi^*; \Phi - \Phi^*)$. As $I_{CSD}(\Phi^*; \Phi - \Phi^*)$ signifies how compact the codebook Φ^* is, a higher value of $I_{CSD}(\Phi^*; \Phi - \Phi^*)$ means a more compact codebook. However, that codebook may not be discriminative, because it does not give any information regarding the new CKD Ψ^* from their class label L. Therefore, we also need to maximize the CSQMI between Ψ^* and L, i.e. $I_{CSD}(\Psi^*; L)$, which provides the discriminability of the new CKD generated from the codebook Φ^*. To this end, the codebook learning problem can be mathematically formulated as

$$\underset{\mathbf{\Phi}^*}{\arg\max}[I_{CSD}(\mathbf{\Phi}^*; \mathbf{\Phi} - \mathbf{\Phi}^*) + \lambda I_{CSD}(\mathbf{\Psi}^*; L)] \tag{12}$$

where λ is the weight parameter to make a tradeoff between the compactness and discriminability terms. We use a two-step strategy to optimize the compactness and discriminability of the codebook simultaneously. In the first step (*Codebook Selection*), the codebook that maximizes Eq. (12) is selected from the initial dictionary in a greedy search manner. In the second step (*Codebook Refinement*), the selected codebook is refined via a gradient ascent method to further maximize the discriminability term $I_{CSD}(\mathbf{\Psi}^*; L)$ while keeping the approximation error as low as possible.

3.2.1 Codebook Selection

The first term in Eq. (12), i.e. $I_{CSD}(\mathbf{\Phi}^*; \mathbf{\Phi} - \mathbf{\Phi}^*)$, is a compactness term which measures the compactness of the codebook $\mathbf{\Phi}^*$. The second term, i.e. $I_{CSD}(\mathbf{\Psi}^*; L)$, measures the discriminability of the codebook $\mathbf{\Phi}^*$. Based on [34], the probability of Bayes classification error resulted from the final CKD $\mathbf{\Psi}^*$, i.e. $P(e^{\mathbf{\Psi}^*})$, has its upper bound given by $P(e^{\mathbf{\Psi}^*}) \leq \frac{1}{2}(H_2(L) - I_{CSD}(\mathbf{\Psi}^*; L))$. Thus, the selected discriminative codebook $\mathbf{\Phi}^*$ corresponding to the minimal Bayes classification error bound should maximize the $I_{CSD}(\mathbf{\Psi}^*; L)$.

During the codebook selection, we start with an empty set of $\mathbf{\Phi}^*$ and iteratively select the next best basis vector ϕ^* out of the remaining set $\mathbf{\Phi} - \mathbf{\Phi}^*$, such that the mutual information gain between the new codebook $\mathbf{\Phi}^* \cup \phi^*$ and the remaining set, as well as the mutual information gain between the CKD derived from the new codebook and the class label, are maximized, i.e.

$$\underset{\phi^* \in \mathbf{\Phi} - \mathbf{\Phi}^*}{\arg\max}\left\{[I_{CSD}(\mathbf{\Phi}^* \cup \phi^*; \mathbf{\Phi} - (\mathbf{\Phi}^* \cup \phi^*)) - I_{CSD}(\mathbf{\Phi}^*; \mathbf{\Phi} - \mathbf{\Phi}^*)] + [I_{CSD}(\mathbf{\Psi}^{\mathbf{\Phi}^* \cup \phi^*}; L) - I_{CSD}(\mathbf{\Psi}^{\mathbf{\Phi}^*}; L)]\right\} \tag{13}$$

3.2.2 Codebook Refinement

Once the initial codebook $\mathbf{\Phi}^*$ is achieved, we refine $\mathbf{\Phi}^*$ to further enhance its discriminability by maximizing the discriminability term in Eq. (12), i.e. $\underset{\mathbf{\Phi}^*}{\max} \lambda I_{CSD}(\mathbf{\Psi}^*; L)$. To guarantee a compact codebook, we assume that *cardinality* ($\mathbf{\Phi}^*$) $\ll$ *cardinality* ($\mathbf{\Phi}$). Under such an assumption, the projection coefficient is solved by $F^* = \mathbf{\Phi}^\dagger \mathbf{\Psi}$ which minimizes the approximation error $e = \| \mathbf{\Psi} - \mathbf{\Phi}^* F^* \|^2$, where $\mathbf{\Phi}^\dagger = pinv(\mathbf{\Phi}^*) = (\mathbf{\Phi}^{*T}\mathbf{\Phi}^*)^{-1}\mathbf{\Phi}^{*T}$ is the pseudo-inverse of $\mathbf{\Phi}^*$. Thus, the problem of refining $\mathbf{\Phi}^*$ for improving the discriminability of codebook while keeping its approximation accuracy is converted to the following constraint optimization problem.

$$\underset{\mathbf{\Phi}^*}{\max} I_{CSD}(\mathbf{\Psi}^*; L), \text{ subject to } F^* = \mathbf{\Phi}^\dagger \mathbf{\Psi} \tag{14}$$

Since $I_{CSD}(\bullet; \bullet)$ is a quadratic symmetric measure, the objective function $I_{CSD}(\mathbf{\Psi}^*; L)$ is differentiable. We use the gradient ascend method to iteratively refine $\mathbf{\Phi}^*$ such that $I_{CSD}(\mathbf{\Psi}^*; L)$ is maximized. In each iteration, $\mathbf{\Phi}^*$ is updated with a step size v. After k-th iteration, $\mathbf{\Phi}_k^*$ becomes

$$\Phi_k^* = \Phi_{k-1}^* + \upsilon \frac{\partial I_{CSD}(\Psi^*; L)}{\partial \Phi^*}\Big|_{\Phi^* = \Phi_{k-1}^*}$$

$$\frac{\partial I_{CSD}(\Psi^*; L)}{\partial \Phi^*} = \sum_{c=1}^{C} \sum_{i=1}^{M_c} \frac{\partial I_{CSD}(\Psi^*; L)}{\partial \psi_c^{*i}} \frac{\partial \psi_c^{*i}}{\partial \Phi^*} = \sum_{c=1}^{C} \sum_{i=1}^{M_c} \left(F_c^i\right)^T \frac{\partial I_{CSD}(\Psi^*; L)}{\partial \psi_c^{*i}} \tag{15}$$

Once Φ^* is refined, we update the projection coefficients F^* and the low-dimensional discriminative CKD Ψ^* according to $F^* = \Phi^{\dagger}\Psi$ and $\Psi^* = \Phi^* F^*$, respectively. The bound of $I_{CSD}(\Psi^*; L)$ guarantees the convergence of codebook refinement.

4 Experiments

To verify the effectiveness of our method in the context of object recognition and detection, we first investigate the performance of CSQMI based codebook learning on the extended YaleB face dataset [35], then we test our model on Caltech-101 [36] and CIFAR-10 [37] for recognition and on our own chicken feet dataset for detection. We also compare our results with other state-of-the-art works, including the original KDES [1], supervised kernel descriptors [4], handcrafted dense SIFT features [7, 8], and the popular deep feature learning approaches [44, 51–53].

4.1 Parameter Configuration

We adopt the code provided from www.cs.washington.edu/robotics/projects/kdes/ to implement the original KDES. To make a fair comparison, in all experiments, except for the final feature dimensionality, we follow the setting of [1] for common parameters used in our method. Namely, basis vectors for κ_o, κ_c, and κ_s are sampled using 25, $5 \times 5 \times 5$, and 5×5 uniform grids, respectively. For κ_{lbp}, we choose all 256 basis vectors. For all CKD, κ_{con} shares the same basis vectors with their attribute kernels κ_a. We use a 3-level spatial pyramid for pooling CKD at different levels. The pyramid level is set as 1×1, 2×2 and 4×4. Gaussian Parzen windows are used to estimate the CSQMI, and the width parameter σ is tuned using a grid search in the range $[0.01\sigma_d, 100\sigma_d]$, where σ_d is the median distance of all training samples. The best window width is selected by cross-validation. The optimal neighborhood distance parameter, k, is decided using a grid search between 1 and 8. Linear SVM classifiers used in all experiments are implemented with the LIBlinear, downloaded from www.csie.ntu.edu. tw/~cjlin/liblinear/.

4.2 Evaluation of Codebook Learning

We first evaluate the discriminability of our CSQMI based codebook learning method by comparing it with other popular kernel based dimensionality reduction methods on the extended YaleB face dataset [35] that contains 16128 face images from 28 individuals. This dataset is challenging due to varying illumination conditions and expressions.

For each individual, half of the frontal face images are used to train the relevant codebook and feature subset. The remaining frontal face samples are used to test the distinctiveness of the learned codebook. *LBP_CKD* is applied to extract the face features. KPCA [14, 15], Kernel Fisher Discriminant Analysis (KFDA) [38], and Kernel Locality Preserving Projections (KLPP) [39] are compared with our codebook learning method. For each method, as suggested in [1], a reduced 200-dimensional feature subset is learned. To visualize the results, we randomly select five subjects and plot the distributions of projected samples onto the leading three most significant feature subsets yielded from each method in Fig. 1. As shown in Fig. 1, the clusters of the face samples resulted from our codebook represents a significant improvement on the class separation over that obtained from the alternative kernel based dimensionality reduction methods. This is because that the feature subset derived from CSQMI captures the angular pattern of the cluster distribution of the analyzing face patterns. Consequently, it is more discriminative than the feature subset selected from principal component vectors based only on magnitude of eigenvalues, such as KPCA.

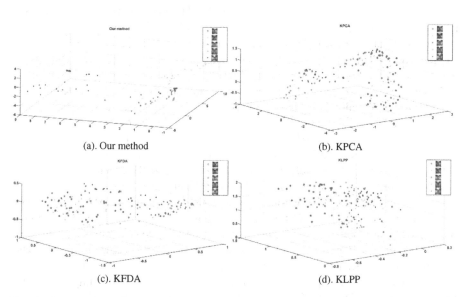

(a). Our method   (b). KPCA

(c). KFDA   (d). KLPP

Fig. 1. Visualization of the leading 3-dimensional *LBP_CKD* features from different methods.

4.3   Evaluation of Object Recognition

Caltech-101: This dataset is one of the most popular benchmarks for multiclass image recognition. It collects 9144 images from 101 object categories and a background category. Each category has 31 to 800 images with significant color, pose and lighting variations. We use this dataset for a comprehensive comparison on the recognition performance of the original KDES, supervised kernel descriptor (SKDES) [4] and our CKD. A 4-neighborhood which achieves the best performance is used to evaluate the context information for CKD. For each category, following the experimental setup of

Table 1. Comparison of mean recognition accuracy (%) and standard deviation of KDES, SKDES and CKD on Caltech-101.

Features	KDES[1]	SKDES [4]	CKD
gradient	75.2±0.4	77.3±0.7	**77.8±0.6**
color	42.4±0.5	68.4±1.4	**69.1±0.9**
texture(*lbp*)	70.3±0.6	71.6±1.3	**74.3±0.8**
combination	76.4±0.7	79.2±0.6	**83.3±0.6**
Method	Accuracy	Method	Accuracy
Jia et al. [40]	75.3±0.7	Feng et al. [45]	82.60
SLC [47]	81±0.2	SDL [41]	75.3±0.4
Adaptive deconvolutional net [44]	71.0±1.0	SSC [46]	80.02±0.36
Boureau et al. [42]	77.3±0.6	M-HMP [48]	82.5±0.5
LSAQ [43]	74.21±0.8	SPM_SIFT [7]	64.6±0.8
Pyramid SIFT (P-SIFT) [49]	80.13	PHOW [50]	81.3±0.8

original KDES [1], we train one-vs-all linear SVM classifiers on 30 images and test on no more than 80 images for KDES and our method. We run five rounds of testing for a confident evaluation. Results of SKDES are quoted from the original papers. Table 1 lists the average recognition accuracy and standard deviation of different options of kernel descriptors. Some recently reported results are also provided for comparison.

From Table 1, we observe that our CKD consistently outperforms KDES and SKDES, for both individual and combined version. Except for the gradient CKD (G_CKD), both color CKD (C_CKD) and texture CKD (LBP_CKD) are significantly better than their original KDES. In particular, compared with the original color and texture KDES, the recognition accuracy of C_CKD and LBP_CKD is increased by 62.97 % and 5.69 %, respectively. For the combined version, the accuracy of combined *CKD* is 83.3 %, which is 6.9 % higher than the original KDES combination and 4.1 % higher than the SKDES combination. We also notice the smaller standard deviation of recognition accuracy in our results compared with that of the SKDES. It means CKD is more robust than SKDES, thanks to the spatial co-occurrence constraints embedded in the CKD. We argue that the performance improvement of CKD comes from two facts: (1) compared with KDES and SKDES, the additional spatial co-occurrence constraint defined in CKD further improves its robustness to the semantic ambiguity, caused by the lack of features in case of partial occlusion; (2) KDES applies KPCA to reduce feature dimensionality, whereas we use CSQMI to learn low-dimensional CKD. KPCA only keeps KDES components that contribute most significantly to image reconstruction. In contrast, our CSQMI criterion selects the CKD that minimize the information redundancy and approximation error while maximize the mutual information between the CKD and its class label in terms of the 'angle distance'. Therefore, the resulting low-dimensional CKD are more discriminative than KDES in that they reveal the cluster structure of density distribution of pixel attributes and relate to the angular manifold of the object category.

To investigate the impact of codebook size on the recognition performance, we train classifiers using different codebook sizes and compare the recognition accuracy of

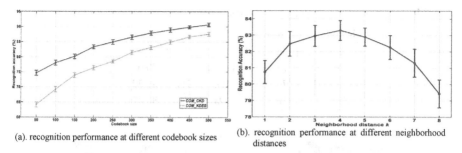

(a). recognition performance at different codebook sizes

(b). recognition performance at different neighborhood distances

Fig. 2. Performance comparison at different codebook sizes and neighborhood distances on Caltech-101.

the combined CKD (*COM_CKD*) with that of the combined KDES (*COM_KDES*) in Fig. 2(a). As expected, *COM_CKD* outperforms *COM_KDES* consistently over all codebook sizes. We also note a relative small performance drop (14 %) of *COM_CKD* when codebook size decreases from 500 to 50, whereas for *COM_KDES* the accuracy drop is 26 %. This verifies the effectiveness of our codebook learning model, which can select discriminative CKD codebook even in low-dimensional situations. We also compare the recognition performance of CKD yielded under different neighborhood distances. As shown in Fig. 2(b), neighborhoods with moderate distances perform better than neighborhoods with small distances, and recognition accuracy tends to decrease for neighborhoods with large distances. This can be understood by the fact that the discriminability of CKD tends to be smoothed, as more noises and outlier data may be included when the neighborhood distance becomes larger.

CIFAR-10: This dataset consists of 60000 tiny images with a size of 32 × 32 pixels. It has 10 categories, with 5000 training images and 1000 test images per category. We choose this dataset to test the performance of our method on recognition of tiny objects. Similar to [1], we calculate CKD around 8 × 8 image patches on a dense grid with a spacing of 2 pixels. A 3-neighborhood which gives the best performance is applied to calculate CKD. The whole training images are split into 10,000/40,000 training/validation set, and the validation set is used to optimize the kernel parameters of γ_s, γ_o, γ_c, and γ_{lbp} using a grid search. Finally, a linear SVM classifier is trained on the whole training set using the optimized kernel parameters.

We compare the performance of *COM_CKD* with several recent feature learning approaches using deep learning (stochastic pooling based Deep Convolutional Neural Network − spDCNN [52], tiled Convolutional Neural Networks − tCNN [53], Multi-column Deep Neural Networks − MDNN [51]), sparse coding (improved local Coordinate Coding − iLCC [54], spike-and-slab Sparse Coding − ssSC [55]), hierarchical kernel descriptor (HKDES) [2] and spatial pyramid dense SIFT (SPM_SIFT) [7]. For SPM_SIFT, we use a 3-layer spatial pyramid structure and calculate dense SIFT feature in an 8 × 8 patch over a regular grid with a spacing of 2 pixels. Table 2 reports the recognition accuracy of various methods. As we see, *COM_CKD* and MDNN defeat other methods by a large margin. Compared with MDNN, *COM_CKD* achieves a comparable performance with only a 0.37 % deficit in classification rate. However,

Table 2. Comparison of recognition accuracy (%) of various methods on CIFAR-10.

Method	Accuracy	Method	Accuracy
spDCNN [52]	84.88	SPM_SIFT [7]	65.60
tCNN [53]	73.10	HKDES [2]	80.00
iLCC [54]	74.50	MDNN [51]	**88.79**
ssSC [55]	78.80	*COM_CKD*	88.42

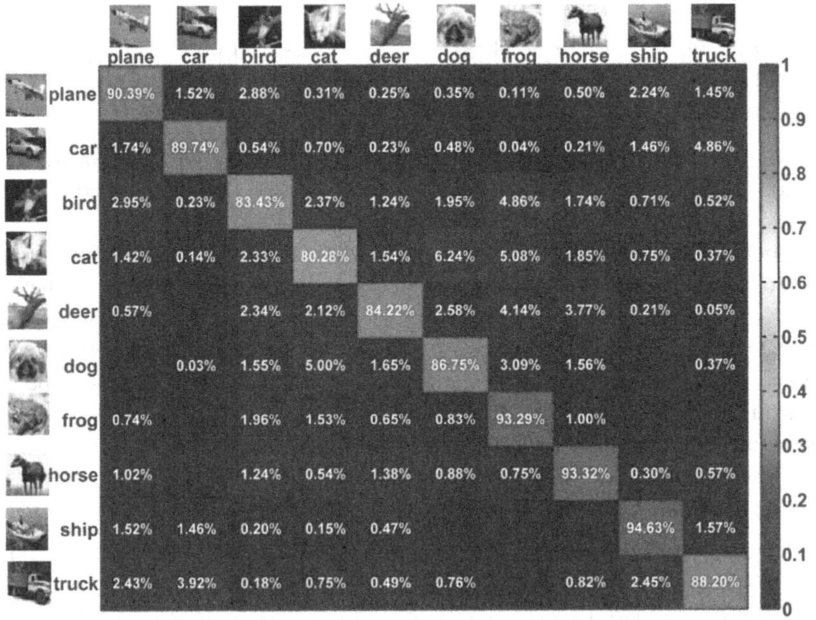

Fig. 3. Confusion matrix for CIFAR-10 using *COM_CKD*. Vertical axis shows the ground truth labels and the predicted labels go along the horizontal axis.

our method is much more simple and efficient than MDNN model. For example, for a 32 × 32 pixel image, our method takes 224.63 ms to calculate the full-dimensional 3-neighborhood *COM_CKD* and 320.21 s to learn a 200-dimensional discriminative codebook using CSQMI on average on a platform with Intel Core i7 2.7 GHz CPU and 16G RAM. Merging different pixel attributes in the kernel space, CKD tune low-level complementary cues into image-level discriminative descriptors. Even coupled with simple linear SVM classifier, our method still achieves superior performance compared with other sophisticated models.

To further analyze the classification performance of our method, we visualize the confusion matrix in Fig. 3. The confusion matrix shows that our *COM_CKD* is able to clearly distinguish animals from rigid artifacts, except for planes and birds. It is understandable because flying birds look very similar to planes (as shown in Fig. 4), especially in low-resolution images. Due to the non-rigid and deformable property of articulated objects, we also observe many confusions between different animals.

(a) Images from plane wrongly classified as bird

(b) Images from bird wrongly classified as plane

Fig. 4. Some wrongly classified samples between plane and bird.

Among all animal classes, the frog class obtains the highest false positive rate of 18.07 % from other animal classes, but it has with very few false negatives. As expected, car and truck are the most confusing artifact classes, which collectively cause a classification error rate of 8.78 %. Whereas, cat and dog are the most confusing animal classes, which collectively cause a classification error rate of 11.24 %.

4.4   Evaluation of Object Detection

To adapt our method for object detection, we train a two-class linear SVM classifier as the detector using *COM_CKD* features. For an instance image, we decompose it into several scales and detect possible locations of all candidate objects using a sliding window at each scale. Finally, we merge detection results at different scales and remove the duplicate detections at the same location. We test our detector on a chicken feet dataset collected in a chicken slaughter house. The aim of our detector is to find and localize chicken feet. As illustrated in Fig. 6, this chicken feet dataset is very challenging due to the following facts: chicken feet themselves are very small compared with other parts of the body, usually more than forty chickens are squeezed in a box, multiple chicken feet may appear in one image, in many cases feet are severely occluded (most part of feet are hidden under feather), the appearance of feet changes drastically due to different poses, and finally the color of the feet is very similar to feather and chest.

We crop a total of 717 image patches containing chicken feet as positive training examples, and 2000 patches without chicken feet as negative training examples. Another set of 318 images containing chicken feet patches never occurred in the training set are used as test set. Since chicken feet are also tiny, we use the same patch size and sampling grid for the CIFAR-10 dataset to evaluate CKD. The parameters of CKD and SVM are tuned by a 10-fold cross-validation on the training set. To judge the correctness of detections, we adopt standards of the PASCAL Challenge criterion [56], i.e. a detection is considered as correct only if the predicted bounding box overlaps at least 50 % area with the ground-truth bounding box. All other detections of the same object are counted as false positives. We compare the detection performance of our model with that of the HKDES model [2] and a 3-level SPM_SIFT [7] in terms of the Equal Error Rate (EER) on the Precision-Recall (PR) curves, i.e. PR-EER. PR-EER defines the point on the PR curve, where the recall rate equals the precision rate.

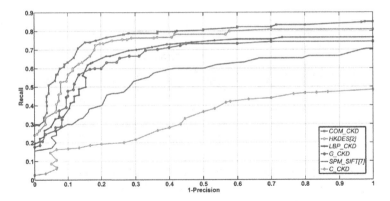

Fig. 5. Precision-Recall curves of all methods tested on chicken feet dataset.

Figure 5 plots the Precision-Recall curves for all methods. As we see, among all tested models, *COM_CKD* achieves the best overall performance (EER = 78.53 %), followed by the HKDES model (EER = 75.61 %) that combines gradient, color and shape cues into KDES. This further confirms that merging different visual cues into object representation can significantly boost the performance of the classifier. One interesting observation is that, expect for *C_CKD*, results from our single CKD models are better than the sophisticated SIFT method. In particular, EERs of *LBP_CKD* and

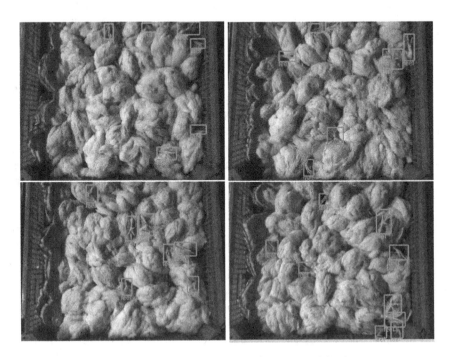

Fig. 6. Detection examples resulting from *COM_CKD* feature.

G_CKD model are 71.23 % and 69.55 %, respectively, whereas EER of SPM_SIFT is only 59.41 %. Considering individual CKD, *C_CKD* gives the worst result with EER = 44.10 %. Both *LBP_CKD* and *G_CKD* perform well, with *LBP_CKD* achieving a slightly better average accuracy. This is not surprising. Color difference between chicken feet and other parts (feather and chest) is marginal (refer to Fig. 6). Color distributions of chicken feet and other parts overlap quite much. In particular, the color distribution of feet and chest can hardly allow an acceptable separation based on color cue alone. In contrast, feet show a moderate difference in texture structures from feature and chest. Hence, texture based *LBP_CKD* outperforms other single feature for this dataset. Figure 6 shows some detection examples resulting from the best *COM_CKD* feature. Due to the influence of shadow caused by the box boundary and severe occlusions, some small chicken feet under the box shadow (in left images) or hidden by the feather (in right images) are missed by the detector, which give the false negative detections. But for these images no false positive detections appear.

5 Conclusion

Based on the context cue and Rényi quadratic entropy based CSQMI, we propose a set of novel kernel descriptors called context kernel descriptors and an information theoretic measure to select a compact and discriminative codebook for object representation in kernel feature space. We evaluate the performance of our algorithm in applications of object recognition and detection. The highlights of our work lie in: (1) the new CKD enhances the original KDES by adding extra spatial co-occurrence constraints to reduce the mismatch of image attributes (features) in kernel space; (2) instead of applying the traditional KPCA for feature dimensionality reduction, CSQMI criterion is employed in our method to learn a subset of low-dimensional discriminative CKD that correspond to the cluster structure of the density distribution of CKD. Evaluation results on both popular benchmark and our own datasets show the effectiveness of our method for generic (especially tiny) object recognition and detection.

Acknowledgements.   This work is supported by The Danish Agency for Science, Technology and Innovation, project "Real-time controlled robots for the meat industry", and partly supported by Jiangsu Natural Science Foundation (JSNSF) under Grant BK20131296, and National Nature Science Foundation of China (NSFC) under Grant 61101165. The authors thank Lantmännen Danpo A/S for providing the chicken images.

References

1. Bo, L., Ren, X., Fox, D.: Kernel descriptors for visual recognition. In: NIPS, pp. 244–252 (2010)
2. Bo, L., Lai, K., Ren, X., Fox, D.: Object recognition with hierarchical kernel descriptors. In: CVPR, vol. 1, pp. 1729–1736 (2011)
3. Bo, L., Sminchisescu, C.: Efficient match kernel between sets of features for visual recognition. In: NIPS, vol. 1, pp. 135–143 (2009)

4. Wang, P., et al.: Supervised kernel descriptor for visual recognition. In: CVPR, vol. 1, pp. 2858–2865 (2013)
5. Jégou, H., Douze, M., Schmid, C.: Packing bag-of-features. In: ICCV, vol. 1, pp. 2357–2364 (2009)
6. Cao, Y. et al.: Spatial-bag-of-features. In: CVPR, vol. 1, pp. 3352–3359 (2010)
7. Lazebnik, S., Schmid, C., Ponce, J.: Beyond bags of features: Spatial pyramid matching for recognizing natural scene categories. In: CVPR, vol. 1, pp. 2169–2178 (2006)
8. Lowe, D.: Distinctive image features from scale-invariant keypoints. IJCV 60(2), 91–110 (2004)
9. Bay, H., Ess, A., Tuytelaars, T., Gool, L.: Van.: SURF: speeded up robust features. Comput. Vis. Image Underst. 110(3), 346–359 (2008)
10. Ojala, T., Pietikäinen, M., Mäenpää, T.: Multiresolution gray-scale and rotation invariant texture classification with local binary patterns. IEEE Trans. PAMI 24(7), 971–987 (2002)
11. Dalal, N., Triggs, B.: Histograms of oriented gradients for human detection. In: CVPR, vol. 1, pp. 886–893 (2005)
12. Pedersen, K., Smidt, K., Ziem, A., Igel, C.: Shape index descriptors applied to texture-based galaxy analysis. In: ICCV, vol. 1, pp. 2240–2447 (2013)
13. Alcantarilla, P.F., Bartoli, A., Davison, A.J.: KAZE features. In: Fitzgibbon, A., Lazebnik, S., Perona, P., Sato, Y., Schmid, C. (eds.) ECCV 2012, Part VI. LNCS, vol. 7577, pp. 214–227. Springer, Heidelberg (2012)
14. Scholkopf, B., Smola, A., Mulle, K.: Kernel principal component analysis. In: ICANN, vol. 1327, pp. 583–588 (1997)
15. Scholkopf, B., Smola, A., Mulle, K.: Nonlinear component analysis as a kernel eigenvalue problem. Neural Comput. 10(5), 1299–1319 (1998)
16. Battiti, R.: Using mutual information for selecting features in supervised neural net learning. IEEE Trans. Neural Netw. 5(4), 537–550 (1994)
17. Peng, H., Long, F., Ding, C.: Feature selection based on mutual information: criteria of max-dependency, max-relevance, and min-redundancy. IEEE Trans. PAMI 27(8), 1226–1238 (2005)
18. Yang, H., Moody, J.: Feature selection based on joint mutual information. Int. ICSC Symp. Adv. Intell. Data Anal. vol. 1, pp. 22–25 (1999)
19. Kwak, N., Choi, C.: Input feature selection by mutual information based on parzen window. IEEE Trans. PAMI 24(12), 1667–1671 (2002)
20. Zhang, Z., Hancock, E.R.: A graph-based approach to feature selection. In: Jiang, X., Ferrer, M., Torsello, A. (eds.) GbRPR 2011. LNCS, vol. 6658, pp. 205–214. Springer, Heidelberg (2011)
21. Liu, C., Shum, H.: Kullback-Leibler boosting. In: CVPR, vol. 1, pp. 587–594 (2003)
22. Qiu, Q., Patel, V., Chellappa, R.: Information-theoretic dictionary learning for image classification. IEEE Trans. PAMI 36(11), 2173–2184 (2014)
23. Brown, G., Pocock, A., Zhao, M., Luján, M.: Conditional likelihood maximisation: a unifying framework for information theoretic feature selection. J. Mach. Learn. Res. 13(1), 27–66 (2012)
24. Leiva, J., Artes, A.: Information-theoretic linear feature extraction based on kernel density estimators: a review. IEEE Trans. Syst. Man Cybern. Part C Appl. Rev. 42(6), 1180–1189 (2012)
25. Hild II, K., Erdogmus, D., Principe, J.: An analysis of entropy estimators for blind source separation. Sign. Proces. 86(1), 182–194 (2006)
26. Hild II, K., Erdogmus, D., Torkkola, K., Principe, J.: Feature extraction using information-theoretic learning. IEEE Trans. PAMI 28(9), 1385–1392 (2006)

27. Rényi, A.: On measures of entropy and information. In: Fourth Berkeley Symposium on Mathematical Statistics and Probability, pp. 547–561 (1961)
28. Principe, J.: Information theoretic learning: Renyi's entropy and kernel perspectives. Springer, Heidelberg (2010)
29. Parzen, E.: On the estimation of a probability density function and the mode. Ann. Math. Statist. **33**(3), 1065–1076 (1962)
30. Jenssen, R.: Kernel entropy component analysis. IEEE Trans. PAMI **32**(5), 847–860 (2010)
31. Jenssen, R., Eltoft, T.: A new information theoretic analysis of sum-of-squared-error kernel clustering. Neurocomputing **72**(1–3), 23–31 (2008)
32. Gómez, L., Jenssen, R., Camps-Valls, G.: Kernel entropy component analysis for remote sensing image clustering. IEEE Geosci. Remote Sens. Lett. **9**(2), 312–316 (2012)
33. Zhong, Z., Hancock, E.: Kernel entropy-based unsupervised spectral feature selection. Int. J. Pattern Recogn. Artif. Intell. **26**(5), 126002-1-18 (2012)
34. Hellman, M., Raviv, J.: Probability of error, equivocation, and the Chernoff bound. IEEE Trans. Inf. Theor. **16**(4), 368–372 (1979)
35. Georghiades, A., Belhumeur, P., Kriegman, D.: From few to many: Ilumination cone models for face recognition under variable lighting and pose. IEEE Trans. PAMI **23**, 643–660 (2001)
36. Li, F., Fergus, R., Perona, P.: One-shot learning of object categories. IEEE Trans. PAMI **28**(4), 594–611 (2006)
37. Torralba, A., Fergus, R., Freeman, W.: 80 million tiny images: A large data set for nonparametric object and scene recognition. IEEE Trans. PAMI **30**(11), 1958–1970 (2008)
38. Mika, S., et al.: Fisher discriminant analysis with kernels. In: IEEE Neural Networks for Signal Processing Workshop, pp. 41–48 (1999)
39. He, X., et al.: Face recognition using laplacianfaces. IEEE Trans. PAMI **27**(3), 328–340 (2005)
40. Jia, Y., Huang, C., Darrell, T.: Beyond spatial pyramids: Receptive field learning for pooled image features. In: CVPR, vol. 1, pp. 3370–3377 (2012)
41. Jiang, Z., Zhang, G., Davis, L.: Submodular dictionary learning for sparse coding. In: CVPR, vol. 1, pp. 3418–3425 (2012)
42. Boureau, Y., et al.: Ask the locals: Multi-way local pooling for image recognition. In: ICCV, vol. 1, pp. 2651–2658 (2011)
43. Liu, L., et al.: In defense of soft-assignment coding. In: ICCV, pp. 2486–2493 (2011)
44. Zeiler, M., Taylor, W., Fergus, R.: Adaptive deconvolutional networks for mid and high level feature learning. In: ICCV, vol. 1, pp. 2018–2025 (2011)
45. Feng, J., Ni, B., Tian, Q., Yan, S.: Geometric p-norm feature pooling for image classification. In: CVPR, vol. 1, pp. 2697–2704 (2011)
46. Oliveira, G., Nascimento, E., Vieira, A.: Sparse spatial coding: a novel approach for efficient and accurate object recognition. In: ICRA, pp. 2592–2598 (2012)
47. McCann, S., Lowe, D.G.: Spatially local coding for object recognition. In: Lee, K.M., Matsushita, Y., Rehg, J.M., Hu, Z. (eds.) ACCV 2012, Part I. LNCS, vol. 7724, pp. 204–217. Springer, Heidelberg (2013)
48. Bo, L., Ren, X., Fox, D.: Multipath sparse coding using hierarchical matching pursuit. In: CVPR, vol. 1, pp. 660–667 (2013)
49. Seidenari, L., Serra, G., Bagdanov, A., Del Bimbo, A.: Local pyramidal descriptors for image recognition. IEEE Trans. PAMI **36**(5), 1033–1040 (2014)
50. Bosch, A., Zisserman, A., Munoz, X.: Image classification using random forests and ferns. In: ICCV, vol. 1, pp. 1–8 (2007)
51. Ciresan, D., Meier, U., Schmidhuber, J.: Multi-column deep neural networks for image classification. In: CVPR, pp. 3642–3649 (2012)

52. Zeiler, M., Fergus, R.: Stochastic pooling for regularization of deep convolutional neural networks. In: ICLR (2013)
53. Le, Q., et al.: Tiled convolutional neural networks. In: NIPS, vol. 1, pp. 1279–1287 (2010)
54. Yu, K., Zhang, T.: Improved local coordinate coding using local tangents. In: ICML, vol. 1, pp. 1215–1222 (2010)
55. Goodfellow, I., Courville, A., Bengio, Y.: Spike-and-slab sparse coding for unsupervised feature discovery. In: NIPS Workshop on Challenges in Learning Hierarchical Models (2011)
56. Everingham, M., et al.: The pascal visual object classes (VOC) challenge. Int. J. Comput. Vis. **88**(2), 303–338 (2010)

Detection of Abrupt Changes in Spatial Relationships in Video Sequences

Abdalbassir Abou-Elailah[1,2]([✉]), Valerie Gouet-Brunet[2], and Isabelle Bloch[1]

[1] LTCI, CNRS, Télécom ParisTech, Université Paris-Saclay, Paris, France
{elailah,isabelle.bloch}@telecom-paristech.fr
[2] Université Paris-Est, IGN, SRIG, MATIS, 73 avenue de Paris,
94160 Saint Mandé, France
valerie.gouet@ign.fr

Abstract. Detecting unusual events in video sequences is very challenging due to cluttered background, the difficulties of accurate extraction and tracking of moving objects, illumination change, etc. In this work, we focus on detecting strong changes in spatial relationships between moving objects in video sequences, with a limited knowledge of the objects. In this approach, the spatial relationships between two objects of interest are modeled using angle and distance histograms as examples. To evaluate the evolution of the spatial relationships during time, the distances between two angle or distance histograms at two different instants in time are estimated. In addition, a combination approach is proposed to combine the evolution of directional (angle) and metric (distance) relationships. Studying the evolution of the spatial relationships during time allows us to detect the ruptures in such spatial relationships. This study can constitute a promising step toward event detection in video sequences, with few a priori models on the objects.

Keywords: Spatial relationships · Angle histogram · Distances · Fuzzy object representation · Detection of ruptures · Fusion

1 Introduction

In the literature, there are many intelligent video surveillance systems, and each system is dedicated to a specific application, such as sport match analysis, people counting, analysis of personal movements in public shops, behavior recognition in urban environments, drowning detection in swimming pools, etc[1]. The VSAM project [35] was probably one of the first projects dedicated to surveillance from video sequences. The goal of ICONS project [18] was to recognize the incidents in video surveillance sequences. The goal of the three projects ADVISOR [2], ETISEO [11] and CareTracker [5] was to analyze record streaming video, in order to recognize events in urban areas and to evaluate scene understanding.

---

[1] See http://www.cs.ubc.ca/~lowe/vision.html for examples of companies and projects on these topics.

© Springer International Publishing Switzerland 2015
A. Fred et al. (Eds.): ICPRAM 2015, LNCS 9493, pp. 89–106, 2015.
DOI: 10.1007/978-3-319-27677-9_6

The AVITRACK project [3] was applied to the monitoring of airport runways, while the BEWARE project [4] aimed to use dense camera networks for monitoring transport areas (railway stations, metro).

In this context, an increasing attention is paid to "event" detection. In [28], an approach is proposed to detect anomalous events based on learning 2-D trajectories. In [30], a probabilistic model of scene dynamics is proposed for applications such as anomaly detection and improvement of foreground detection. For crowded scenes, tracking moving objects becomes very difficult due to the large number of persons and background clutter. There are many approaches proposed in the literature for abnormal event detection, based on spatio-temporal features. In [19], an unsupervised approach is proposed based on motion contextual anomaly of crowd scenes. In [23], a social force model is used for abnormal crowd behavior detection. In [9], an abnormal event detection framework in crowded scenes is proposed based on spatial and temporal contexts. The same authors proposed in [8] a similar approach based on sparse representations over normal bases. Recently, Hu et al. [16] proposed a local nearest neighbor distance descriptor to detect anomaly regions in video sequences. More recently, the authors in [32] have proposed a video event detection approach based on spatio-temporal path search. It is also applied for walking and running detection.

We adopt a different point of view. We address the question of detecting structural changes or ruptures, which can be seen as a first step for event detection. We propose to use low-level generic primitives and their spatial relationships, and we do not assume a known set of normal situations or behaviors. To our knowledge, the proposed approach is the first one that exploits low-level primitives and spatial relationships in an unsupervised manner to detect ruptures in video. In order to illustrate the interest of spatial relationships, let us consider a person leaving a luggage unattended on the ground. For human beings, it is easy to detect and recognize this kind of event. To learn an intelligent system to detect and recognize this event, one solution is to break down this event into the spatial relationships between the luggage and the person at many points in time. For example, the person holds the luggage at the beginning. If the person leaves the luggage unattended, the spatial relationships between the person and the luggage rapidly changes from very close state to far away state. Thus, detecting ruptures in spatial relationships can be important in detecting and recognizing actions or events in video sequences.

We propose to detect in an unsupervised way strong changes (or ruptures) in spatial relationships in video sequences. This rules out supervised learning-based algorithms which require specific training data. This is useful in all situations where an action or an event can be detected based on such changes or ruptures. Here, we use Harris detector [15], and/or SIFT detector [22] to extract low-level primitives, which are suitable to efficiently detect and track moving objects during time in video sequences [31,36]. In order to associate features points to objects (to compute the fuzzy representation), the algorithm proposed in [31,36] can be used. The work presented is considered as a further analysis step after tracking the objects using feature points. Furthermore, we propose a fuzzy representation of the objects, based on their feature points, to improve the representation of the objects and of the spatial relationships. Then, the structure of the scene is modeled by spatial relationships between different objects using their fuzzy representation. There are several types of spatial relationships: topological relations, metric relations, directional relations, etc. We use directional and metric relationships as an example. More specifically, we consider the angle histogram [24] for its simplicity and reliability, and similarly the distance histogram. In order to study the evolution of the spatial relationships over time and to detect strong changes in the video sequences, we need to measure the changes in the angle or distance histograms during time. Note that this approach differs from methods based on motion detection and analysis, since it considers structural information and the evolving spatial arrangement of the objects in the observed scene. In the literature, many measures have been proposed to measure the distance between two normalized histograms. Here, we propose to adapt these measures to angle histograms, in order to use them in our method. Finally, a criterion is proposed to detect ruptures in the spatial relationships based on distances between angle or distance histograms over time. In addition, a new approach is proposed for combining the distances between angle and distance histograms. The fusion consists in creating a summarized information that represents both the directional and metric spatial relationships. This is a new feature with respect to our preliminary work in [1].

The proposed methods for the fuzzy representation and detection of ruptures in the spatial relationships are described in Sect. 2. Experimental results are shown in Sect. 3 in order to evaluate the performance of the proposed approach. Finally, conclusions and future work are presented in Sect. 4.

2 Rupture Detection Approach

The proposed approach is divided into two main parts. In the first part, our goal is to estimate a fuzzy representation of the objects exploiting only feature points. In the second one, spatial relationships between objects are investigated, using this representation of the objects. Based on the evolution of the spatial relationships during time, strong changes in video sequences are detected.

The fuzzy representation of the objects using the features points is described in Sect. 2.1. Specifically, we study the spatial distribution of the feature points that are extracted using a detector such as Harris or SIFT, for a given object. Feature points can be used to isolate and track objects in video sequences [31,36]. Thus, we suppose that each moving object is represented by a set of interest points isolated from others with the help of such techniques. Here, we propose two different criteria to represent the objects as regions, exploiting only the feature points. The first one is based on the **depth** of the feature points, by assigning a value to each point based on its centrality with respect to the feature points. The second one assigns a value to each point depending on the **density** of its closest feature points. Finally, the depth and density estimations are combined together, to form a fuzzy representation of the object, where the combined value at each pixel represents the membership degree of this pixel to the object.

This allows reasoning on the feature points or on the fuzzy regions derived from them, without needing a precise segmentation of the objects.

In Sect. 2.2, the computation of the spatial relationships is discussed based on the fuzzy representation of the objects. As an example, we illustrate the concept with the computation of the angle and distance histograms. Then, the existing distances between two normalized histograms are detailed, and the adaptation of these distances to angle histograms is also discussed. Finally, a criterion is defined as the distance between the angle or distance histograms during time, in order to detect ruptures in the spatial relationships.

2.1 Fuzzy Object Representation

In this section, we detail the estimation of the fuzzy representation based on the feature points.

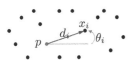

Fig. 1. Feature point distribution for a given object.

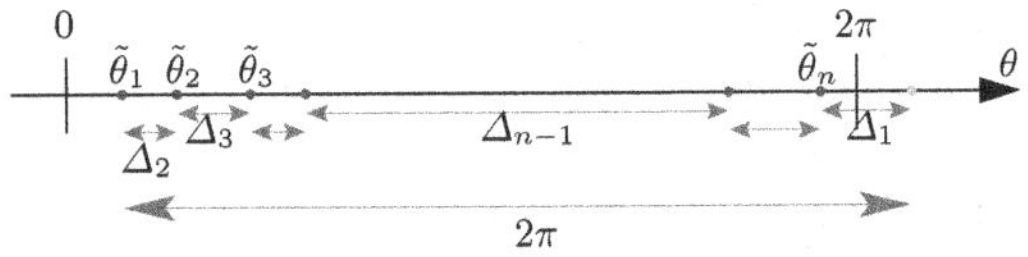

Fig. 2. Sorted angles.

Feature Detection. For a given object, let x_k $(k = 1, 2, ..., n)$ be the detected feature points. For a given pixel p of the object, let px_i denote the line connecting the pixel p and x_i $(i \in [1...n])$, d_i the distance between p and x_i, and θ_i the angle between $\vec{px_i}$ and the horizontal line as shown in Fig. 1 $(\theta_i \in [0, 2\pi])$.

Distances d_i and angles θ_i are used to estimate depth and density weights for each object based on the x_i. The depth weight is computed using the angles θ_i, and is denoted by dh. The second weight is computed using the distances d_i, and is denoted by dy. Hereafter, their estimations are described, as well as their fusion.

Depth Estimation. In the depth estimation (i.e. centrality), all the feature points are taken into account. Several approaches have been proposed in the literature for depth measures [17], such as simplicial estimation [20], half-space estimation [33], convex-hull peeling estimation [10], L1-depth [34], etc. We propose a new depth measure which is based on the entropy. For each pixel p, the computed angles θ_i are sorted in ascending order as shown in Fig. 2. Let $\tilde{\theta}_i$ $(\tilde{\theta}_j \geq \tilde{\theta}_i$ if $j > i)$ be the sorted angles. We define Δ_i as follows:

$$\Delta_i = \begin{cases} (2\pi + \tilde{\theta}_1) - \tilde{\theta}_n & \text{if } i = 1 \\ \tilde{\theta}_i - \tilde{\theta}_{i-1} & \text{if } i \in [2...n] \end{cases} \tag{1}$$

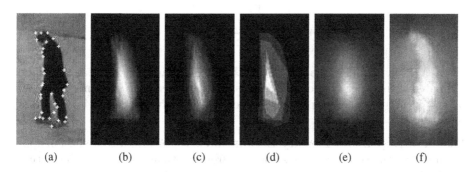

(a)   (b)   (c)   (d)   (e)   (f)

Fig. 3. Depth measures: original object with feature points (a), simplicial estimation [20] (b), half-space estimation [33] (c), convex-hull peeling estimation [10] (d), L1-depth [34] (e), and the proposed depth (f) (image from PETS 2009 database [27]).

Let $p_i = \frac{\Delta_i}{2\pi}$, p_i has two properties: $0 \leq p_i \leq 1$ and $\sum_{i=1}^{n} p_i = 1$. Thus, p_i can be seen as a discrete probability distribution of the angles. Then, the depth weight is defined as the entropy of this probability distribution:

$$\mathrm{dh}(p) = \frac{1}{n} \sum_{i=1}^{n} -p_i \log_2 p_i \tag{2}$$

This depth measure can be explained as follows: let us consider a point q inside the object with feature points distributed equitably around it in terms of directions. In this case, we obtain $p_0 = p_1 = ... = p_n$, and the depth weight of point q is equal to 1 (the highest weight). Otherwise, if the point q is outside the object, the depth weight depends on the angle view (Δ_1 can represent the angle view) and the distribution of the feature points inside the object ($p_2, p_3, ..., p_n$). If the angle view becomes smaller and smaller (e.g. the point q is moving away from the object), the depth weight of the point q becomes also smaller accordingly.

Figure 3 shows the representation of several state of the art depth estimations for an object, including our proposal. As we can see, the entropy depth can better represent the shape of the object than the existing depth measures. In terms of computation time, the L1-depth and the proposed depth are the most efficient ones compared to other measures. Our experimental tests showed that the choice of a particular depth measure has a limited impact on the detection of the rupture. However, the entropy depth measure may present a significant enhancement compared to other depth measures, in the applications that need a precise shape estimation, to describe fine relationships, for example when objects meet.

Density Estimation. For density estimation, for a given pixel inside the object, only the neighbor feature points are taken into consideration (feature points within a certain distance r, or k closest feature points). Thus, the distances d_i that are lower than a certain distance r are taken into account to compute the density weight for the pixel p as follows:

$$dy(p) = \sum_{i=1}^{M}(1 - \frac{d_i}{r}), \text{ where } d_i \leq r \qquad (3)$$

where M is the number of points inside the circle of radius r. This radius can be estimated automatically and online, based on statistics on the distances between points, in order to be adapted to the scale of the object. Figure 4(c) shows a representation of the density estimation.

Fusion of Depth and Density Estimations. We present a combination approach to fuse the two estimations obtained from depth and density of the feature points. For the sake of optimization, the pixels q that are taken into consideration for the fusion are defined as follows: $dy(q) > 0$ or $dh(q) > th$, where th is a given threshold. The obtained estimation of the object is referred to as "fuzzy representation".

Here, the z-score [6] is applied on the two estimations, in order to make them comparable. The z-score is the most commonly used normalization process. It converts all estimations to a common scale with an average of **zero** and a standard deviation of **one**. It is defined as follows: $Z = (X - \overline{M})/(\sigma)$, where $\overline{M}$ and σ represent the average and the standard deviation of the X estimation, respectively. Let Z^{dh} and Z^{dy} be the depth and density estimations respectively, after applying the z-score normalization.

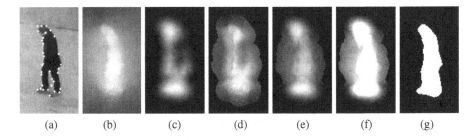

Fig. 4. Original object with the feature points (a), depth estimation (b), density estimation (c), fusion using min operator (d), fusion using max operator (e), fusion using Eq. 4 (f), and the object segmented precisely GT (g).

The obtained fuzzy representation, using different fusion operators, are compared with a Ground Truth (GT) where the objects are segmented precisely (see Sect. 3 for details, and an example in Fig. 4(g)). The combination approach which gives the best performance consists in using the two operators min and max together as defined in the following expression:

$$F(p) = \min\left(\max\left(Z^{dh}(p), Z^{dy}(p)\right), \hat{\sigma}\right) \qquad (4)$$

where $\hat{\sigma} = \frac{1}{2th}$. Then, F is normalized using Min-Max scaling [14] to obtain the membership function μ_F which varies in $[0,1]$. This fusion can be explained as follows: when Z^{dh} (or Z^{dy}) is greater than $\hat{\sigma}$, the membership value $\mu_F(p)$ is equal to 1. Otherwise, $\mu_F(p)$ is less than 1 according to the maximum between them. As an example, Fig. 4 shows different fuzzy representations of the object using min operator, max operator, and Eq. 4 for the fusion. As we can see, the last fusion approach shows the best fuzzy representation of the object according to the ground truth. The obtained fuzzy representations are used to compute the spatial relationships.

2.2 Spatial Relationships and Rupture Detection

Here, the goal is to estimate the spatial relationships between two objects based on their fuzzy representation. The angle [24] and distance histograms are selected as examples to model the spatial relationships. It is important to note that the proposed method also applies to other types of spatial relationships.

Angle Histogram. Given two fuzzy regions $A = \{(a_i, \mu_A(a_i)), i = 1, ..., n\}$ and $B = \{(b_j, \mu_B(b_j)), j = 1, ..., m\}$, where a_i and b_j are the elements of A and B, and μ_A and μ_B represent their membership functions respectively, for all possible pairs $\{(a_i, b_j), a_i \in A$ and $b_j \in B\}$, the angle θ_{ij} between a_i and b_j is computed, and a coefficient $\mu_\Theta(\theta_{ij}) = \mu_A(a_i) \times \mu_B(b_j)$ is derived. For a given direction α, all the coefficients of the angles that are equal to α are accumulated as follows:

$$h^\alpha = \sum_{\theta_{ij}=\alpha, i=1,..,n, j=1,..,m} \mu_\Theta(\theta_{ij}) \tag{5}$$

Finally, $h = \{(\alpha, h^\alpha), \alpha \in [0, 2\pi]\}$ is the angle histogram. In our case, the histogram can be seen as an estimate of the probability distribution of the angles. Thus, the obtained histogram is normalized to display frequencies of the existed angles with the total area equaling 1. It is normalized by dividing each value by the sum $R_h = \sum_{\alpha \in [0,2\pi]} h^\alpha$, instead of normalizing by the maximum value (which would correspond to a possibilistic interpretation).

When the objects are represented sparsely by feature points, then $\mu_A(a_i) = 1$ and $\mu_B(b_j) = 1$ (where a_i and b_j represent the feature points on the objects A and B respectively), and the same approach is used to compute the angle histogram between the two sparse objects A and B.

Distance Histogram. In this case, all the distances d_{ij} between a_i ($i = 1, ..., n$) and b_j ($j = 1, ..., m$) are computed. Based on these distances, the distance histogram is formulated in the same way as the angle histogram:

$$h^l = \sum_{d_{ij}=l, i=1,..,n, j=1,..,m} \mu_L(d_{ij}) \tag{6}$$

where $\mu_L(d_{ij}) = \mu_A(a_i) \times \mu_B(b_j)$ and l represents a given distance value. The obtained histogram is normalized such that the sum of all bins is equal to 1.

Comparison of Spatial Relationships. There are two main approaches to estimate distances between histograms. The first approach is known as bin-to-bin distances such as L_1 and L_2 norms. The second one is called cross-bin distances; it is more robust and discriminative since it takes the distance on the support of the distributions into account. Note that the bin-to-bin distances may be seen as particular cases of the cross-bin distances. Several distances based on cross-bin distances, such as Quadratic-Form (QF) distance [13], Earth Mover's Distance (EMD) [29], Quadratic-Chi (QC) histogram distance [25], have been proposed in the literature. We have tested these three distances on different examples, and experiments showed that they were well adapted to angle histograms. Finally, the QF distance was used in our experiments to assess the distance between the angle or distance histograms during time, because of its simplicity. It is defined as follows: $d(h_1, h_2) = \sqrt{ZSZ^T}$, where $Z = h_1 - h_2$ and $S = \{s_{ij}\}$ is the bin-similarity matrix. This distance is commonly used for normalized histograms (the distance histogram for example). Here, we propose an approach to adapt it to the case of angle histograms just by adjusting the elements of the similarity matrix S. We consider that the two histograms h_1 and h_2 defined on $[0, 2\pi]$ consist of k bins B_i. Usually, for a distribution on the real line, the distance between B_i and B_j is defined as follows: $x_{ij} = |B_i - B_j|$, where $1 \le i \le k$ and $1 \le j \le k$. However, in the case of angle histograms, the distance between B_i and B_j is defined as follows: $x_{ij}^c = \min(x_{ij}, 2\pi - x_{ij})$ to account for the periodicity on $[0, 2\pi]$. Thus, the elements of the matrix S are simply defined, in the case of angle histograms, using x_{ij}^c instead of x_{ij} as follows:

$$s_{ij} = 1 - \frac{x_{ij}^c}{\max_{i,j}(x_{ij}^c)} \tag{7}$$

Criterion for Rupture Detection. Based on the fuzzy representation of the objects exploiting only the feature points, the angle or distance histogram h between two different objects is computed. Let f_i $(i = 0, 1, ..., N - 1)$ be the frames of the video sequences, and h_i be the computed angle or distance histogram between the objects A and B in frame f_i. We define $y(i) = d(h_i, h_{i+1})$ for each $i = 0, 1, ..., N - 1$. This function describes the evolution of the angle or distance histograms over time. If a strong change in the spatial relationships occurs at instant R $(R < N)$, where R denotes the instant of rupture, this means that the angle or distance histogram h_R effectively changes compared to previous angle or distance histograms $(h_i, i < R)$. A rupture is detected according to the following criterion W: $\forall i < R - 1$, $y(R - 1) - y(i) > t$, and t is a threshold value. Thus, the instant of rupture R can be effectively detected from the analysis of the function y.

Here, in order to clearly show the instant of ruptures in the spatial relationships and remove noise, we also show the evolution of the function y filtered by a Gaussian derivative, denoted by g, instead of a simple finite difference. This filter can remove noise and the function g effectively exhibits the instant of the strong changes in the spatial relationships using a threshold approach. This approach is particularly well suited for abrupt changes, leading to clear peaks in the function g, that are then easy to detect (a simple threshold can be sufficient). For slower changes, a multiscale approach can be useful to detect more spread peaks.

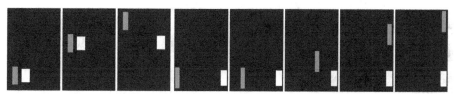

(a) Frames number 1, 30, and 50 of (b) Frames number 45, 55, 74, 95, and 105 of SE 2.
SE 1.

(c) Frames number 450, 462, and 468 (d) Frames number 595, 630, 670, and 700 of RE 2 selected
of RE 1 selected from PETS 2009.   from PETS 2009.

Fig. 5. Events SE 1 (a), SE 2 (b), RE 1 (c) and RE 2 (d).

Fusion of Directional (angle) and Metric (distance) Evolutions. To distinguish between two functions that are derived from angle and distance histogram, let y^θ and y^d be the functions that represent the evolution of directional (angle) and metric (distance) spatial relationships during time respectively. The goal of this study is to combine the two functions y^θ and y^d in an efficient way, in order to produce a unique function y^u, which allows us to detect the strong changes in both directional and metric relationships, at the same time. Thus, if a rupture occurs in at least one of them (y^θ and y^d), this rupture must be efficiently detected using the function y^u.

To combine the two evolutions, at each instant time k, the two distances $y^\theta(k)$ and $y^d(k)$ are extracted and used to provide a single point $p_k(x_k, y_k)$ in $\mathbf{R}^2$, defined as follows :

$$\begin{cases} x_k = (d_0 + y^d(k)) \cos(y^\theta(k)) \\ y_k = (d_0 + y^d(k)) \sin(y^\theta(k)) \end{cases} \tag{8}$$

where d_0 is a constant, to account for the variation of the distance $y^\theta(k)$ when the value of $y^d(k)$ is very small (e.g. close to 0). Furthermore, we define a single function y^u using the points p_k ($k = 1, ..., n$), by computing the distances between two consecutive points p_k and p_{k+1}, over time. The value of the function y^u at instant time k is computed as follows:

$$y^u(k) = \sqrt{(x_{k+1} - x_k)^2 + (y_{k+1} - y_k)^2} \tag{9}$$

this function y^u can be used to detect the ruptures in both directional and metric relationships, using the approach described above.

(a) Frames number 1, 5, 10, and (b) Frames number 1955, 2010, 2060, and 2100 of RE 3 selected
50 of SE 3.   from PETS 2006 [26].

Fig. 6. Events SE 3 (a) and RE 3 (b).

3   Experiments and Evaluations

To evaluate the performance of the proposed approach, we created some synthetic events (illustrated in Fig. 5(a) and (b)), and also used a variety of events selected from the PETS 2009 datasets [27] (illustrated in Fig. 5(c) and (d)). Here, we call "event", some frames that contain a rupture in the spatial behavior. The results of the proposed fuzzy representation are also compared to classical segmentation approaches: a binary segmentation approach [7] and an approach using differences between the background and the actual frame. Then, morphological operations are carried out to remove small objects and fill holes. The last one is used as ground truth (GT) because it produces very precise segmentations.

A synthetic event and an event selected from PETS 2006 dataset [26], displayed in Fig. 6, are used to illustrate the proposed approach using the distance histogram. To associate feature points to objects, here we simply consider the points included in the bounding boxes associated with objects available in the PETS 2009 dataset.

3.1   Parameters Tuning

In this section, some results are detailed concerning the tuning of the parameters that are used in the proposed approach. Specifically, we discuss the estimation of the radius r, which is used in the computation of the density estimation. Then, some results are shown for different values of the threshold th, which is used in the combination of depth and density estimations. Finally, we show the effect of the number of bins on the computation of the distance between two angle histograms.

r **Parameter.** Fig. 7 shows different estimations of the radius r (normalized) during time. First, all the possible distances d_{ij} among the feature points are computed. The mean, median, and maximum of these distances are computed, as shown in the figure (three first curves). Then, Delaunay triangulation is applied on the feature points, and two other estimations of the radius r are computed, as the mean and median of the lengths of the triangle edges (fourth and fifth curves). Finally, as in [21], the median of all radius of the circumscribed circle around the Delaunay triangles provides the last estimation (last curve). As we can see, the maximum of the distances (third curve) gives the most robust and stable estimation during time. Other experiments on different objects show the same result. Thus, the expression

$$r = \max_{i=1,..,n, j=i,..,m} \frac{d_{ij}}{6} \tag{10}$$

is adopted to estimate the radius r for the density estimation.

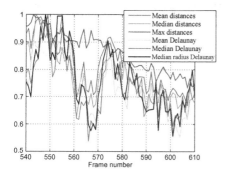

Fig. 7. Different estimations of the radius r based on the feature points.

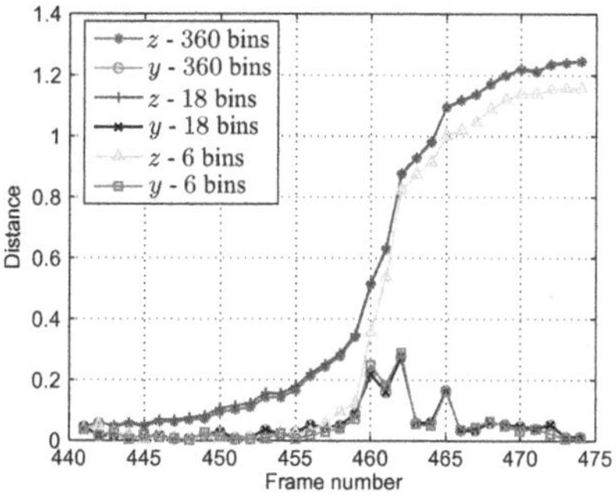

Fig. 8. The functions y and z over time using various number of bins.

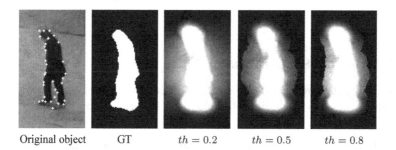

Original object   GT $th = 0.2$ $th = 0.5$ $th = 0.8$

Fig. 9. Original object with the feature points, GT of the object, and fuzzy representations of the object for th equal to 0.2, 0.5, and 0.8 respectively.

th **Parameter.** In the fusion of depth and density estimations, a threshold th is used. Figure 9 shows the original object with the feature points, the ground truth (GT) of the object, and the fuzzy representation (FR) of the object for different values of th. As we can see, the proposed fusion approach is quite robust to the variation of the used threshold th. In the paper, a value of th equal to 0.5 is used in the combination of depth and density estimations.

Number of Bins. In this section, we study the effect of the number of bins (quantification) on the distance between two angle histograms. We defined the function y as the distance between two successive angle histograms in frames f_i and f_{i+1}. Here, we also define $z(i) = d(h_0, h_i)$ for $i = 0, 1, ..., N - 1$, i.e. the distance to the histogram in the initial frame, to consider strong changes in the angle histograms. Figure 8 shows the evolution of the two functions y and z, for numbers of bins of 360, 18, and 6. As we can see, there is almost no difference between 360 and 18 bins, for the two functions. For a number of bins equal to 6, there is a difference compared to 360 and 16 bins for the function z. For the function y, the three curves are almost the same. Thus, the used distance between two angle histograms is robust to the variation of the number of bins.

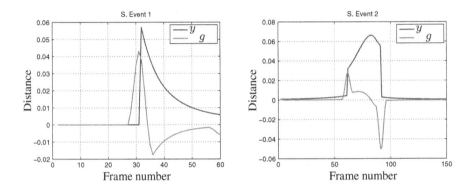

Fig. 10. Functions y and g for events SE 1 (left) and 2 (right), computed from angle histograms.

3.2   Ruptures in Spatial Relationships

We now illustrate how the analysis of the distances between histograms allows us to detect ruptures in spatial relations, both for orientation and distances.

Angle Histogram. Three snapshots of the first synthetic event (SE) are shown in Fig. 5(a) (two objects moving together and then separately). In this case, there is a rupture in the directional spatial relationships, when the two objects diverge. Figure 5(b) shows five snapshots of the second SE. In this event, the object B moves towards the object A (fixed) from the left to the right. Then, the object B changes of direction (frame 74), and when the object B becomes above the object A, it goes towards the top.

Figure 10 shows the functions y and g during time for the two events SE 1 and 2. For the event SE 1, the function y shows a strong variation at frame number 31. At this instant, there is the rupture in the spatial relationships (the two objects begin to separate). Using the evolution of g over time, the instant of the rupture can be detected by applying a threshold (a threshold of 0.02 can be used to detect the instants of rupture for the SE). For the second SE, we can see two strong variations in the function y; the first strong variation (frame 60) occurs when B

changes of direction with respect to A, the second strong variation (frame 90) occurs when B becomes above A and changes its direction towards the top. The function g clearly shows the two strong variations. Thus, the proposed method can efficiently detect the instants of ruptures in the spatial relationships. Other SE were created and tested using the proposed approach, and similar results were obtained.

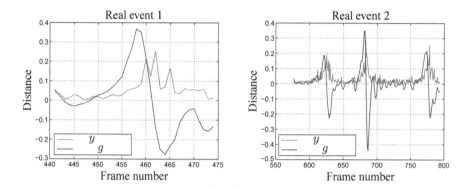

Fig. 11. Functions y and g over time, using the proposed fuzzy representation, for the events RE 1 (left) and RE 2 (right), computed from angle histograms.

Let us now evaluate the proposed detection of ruptures in the spatial relationships in the presence of noise (deformation of objects, etc.) in real events. For the real event (RE) 1 (Fig. 5(c)), the two persons converge then diverge. Figure 11 (left) shows the functions y and g over time using the proposed fuzzy representation, for the event RE 1. Two ruptures in the directional spatial relationships exist in this event. The first one is when the two persons meet, and the second rupture when the two persons separate. It is clear that the two instants of the ruptures can be efficiently detected using the evolution of g (a threshold of 0.2 can be used to detect the instants of ruptures for the RE). In the event RE 2 (Fig. 5(d)), the two persons (surrounded by white and blue bounding boxes) converge and diverge several times. In Fig. 11 (right), we show the functions y and g over time, using the fuzzy representation of the objects, for the event RE 2. All the ruptures in the directional spatial relationships can be efficiently detected using the function g.

Distance Histogram. Four snapshots of the third synthetic event are shown in Fig. 6(a). At the beginning of this event, the two objects diverge at a speed of 5 pixels/frame, and at a given instant (precisely at frame 10), the speed of the two objects becomes 10 pixels/frame. Thus, the velocity of the objects is suddenly increased. Figure 6(b) shows four snapshots of the third real event selected from PETS 2006. In this event, the luggage is attended to by the owner for a moment, and then the person leaves the place and goes away.

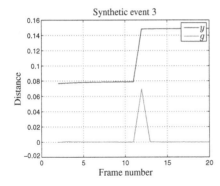

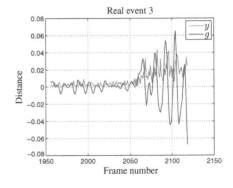

Fig. 12. Functions y and g over time, using the proposed fuzzy representation, for the event SE 3, computed from distance histograms.

Fig. 13. Functions y and g over time, using the proposed fuzzy representation, for the event RE 3, computed from distance histograms.

In Fig. 12, the functions y and g during time for the event SE 3 are shown. As we can see, the function y shows a strong variation at frame number 10, when the velocity of the objects changes. At this instant, a rupture in the metric spatial relationships is detected, using the evolution of g over time.

In the presence of noise, we show in Fig. 13 the functions y and g during time for the third real event. When the person leaves the place and goes away, we can see a strong change in the function y. By nalyzing the obtained results, the instant of rupture in the metric spatial relationships can be detected. These results can be used to indicate events occurring in the video sequences, such as escaping in Fig. 6(a) and Left-Luggage in Fig. 6(b).

Fusion of Angle and Distance Histograms. To evaluate the performance of the proposed approach for the fusion of directional and metric information, we create a synthetic event that contains many ruptures, in directional and metric spatial relationships, during time. In Fig. 14, we show the obtained functions y^θ and y^d over time. Note that y^θ and y^d represent the evolution of spatial relationships using the angle histogram and the distance histogram respectively. As we can see, there are many directional and metric ruptures in these functions. On the right side, we show the results of the combination of the two functions y^θ and y^d using a naive method (in this approach, the fusion consists in averaging the two functions) and the proposed approach. As we can observe, the proposed approach shows clearly the instant of ruptures in both directional and metric spatial relationships, and provides higher values than the naive fusion for most of the ruptures.

As we can see, when a rupture occurs in both functions y^θ and y^d, it is clearly shown in the function y^u (see instants 10 and 94). In addition, the last rupture in the function y^d (at instant 100) can be efficiently detected using the function

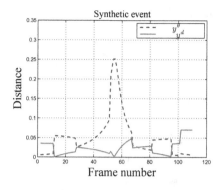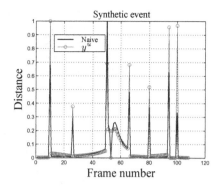

Fig. 14. Functions y^θ and y^d over time, for a synthetic event (left) and the combination of the functions y^θ and y^d using the naive fusion and the proposed combination (right).

y^u, even if there is no strong change in the function y^θ at this instant. Similarly, the rupture at instant 50 in the function y^θ can be efficiently detected using y^u, even if at this instant, no rupture is shown in the function y^d. Thus, the function y^u can show clearly the ruptures that occur in at least one of the functions y^θ and y^d.

3.3 Impact of Object Representation

Here, we show the importance of the fuzzy representation based on a simple feature points representation. Two feature detectors, Harris and SIFT, are tested. Figure 15 illustrates the function y during time (computed here from angle histograms) for different representations of the objects, for RE 1. The Harris and SIFT features are directly used to estimate the spatial relationships between the two objects and to compute the function y (red and green curves in the figure). In addition, we show in the same figure the evolution of the function y computed on the fuzzy representation of the objects using the Harris and SIFT features (blue and black curves in the figure). As we can see, the evolution of the function y obtained from the fuzzy representation of the objects using the SIFT features (black curve) can significantly reduce the variation of the distance (i.e. less amplitude of the curve) on areas when there is no rupture in the spatial relationships (see Fig. 15, frames 440 to 456) with respect to the SIFT features without computing the fuzzy representation. Thus, the proposed fuzzy representation of the objects before computing the spatial relationships can improve the robustness of the detection of ruptures, based on the observation that SIFT features are more noisy across frames than Harris features in this sequence.

However, noise is present in the function y for all object representations. Assuming that the function y has additive Gaussian noise, the algorithm proposed by Garcia [12] is used to estimate the variance of the noise of the function y, for the different object representations: Harris features, fuzzy representation

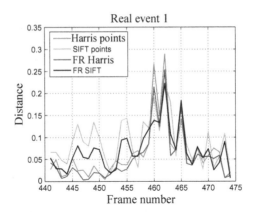

Fig. 15. Function y over time, computed from angle histograms, for different estimations of the objects: Harris features, SIFT features, fuzzy representation (FR) of the objects using Harris features (FR Harris) and SIFT features (FR SIFT), for RE 1.

of the objects using Harris features (FR Harris), SIFT features, fuzzy representation of the objects using SIFT features (FR SIFT), the binary segmentation using Mean-Shift algorithm [7] and GT.

Table 1. Estimated variance of the noise $(\times 10^{-4})$ [12] in the function y, for different object representations, for RE 1 and 2.

Event	Harris	FR Harris	SIFT	FR SIFT	Mean-Shift	GT
RE 1	13	12	27	10	31	12
RE 2	7.7	5.48	8.9	7	31	5.4

Table 1 shows the variance of the noise in the function y, for the different object representations, for the two events RE 1 and 2. It is clear that the proposed fuzzy representation significantly reduces the variance of the noise, which becomes close to the one of the GT. Especially, for SIFT features, the variance of the noise reduces from 27 to 10 for RE 1, and from 8.9 to 7 for RE 2. In addition, the variance of the noise of the proposed object representation is significantly less than the one of the binary segmentation using Mean-Shift algorithm.

4   Conclusion

In this paper, a new method was proposed to detect strong changes in spatial relationships in video sequences. Specifically, new approaches have been proposed to compute depth and density estimations, based on feature points, as well as fuzzy representations of the objects by combining depth and density estimations. Exploiting the fuzzy representations of the objects, the angle and distance histograms are computed. Then, the distance between the angle or distance histograms is estimated during time. Based on these distances, a criterion is defined in order to detect the significant changes in the spatial relationships. A new approach has been also proposed to combine directional and metric spatial relationships. The proposed method shows good performances in detecting ruptures in the spatial relationships for both synthetic and real video sequences.

Future work will focus on further improvement of the proposed method in order to detect other kinds of ruptures, and investigating the use of spatio-temporal relationships. Besides, we will investigate multi-time scale analysis, in order to better detect events that take more time to happen. In addition, proposing a complete event detection framework based on spatial relationships as discriminative features seems to be promising.

Acknowledgements. This research is part of French ANR project DESCRIBE "Online event detection in video sequences using structural and Bayesian approaches".

References

1. Abou-Elailah, A., Gouet-Brunet, V., Bloch, I.: Detection of ruptures in spatial relationships in video sequences. In: International Conference on Pattern Recognition Applications and Methods (ICPRAM), pp. 110–120 (2015)
2. Advisor. Advisor Project (2000). http://www-sop.inria.fr/orion/ADVISOR/
3. Avitrackr. Avitrackr Project (2004). http://www-sop.inria.fr/members/Francois.Bremond/topicsText/avitrackProject.html
4. Beware. Beware Project (2007). http://www.eecs.qmul.ac.uk/~sgg/BEWARE/
5. Caretaker. Caretaker Project (2006). http://www-sop.inria.fr/members/Francois.Bremond/topicsText/caretakerProject.htm
6. Carroll, S.R., Carroll, D.J.: Statistics made simple for school leaders: data-driven decision making. R&L Education (2002)
7. Comanicu, D., Meer, P.: Mean shift: a robust approach toward feature space analysis. IEEE Trans. Pattern Anal. Mach. Intell. **24**(5), 603–619 (2002)
8. Cong, Y., Yuan, J., Liu, J.: Abnormal event detection in crowded scenes using sparse representation. Pattern Recogn. **46**(7), 1851–1864 (2013)
9. Cong, Y., Yuan, J., Tang, Y.: Video anomaly search in crowded scenes via spatio-temporal motion context. IEEE Trans. Inf. Forensics Secur. **8**(10), 1590–1599 (2013)
10. Eddy, W.: Convex hull peeling. In: COMPSTAT, pp. 42–47 (1982)
11. Etiseo (2004). http://www-sop.inria.fr/orion/ETISEO/
12. Garcia, D.: Robust smoothing of gridded data in one and higher dimensions with missing values. Comput. Stat. Data Anal. **54**(4), 1167–1178 (2010)
13. Hafner, J., Sawhney, H., Equitz, W., Flickner, M., Niblack, W.: Efficient color histogram indexing for quadratic form distance functions. IEEE Trans. Pattern Anal. Mach. Intell. **17**(7), 729–736 (1995)
14. Han, J., Kamber, M., Pei, J.: Data Mining: Concepts and Techniques. Morgan Kaufmann, San Francisco (2006)
15. Harris, C.G., Stephens, M.J.: A combined corner and edge detector. In: Fourth Alvey Vision Conference, pp. 147–151 (1988)

16. Hu, X., Hu, S., Zhang, X., Zhang, H., Luo, L.: Anomaly detection based on local nearest neighbor distance descriptor in crowded scenes. The Sci. World J. **2014**, 12 pages (2014)
17. Hugg, J., Rafalin, E., Seyboth, K., Souvaine, D.: An experimental study of old and new depth measures. In: Workshop on Algorithm Engineering and Experiments (ALENEX), pp. 51–64 (2006)
18. Icons. Icons Project (2000). http://www.dcs.qmul.ac.uk/research/vision/projects/ICONS/
19. Jiang, F., Wu, Y., Katsaggelos, A.K.: Detecting contextual anomalies of crowd motion in surveillance video. In: 16th IEEE International Conference on Image Processing, pp. 1117–1120 (2009)
20. Liu, R.: On a notion of data depth based on random simplices. The Ann. Stat. **18**(1), 405–414 (1990)
21. Loménie, N., Stamon, G.: Morphological mesh filtering and α-objects. Pattern Recogn. Lett. **29**(10), 1571–1579 (2008)
22. Lowe, D.G.: Distinctive image features from scale-invariant keypoints. Int. J. Comput. Vis. **60**(2), 91–110 (2004)
23. Mehran, R., Oyama, A., Shah, M.: Abnormal crowd behavior detection using social force model. In: IEEE Conference on Computer Vision and Pattern Recognition, pp. 935–942 (2009)
24. Miyajima, K., Ralescu, A.: Spatial organization in 2D images. In: Third IEEE Conference on Fuzzy Systems, pp. 100–105 (1994)
25. Pele, O., Werman, M.: The quadratic-chi histogram distance family. In: Daniilidis, K., Maragos, P., Paragios, N. (eds.) ECCV 2010, Part II. LNCS, vol. 6312, pp. 749–762. Springer, Heidelberg (2010)
26. PETS (2006). http://www.cvg.rdg.ac.uk/PETS2006/data.html
27. PETS (2009). http://www.cvg.rdg.ac.uk/PETS2009/a.html
28. Piciarelli, C., Micheloni, C., Foresti, G.L.: Trajectory-based anomalous event detection. IEEE Trans. Circ. Syst. Video Technol. **18**(11), 1544–1554 (2008)
29. Rubner, Y., Tomasi, C., Guibas, L.J.: The earth mover's distance as a metric for image retrieval. Int. J. Comput. Vis. **40**(2), 99–121 (2000)
30. Saleemi, I., Shafique, K., Shah, M.: Probabilistic modeling of scene dynamics for applications in visual surveillance. IEEE Trans. Pattern Anal. Mach. Intell. **31**(8), 1472–1485 (2009)
31. Tissainayagam, P., Suter, D.: Object tracking in image sequences using point features. Pattern Recogn. **38**(1), 105–113 (2005)
32. Tran, D., Yuan, J., Forsyth, D.: Video event detection: From subvolume localization to spatio-temporal path search. IEEE Trans. Pattern Anal. Mach. Intell. **36**(2), 404–416 (2014)
33. Tukey, J.W.: Mathematics and the picturing of data. In: International Congress of Mathematicians, vol. 2, pp. 523–531 (1975)
34. Vardi, Y., Zhang, C.-H.: The multivariate l1-median and associated data depth. Nat. Acad. Sci. **97**(4), 1423–1426 (2000)
35. Visam. Visam Project (1997). http://www.cs.cmu.edu/~vsam/
36. Zhou, H., Yuan, Y., Shi, C.: Object tracking using SIFT features and mean shift. Comput. Vis. Image Underst. **113**(3), 345–352 (2009)

Diffusion-Based Similarity for Image Analysis

Jan Gaura$^{(\boxtimes)}$ and Eduard Sojka

Faculty of Electrical Engineering and Computer Science,
VŠB - Technical University of Ostrava, 17. listopadu 15, 708 33
Ostrava-Poruba, Czech Republic
{jan.gaura,eduard.sojka}@vsb.cz
http://www.cs.vsb.cz

Abstract. Measuring the distances is a key problem in many image-analysis algorithms. This is especially true for image segmentation. It provides a basis for the decision whether two image points belong to a single or to two different image segments. Many algorithms use the Euclidean distance, which may not be the right choice. The geodesic distance or the k shortest paths measure the distance along the surface that is defined by the image function. The diffusion distance seems to provide better properties since all the paths are taken into account. In this paper, we show that the diffusion distance has the properties that make it difficult to use in some image processing algorithms, mainly in image segmentation, which extends the recent observations of some other authors. We propose a new measure called normalised diffusion cosine similarity that overcomes some problems of diffusion distance. Lastly, we present the necessary theory and the experimental results.

Keywords: Distance measurement · Euclidean distance · Geodesic distance · Diffusion distance · Normalised diffusion similarity

1 Introduction

Measuring the distance is an important problem in clustering and image segmentation. The distance is used as a quantity that makes it possible to decide whether two image pixels belong to one or two different clusters (image segments). The Euclidean distance (i.e. the direct straight-line distance) need not be the best choice. In images, the image points form a certain surface in some space. Measuring the distance along this surface promises better results.

The *geodesic distance* [13,17] measures the length of the shortest path lying entirely on the surface. The problem is that the geodesic distance can be influenced significantly by relatively small disturbances in image since only one (and "thin") path on the surface determines the distance. In [4], the possibility of computing k shortest paths is discussed. This can be viewed as an attempt to take into consideration the connection that is not thin, but has a certain width, which reduces the influence of disturbances and noise.

The *resistance distance* is a metric on graphs [1,8]. The resistance distance between two vertices of graph is equal to the effective resistance between the

© Springer International Publishing Switzerland 2015
A. Fred et al. (Eds.): ICPRAM 2015, LNCS 9493, pp. 107–123, 2015.
DOI: 10.1007/978-3-319-27677-9_7

corresponding nodes in an equivalent electrical network (regular grid in this case). The resistances of edges in the network increase with the increasing local image contrast. Intuitively, the resistance distance explores all the existing paths between two points whereas the geodesic distance explores only the shortest of them.

It was shown that the resistance distance is equivalent to so called *commute-time distance* [5,14,20] which is the distance based on summing the *diffusion distance* in time. Diffusion is a process during which a certain substance, e.g. heat or electric charge diffuses from the places of its greater concentration to the places where the concentration is lower. The mathematical description can be built on the diffusion equation (i.e. can be physically based) or on the Markov matrices describing the random walker technique [6]. The diffusion maps were systematically introduced in [3,12]. Although further papers appear, e.g. [10], almost nothing is reported about successful use of diffusion distance for image segmentation. This can be regarded as surprising since, at a first glance, the method should have the properties that are useful. For measuring every distance, it examines many paths on the image surface.

In this paper, we show that the diffusion distance need not be beneficial for measuring distances in image segmentation. The reason is that the influence of different sizes of image segments may overshadow the influence of the edges between them (i.e. the differences in brightness or colour). This finding extends the observations of some other authors that appeared recently [19]. We introduce a new measure called *normalised diffusion cosine similarity* in which the mentioned problem is significantly reduced. The computational technique (as well as the time complexity) remains similar as is usually presented for the diffusion distance, i.e. it is based on the spectral decomposition of the Laplacian matrix.

The paper is organised as follows. In the following section, we recall the needed theoretical background. In Sect. 3, the problems of diffusion distance are explained. The new similarity is introduced in Sect. 4. Section 5 is devoted to the experimental results. The concluding remarks are given in Sect. 6.

2   Diffusion Distance and Clustering

The diffusion-based methods are usually formulated by making use of the diffusion equation

$$\frac{\partial f(t, x)}{\partial t} = \operatorname{div} \left(g(f(t, x), x) \nabla f(t, x) \right), \tag{1}$$

where $f(t, x)$ is a potential function (e.g. concentration, temperature, charge) evolving in time; $g(\cdot)$ is a diffusion coefficient (generally, it is a function). In some applications, the coefficient does not depend on $f(t, x)$. If $g(\cdot)$ reduces to a constant G, the right-hand side of Eq. (1) reduces to $G\nabla^2 f(t, x)$. In our context, $f(t, x)$ has the meaning of evolving image brightness or colour. The process of evolving starts at $t = 0$; $f(0, x)$ is a given input image.

In the discrete case, the problem is formulated in a graph [15]. The diffusion properties are represented by edge weights that can be understood as proximity between the neighbouring nodes connected by the corresponding edge. The weights may again be considered evolving in time or constant. In this paper, we follow the latter option. The diffusion equation can now be written in the form of

$$\frac{\partial \boldsymbol{f}(t)}{\partial t} = \mathbf{L}\boldsymbol{f}(t), \tag{2}$$

where $\mathbf{L}$ is the Laplacian matrix containing the weights of edges; $\boldsymbol{f}(t)$ is a vector whose entries correspond to the potential in the particular graph nodes, i.e. $\boldsymbol{f}(t) = (f_1(t), \ldots, f_n(t))^\top$ (we suppose the graph with n nodes). The weight, denoted by $w_{i,j}$, of the edge connecting the nodes i and j is often considered according to the formula

$$w_{i,j} = e^{-\frac{\|c_{i,j}\|^2}{2\sigma^2}}, \tag{3}$$

where $c_{i,j}$ denotes the grey-scale or colour contrast between the nodes.

The solution of Eq. (2) can be found in the form of [15]

$$\boldsymbol{f}(t) = \mathbf{H}(t)\boldsymbol{f}(0), \tag{4}$$

where $\mathbf{H}(t)$ is a diffusion matrix. The entry $h_t(p, q)$ of $\mathbf{H}(t)$ expresses the amount of a substance that is transported from the q-th node into the p-th node (or vice versa since $h_t(p, q) = h_t(q, p)$) during the time interval $[0, t]$. It can be shown that the following formula for $\mathbf{H}(t)$ ensures that Eq. (2) is satisfied

$$\mathbf{H}(t) = \sum_{k=1}^{n} e^{-\lambda_k t} \boldsymbol{u}_k \boldsymbol{u}_k^\top, \tag{5}$$

where λ_k and $\boldsymbol{u}_k$, respectively, stand for the k-th eigenvalue and the k-th eigenvector of $\mathbf{L}$. Let $u_{i,k}$ be the i-th entry of the k-th eigenvector. For each graph vertex, the vector of new coordinates can be introduced

$$\boldsymbol{x}_i(t) = \left(e^{-\lambda_1 t} u_{i,1}, e^{-\lambda_2 t} u_{i,2}, \ldots, e^{-\lambda_n t} u_{i,n}\right). \tag{6}$$

If the coordinates are assigned in this way, we call it *diffusion map* [3,9]. This vector can be used for clustering the vertices, which will be discussed later. By making use of this vector, the entries of the diffusion matrix can be expressed as the following dot product

$$h_{2t}(p, q) = \langle \boldsymbol{x}_p(t), \boldsymbol{x}_q(t) \rangle. \tag{7}$$

The square of diffusion distance is defined as a sum of the squared differences of the concentrations caused by putting the unit concentration into the p-th node and into the q-th node, respectively, which corresponds to the formula

$$d_t^2(p, q) = \sum_{i=1}^{n} [h_t(i, p) - h_t(i, q)]^2. \tag{8}$$

After some effort, the following formula can be deduced from Eq. (8)

$$\begin{aligned} d_t^2(p, q) &= h_{2t}(p, p) - 2h_{2t}(p, q) + h_{2t}(q, q) \\ &= \|\boldsymbol{x}_p(t) - \boldsymbol{x}_q(t)\|^2, \end{aligned} \tag{9}$$

which shows that introducing the coordinates according to Eq. (6) may be seen as creating a *diffusion map*, which is a map created in a similar sense as in [18], where the idea was presented that measuring the distance along the data manifold in some space can be done by transforming the problem into a new space in such a way that the Euclidean distance in the new space is equal to the distance measured on the data manifold in the original space.

Diffusion clustering is based on the idea to use the coordinates introduced in Eq. (6) for clustering the graph nodes, i.e. the image pixels [7,9,12]. The time t can be used to set the level of details that is desired. Often, the k-means clustering method is mentioned in this context [7,9]. It is believed that much less than n coordinates are needed in practice.

3   The Problems of Diffusion Distance

In this section, we show that the diffusion distance has the properties that make it difficult to use it for image segmentation. We show that the value of diffusion distance between two image points does not necessarily give a good clue whether or not they belong to one image segment. We note that a certain criticism in a similar sense has already been published for the commute-time distance. In [19], the authors came to the conclusion that the commute-time distance in graph does not reflect its structure correctly if the graph is large. We continue in this direction and show some further problems that are relevant for image segmentation. We also show that the problems appear not only for the commute-time (resistance) distance, but also for the diffusion distance, i.e. they cannot be avoided by a certain suitable choice of time.

Consider two points, denoted by p, q, in image. We study two situations (Fig. 1): (i) Both the points are placed in an image containing one rectangular area with a constant brightness; the size of image is $w \times h$ pixels. (ii) The size of image is $w \times h$ pixels again, but the image area is now split by the vertical line into two halves (areas); the brightness is constant inside each area; the difference of brightness between the areas is equal to 1; each of the points is placed in one area. The Euclidean distance between p and q measured in the xy plane is denoted by a (Fig. 1). We traditionally call these situations as "without edge" and "with edge", respectively. Clearly, from the point of view of image segmentation, these two situations are substantially different. In the second case, we expect two image segments and a big distance between p and q. In the first case, only one image segment and a small distance between p and q are expected.

A simple theoretical consideration might be useful for obtaining the first intuitive overview. We compute the distance $d_t(p, q)$ by making use of the formula from Eq. (8) for both mentioned cases. If we consider all possible sizes of image (from small to infinitely big) and all possible values of time ($0 \le t < \infty$), we can easily see that the values of distance vary between 0 and $\sqrt{2}$ in both cases. (We note that the value of $\sqrt{2}$ is the distance between every two distinct points for $t = 0$). It follows that it is threatening that from the value of diffusion distance itself, it will not be clear whether it was obtained for the case (i) or (ii).

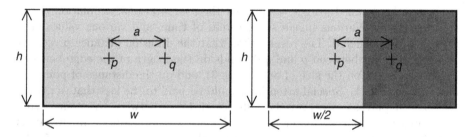

Fig. 1. Two points (p,q) placed into an image containing a single area (*left image*) or two areas (*right image*).

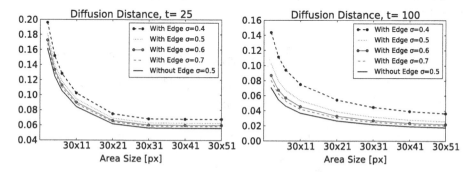

Fig. 2. The dependence of diffusion distance on the length of the edge between the areas: The distance (vertical axis) is computed for the problem from Fig. 1 with/without the edge, for $a = 15$, and for various values of t, σ, and for the increasing value of h (the length of the edge between the areas); the width of the areas remains constant (the value of w). It can be seen that for one value of t and σ, the value of distance depends on h.

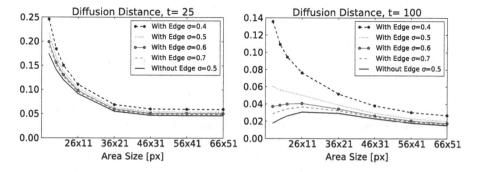

Fig. 3. The dependence of diffusion distance on the area size: The distance (vertical axis) is computed for the problem from Fig. 1 with/without the edge, for $a = 15$, and for various values of t, σ, and for the increasing length of the edge between the areas and for the increasing width of the areas (both w and h are changing in this case). The value of distance depends on the size.

For a more detailed insight, we present the computational simulation of the problem (Fig. 1). Various image sizes, values of time, and various values of σ (Eq. (3)) are considered. The results show that the diffusion distance presented in Figs. 2, 3 and 4 between p and q depends on the length of the edge between the areas (Fig. 2), on the size of areas (Fig. 3), and on the distance of points in the xy plane (Fig. 4). Special attention should be paid to the fact that, for some area sizes, it may happen that the diffusion distance between the points lying in one area (case (i)) is greater than in the case if the points lie in two areas (case (ii)). In Fig. 2, for example, we can see that for $t = 100$ and $\sigma = 0.5$, the distance for $(w = 30, h = 11)$ in the case (i) is greater than the distance for $(w = 30, h = 31)$ in the case (ii). As can be seen, the problem increases with the increasing value of σ. We note that the value of σ must be big enough with respect to the noise intensity that is expected.

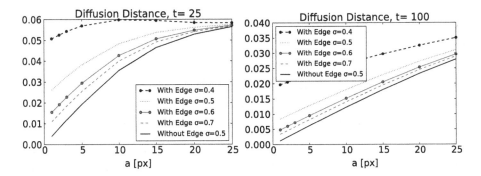

Fig. 4. The dependence of diffusion distance on the distance in the xy plane: The distance (vertical axis) is computed for the problem from Fig. 1 with/without the edge, for a constant image size ($w = 50$, $h = 51$), for various values of t, σ, and for the changing distance in the xy plane (the value of a in pixels that is shown on the horizontal axis). The value of diffusion distance depends on the value of a.

In image segmentation, the neighbouring segments may be of different sizes, which has not been taken into account in the above mentioned simulation (Fig. 1). Therefore, we created another set of test cases to show that the diffusion distance depends on the difference in size and on the mutual position of the segments in which the points are placed. The set is depicted in Fig. 5. We measure the diffusion distance between the points p and q lying in the areas of various shapes. Two cases are considered for each shape: (i) the points are placed in a single area, (ii) the big area is split into two areas by inserting the vertical splitting line (dashed line in Fig. 5); the difference of brightness between both areas is equal to 1. The distance between p, q in the xy plane was $a = 19$ in all cases. Naturally, we would expect that the distances measured in the cases with two areas (with the edge) will always be greater than the distances for the cases with only one area. We could also hope that the distances for all test cases with only one area will remain more or less constant (similarly, for the test cases with two areas). The computational simulation showed that this is not always true. The resulting distances for each case are shown in Fig. 6. It can be seen that the classical diffusion distance does not provide the ordering in the sense that the distances measured between the points lying in one image segment should always be less than the distances measured between the points lying in two different segments. It follows that the value of distance does not give the information that is needed for segmentation if we do not have any apriori knowledge about the size of segments or if the sizes may vary.

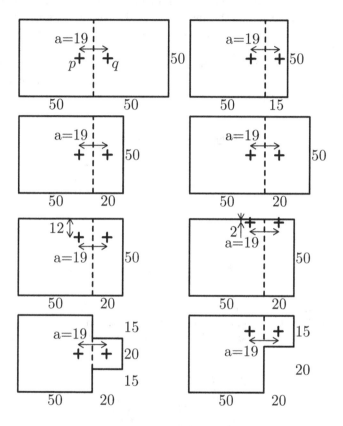

Fig. 5. Various configurations of image segments used for testing the suitability of distance measuring methods. The configurations presented here are referred to as case 1 - 8 in text.

We also use this test set for evaluating the quality of measuring the distance and for comparing the classical diffusion distance with the new measure that is introduced in the next section. We introduce a *discriminative capability* of distance measuring method, which is defined by the following formula

$$\mathcal{D}(t) = \frac{|\mu_e(t) - \mu_s(t)|}{\sqrt{\sigma_e^2(t) + \sigma_s^2(t)}}, \tag{10}$$

where $\mu_s(t)$ and $\sigma_s^2(t)$ stand for the mean value and variance, respectively, of the distance for the cases without edge. Similarly, $\mu_e(t)$, $\sigma_e^2(t)$ stand for the corresponding values for the cases with the edge. (We note that the mentioned values are all dependent on time). The higher is the value of $\mathcal{D}(t)$, the better is the method. The formula in Eq. (10) simply reflects the fact that we would welcome if the distance measured for any case without edge were less than the distance measured for any case with the edge. The computational simulation gave $\mathcal{D}(100) = 0.74$ for the case without noise (Fig. 6). The discriminative capability shows the unsatisfactory behaviour of the classical diffusion distance again. As can be seen, the intervals corresponding to the cases with and without the edge overlap each other, which says again that the diffusion distance cannot distinguish between both cases.

For a certain visual illustration of the behaviour of diffusion distance, we finally present an example in Fig. 7. For a synthetic image with noise, the diffusion distances from the image centerpoint to all other pixels are computed.

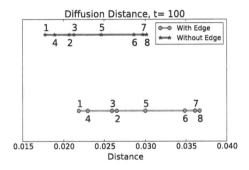

Fig. 6. Diffusion distance for various test cases from Fig. 5 without noise, for the situation with and without the edge, respectively, for $t = 100$, and $\sigma = 0.5$. The value of discriminative capability is $\mathcal{D}(100) = 0.74$ ($\mu_e = 0.0294$, $\sigma_e^2 = 0.31 \times 10^{-4}$, $\mu_s = 0.02398$, $\sigma_s^2 = 0.22 \times 10^{-4}$).

Fig. 7. For an image with noise (*left image*), the diffusion distances from the image centerpoint to all other pixels are computed and depicted by brightness (*right image*); the dark areas correspond to a small distance from the centerpoint. Notice the highly changing distances inside the upper bright object area, especially, in the left part having the shape of vertical strip. The distance step expected along the boundary between the upper and lower object parts can be better seen in the central area of the edge between the parts; in the left area, the distance difference is less convincing.

The parameters were set as follows. The ideal values of brightness were 0.0, 0.6, and 1.0, respectively. Gaussian noise with $\sigma_n = 0.075$ was added. The value of sigma from Eq. (3) was $\sigma = 0.15$. The diffusion distance was computed for $t = 250$.

4   Normalised Diffusion Cosine Similarity

In this section, we propose an improvement that reduces the problems with the diffusion distance that have been mentioned in the previous section. We firstly define our approach. Then we explain why it should be better than the diffusion distance.

For a given image, we introduce the *diffusion cosine similarity* between p, q at the time t as follows

$$s_t(p,q) = \frac{h_{2t}(p,q)}{\sqrt{h_{2t}(p,p)h_{2t}(q,q)}}. \tag{11}$$

By substituting from Eq. (7), it can be easily seen that

$$\begin{aligned}
s_t(p,q) &= \frac{\langle \boldsymbol{x}_p(t), \boldsymbol{x}_q(t)\rangle}{\sqrt{\langle \boldsymbol{x}_p(t), \boldsymbol{x}_p(t)\rangle}\sqrt{\langle \boldsymbol{x}_q(t), \boldsymbol{x}_q(t)\rangle}} \\
&= \frac{\langle \boldsymbol{x}_p(t), \boldsymbol{x}_q(t)\rangle}{\|\boldsymbol{x}_p(t)\|\,\|\boldsymbol{x}_q(t)\|}. \tag{12}
\end{aligned}$$

The value of $s_t(p,q)$ is equal to the value of the cosine of the angle between the vectors $\boldsymbol{x}_p(t)$ and $\boldsymbol{x}_q(t)$. Since the value of $h_t(p,q)$ is always non-negative, the value of $s_t(p,q)$ varies in the range of $[0,1]$.

To obtain a *normalised* cosine similarity, we evaluate the diffusion cosine similarity two times. Firstly, for a given image. Secondly, for the corresponding reference image, which is the image of the same size as is the given input image, but with a constant brightness everywhere. The normalised cosine similarity is now the ratio between the similarity in the given image and the similarity in the reference image. We note that this ratio is only computed if the similarity in the reference image is not close to zero. Otherwise, the normalised cosine similarity is set to zero too, which means that it cannot be computed reliably. Since the diffusion cosine similarity in the given image is not greater than the similarity in the reference image, the maximal possible value of the normalised diffusion cosine similarity is 1.

We should now explain why the normalised diffusion cosine similarity is better than the diffusion distance. The reason is simple. The value of normalised cosine similarity itself tells more clearly whether or not two points are close one to another (i.e. belong to one image segment). No other additional information is needed. The value of 1.0 expresses the maximal possible concordance, decreasing values mean increasing difference. We stress the following properties. The normalised cosine similarity is independent on the length of the edge along which two areas touch (see Fig. 8 and compare it with Fig. 2 for the diffusion distance). The normalised cosine similarity is much less dependent on the total size of area (see Fig. 9 and compare it with Fig. 3). The normalised cosine similarity is much less dependent on the distance between the points in the xy plane (see Fig. 10 and compare it with Fig. 4). Further results will be presented in the next section.

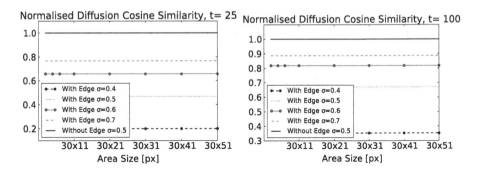

Fig. 8. The dependency of normalised diffusion cosine similarity on the length of the edge between the areas: The similarity (vertical axis) is computed for the problem from Fig. 1 with/without the edge, for $a = 15$, and for various values of t, σ, and for the increasing length of the edge between the areas (the value of h); the width of areas remains constant. In contrast to the diffusion distance, the new similarity does not depend on the edge length in this test environment (compare with Fig. 2).

For completeness, it should be pointed out that the idea of using the cosine similarity in a related area is not completely new. In [2], the author mentions the use of cosine similarity in the context of maximizing satisfaction and profit and in connection with the commute-time distance. The author, however, does not present similar analysis (focused on the use in the area of image segmentation) as we do in this paper. Neither he uses the normalisation.

5   Experimental Results

We start with the tests using the synthetic images. Then the tests with the real-life images are also presented. As a first experiment, we evaluate the discriminative capability introduced in Eq. (10) based on computing the distance or similarity in various area configurations (Fig. 5). For the diffusion distance, the results have already been presented in Fig. 6. For the new method, the values are stated in Fig. 11. Notice that the intervals into which the values of normalised diffusion cosine similarity fall for the cases with and without the edge, respectively, do not overlap, which makes the discriminative capability very good.

Naturally, the behaviour of every method is also important in the presence of noise. We carried out the same test for the noisy images too. Gaussian noise was added to all test images (Fig. 5). For each test image, 1000 samples were used in simulation (see Fig. 12 for further details). Even with a relatively big amount of noise, the results of the new method were better (Fig. 12) then the results obtained for the diffusion distance without noise.

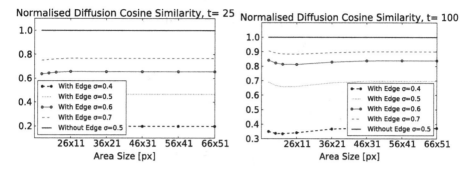

Fig. 9. The dependency of normalised diffusion cosine similarity on the area size: The similarity (vertical axis) is computed for the problem from Fig. 1 with/without the edge, for $a = 15$, and for various values of t, σ, and for the increasing length of the edge between the areas and for the increasing width of the areas (w and h are changing). The dependence of similarity on the area size is much smaller than in the case of diffusion distance (Fig. 3).

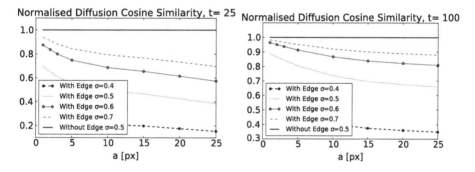

Fig. 10. The dependency of normalised diffusion cosine similarity on the distance in the xy plane: The similarity (vertical axis) is computed for the problem from Fig. 1 with/without the edge, for a constant image size ($w = 50$, $h = 51$), for various values of t, σ, and for the increasing distance in the xy plane (the value of a on the horizontal axis). Due to the normalisation, the similarity depends on a much less than the diffusion distance (Fig. 4).

As a further example, we also present the result for the image from Fig. 7. The normalised diffusion cosine similarity is depicted in Fig. 13. The similarity is measured between the image center point and all remaining pixels and is depicted as brightness. Since big similarity corresponds to a small distance, we also present an inverse image for more convenient comparison with the result for diffusion distance (Fig. 7). Although we do not present any quantitative evaluation in this case, we believe that the result of the new similarity may be regarded as visually better (Fig. 13).

In the rest of this section, we focus on the real-life images and their seeded (interactive) segmentation [16]. For this purpose, the similarity (proximity)

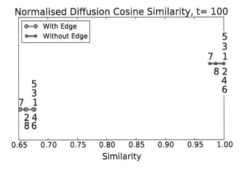

Fig. 11. Normalised diffusion cosine similarity for the test cases from Fig. 5 without noise, for the situation with and without edge, respectively, for $t = 100$, and $\sigma = 0.5$. The value of discriminative capability is $\mathcal{D}(100) = 26.9$ ($\mu_e = 0.669$, $\sigma_e^2 = 0.71 \times 10^{-4}$, $\mu_s = 0.995$, $\sigma_s^2 = 0.75 \times 10^{-4}$), which is a substantial improvement in comparison to the value for the diffusion distance shown in Fig. 6.

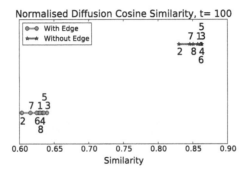

Fig. 12. Normalised diffusion cosine similarity for the same situation as in Fig. 11, but with the Gaussian noise $\sigma_n = 0.2$ added to the test images. The resulting discriminative capability is $\mathcal{D}(100) = 4.164$ ($\mu_e = 0.626$, $\sigma_e^2 = 0.234 \times 10^{-2}$, $\mu_s = 0.854$, $\sigma_s^2 = 0.651 \times 10^{-3}$), which still is better than the value for the diffusion distance without noise.

Fig. 13. The value of the normalised diffusion cosine similarity between the image center point and all remaining pixels is computed under exactly the same conditions as in Fig. 7 for the diffusion distance. In the *left image*, the similarity is depicted by brightness (the bright places indicate a high similarity). For more convenient comparison with the result for the diffusion distance, also the inverse image is presented (*right image*).

between the pixel and area should be defined. Let $\sum_{\text{top}_Q}\{\text{collection}\}$ stand for the sum of the biggest Q elements from a collection of real numbers. The normalised diffusion cosine proximity, denoted by $\tilde{s}_t(p, A)$, between a point p and an area A can be introduced by the formula

$$\tilde{s}_t(p, A) = \frac{1}{Q} \sum_{\text{top}_Q} \{\tilde{s}_t(p, q)\}_{q \in A}. \tag{13}$$

The formula simply reflects the fact that p and A may be regarded as close if at least a certain number of points exist in A that are close to p. The formula can also be easily adapted for the diffusion distance (instead of the Q points with the biggest similarities, Q points with the smallest distances are considered).

We note that the segmentation algorithm itself is not the direct focus of our work, i.e. we do not propose a new algorithm. Instead, by making use of a certain algorithm, we demonstrate the properties of the new similarity measure that might be useful in various known or future algorithms based on measuring the distance or similarity. The algorithm used for testing should make it possible to present the properties of the new measure clearly; understanding the results should not be made more difficult due to the properties of algorithm. For this reason, we use the simple one-step seeded segmentation.

The seeds of the objects and the background are defined manually. Once it is done, the distance to the seeds is computed for all remaining image pixels. If the distance of a pixel is lower to the object seed than to the background seed, the pixel is marked as an object pixel. Otherwise, it is marked as a background pixel. The algorithm can be easily adapted for the use of similarity instead of distance. A certain threshold value of relative similarity may be added that must be fulfilled by object points.

Several real-life images from the Berkeley Segmentation Dataset [11] were used (Fig. 14) and processed as follows. The conversion to greyscale was carried out, which was followed by the normalisation of intensity values into the interval $[0, 1]$. The normalised images were slightly filtered by making use of anisotropic diffusion filtering; the filtered images used for further processing are shown in Fig. 14 too. For all images, we used $\sigma = 0.07$ (Eq. 3), $t = 150$, $Q = 10$, and 750 eigenvectors. The size of images was 160×240 pixels. The relative cosine similarity was computed only if the similarity in the reference image was greater than 0.001; the threshold value for relative similarity was 0.5. The suitable values of t and Q were determined experimentally on the basis of visual evaluation of the results. For the diffusion distance as well as for the new similarity, the best results were obtained for $100 \leq t \leq 200$, $4 \leq Q \leq 50$.

The results of the segmentation using the new normalised diffusion cosine similarity, the diffusion distance, and the cosine similarity mentioned in [2] are shown in Figs. 14 and 15. It can be seen that the new similarity gives visually better results than the remaining mentioned measures. Naturally, all the results could be improved by modifying the position of the seeds. However, we did not do so since we wanted to show the properties of the distance/similarity measures clearly.

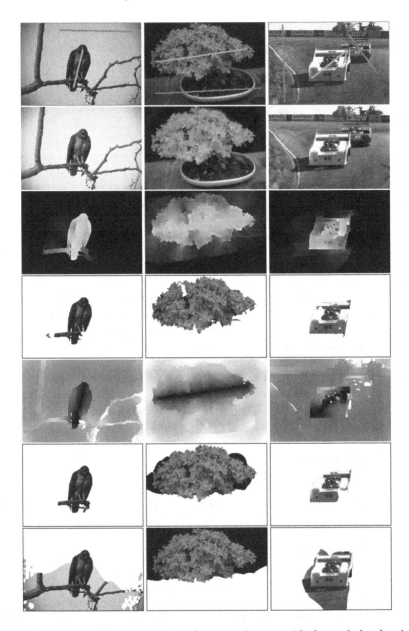

Fig. 14. One-step seeded segmentation: the source images with the seeds for the objects and the background (the *first row*); the filtered images that were used for further processing (the *second row*); the normalised diffusion cosine similarity of pixels to the object seeds, the bright areas correspond to a high similarity (the *third row*); the objects extracted by making use of the new similarity (the *fourth row*); the diffusion distance from the object seeds, the dark areas correspond to a small distance (the *fifth row*); the objects extracted by making use of the diffusion distance (the *sixth row*); the objects extracted by making use of the cosine similarity mentioned in [2] (the *last row*).

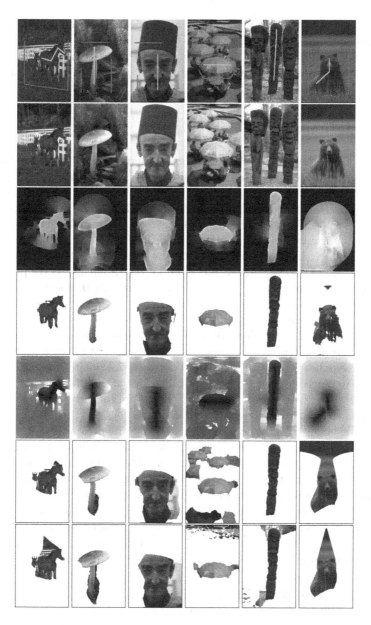

Fig. 15. One-step seeded segmentation: the source images with the seeds for the objects and the background (the *first row*); the filtered images that were used for further processing (the *second row*); the normalised diffusion cosine similarity of pixels to the object seeds, the bright areas correspond to a high similarity (the *third row*); the objects extracted by making use of the new similarity (the *fourth row*); the diffusion distance from the object seeds, the dark areas correspond to a small distance (the *fifth row*); the objects extracted by making use of the diffusion distance (the *sixth row*); the objects extracted by making use of the cosine similarity mentioned in [2] (the *last row*).

Finally, we note that we did not aim at comparing the above segmentation algorithms with all other main state-of-the-art approaches. Instead, we used them to show that the theoretical findings and expectations presented before are correct and useful for the practice (see the conclusion for the discussion about our main goals and contributions).

6   Conclusions

Measuring the distances along the surface that is defined by the image function seems to be useful in more complicated situations. The use of geodesic distance is often mentioned in this context, but its disadvantages are known. One would intuitively say that the diffusion distance should have good properties for the mentioned purpose. We showed (including the computational simulations of the situations that are important for segmentation) that the diffusion distance need not be useful since the presence of edges may be overshadowed by the varying size of image segments (and the size is not often known in advance). We proposed a new measure called *normalised diffusion cosine similarity* that suffers from these problems to a much lesser extent. We have also demonstrated that it can be used in image segmentation algorithms.

We believe that the geodesic distance and diffusion distance (resistance or commute-time distance) are two opposite approaches. While the geodesic distance only searches for the shortest path between the points, the diffusion distance takes into account all possible paths. The idea of simultaneously examining more paths seems to be generally useful. The question, however, remains how it should be exactly done. The diffusion distance does not seem to be the best solution. We believe that a certain gap exists in this area and that the corresponding efficient methods will probably be developed in the future. We intended this paper as a certain step in this direction rather than a paper proposing a new segmentation method for everyday use. That is why we did not aim at comparing the algorithm mentioned in the previous section with various other state-of-the-art algorithms. The goal was to show that some alternatives exist in the area of the diffusion-like distances that may have a chance to be developed into useful and practical tools. We hope that introducing the new normalised diffusion cosine similarity can be regarded as a step in this direction.

Acknowledgements. This work was partially supported by the grant of SGS No. SP2015/141, VŠB - Technical University of Ostrava, Czech Republic.

References

1. Babić, D., Klein, D.J., Lukovits, I., Nikoli, S., Trinajsti, N.: Resistance-distance matrix: a computational algorithm and its application. Int. J. Quantum Chem. **90**(1), 166–176 (2002)
2. Brand, M.: A random walks perspective on maximizing satisfaction and profit. In: Proceedings of the International Conference on Data Mining, pp. 12–19 (2005)

3. Coifman, R., Lafon, S.: Diffusion maps. Appl. Comput. Harmonic Anal. **21**, 5–30 (2006)
4. Eppstein, D.: Finding the k shortest paths. SIAM J. Comput. **28**(2), 652–673 (1999)
5. Fouss, F., Pirotte, A., Renders, J.M., Saerens, M.: Random-walk computation of similarities between nodes of a graph with application to collaborative recommendation. Trans. Knowl. Data Eng. **19**, 355–369 (2007)
6. Grady, L.: Random walks for image segmentation. IEEE Trans. Pattern Anal. Mach. Intell. **28**(11), 1768–1783 (2006)
7. Huang, H., Yoo, S., Qin, H., Yu, D.: A robust clustering algorithm based on aggregated heat kernel mapping. In: International Conference on Data Mining, ICDM 2011, pp. 270–279 (2011)
8. Klein, D.J., Randić, M.: Resistance distance. J. Math. Chem. **12**, 81–95 (1993)
9. Lafon, S., Lee, A.B.: Diffusion maps and coarse-graining: a unified framework for dimensionality reduction, graph partitioning, and data set parameterization. Trans. Pattern Anal. Mach. Intell. **28**(9), 1393–1403 (2006)
10. Lipman, Y., Rustamov, R.M., Funkhouser, T.A.: Biharmonic distance. ACM Trans. Graph. **29**, 1–11 (2010)
11. Martin, D., Fowlkes, C., Tal, D., Malik, J.: A database of human segmented natural images and its application to evaluating segmentation algorithms and measuring ecological statistics. In: Proceedings of the International Conference of Computer Vision, ICCV 2001, pp. 416–423 (2001)
12. Nadler, B., Lafon, S., Coifman, R.R., Kevrekidis, I.G.: Diffusion maps, spectral clustering and reaction coordinates of dynamical systems. Appl. Comput. Harmonic Anal. **21**(1), 113–127 (2006)
13. Papadimitriou, C.H.: An algorithm for shortest-path motion in three dimensions. Inf. Process. Lett. **20**(5), 259–263 (1985)
14. Qiu, H., Hancock, E.R.: Clustering and embedding using commute times. Trans. Pattern Anal. Mach. Intell. **29**(11), 1873–1890 (2007)
15. Sharma, A., Horaud, R., Cech, J., Boyer, E.: Topologically-robust 3D shape matching based on diffusion geometry and seed growing. In: Proceedings of the Conference on Computer Vision and Pattern Recognition, pp. 2481–2488 (2011)
16. Sinop, A.K., Grady, L.: A seeded image segmentation framework unifying graph cuts and random walker which yields a new algorithm. In: Proceedings of the International Conference on Computer Vision, ICCV 2007, pp. 1–8 (2007)
17. Surazhsky, V., Surazhsky, T., Kirsanov, D., Gortler, S.J., Hoppe, H.: Fast exact and approximate geodesics on meshes. ACM Trans. Graph. **24**(3), 553–560 (2005)
18. Tenenbaum, J.B., de Silva, V., Langford, J.C.: A global geometric framework for nonlinear dimensionality reduction. Science **290**, 2319–2323 (2000)
19. von Luxburg, U., Radl, A., Hein, M.: Hitting and commute times in large random neighborhood graphs. J. Mach. Learn. Res. **15**, 1751–1798 (2014)
20. Yen, L., Fouss, F., Decaestecker, C., Francq, P., Saerens, M.: Graph nodes clustering based on the commute-time kernel. In: Zhou, Z.-H., Li, H., Yang, Q. (eds.) PAKDD 2007. LNCS (LNAI), vol. 4426, pp. 1037–1045. Springer, Heidelberg (2007)

Automatic Detection and Recognition of Symbols and Text on the Road Surface

Jack Greenhalgh$^{(\boxtimes)}$ and Majid Mirmehdi

Visual Information Laboratory, University of Bristol, Bristol, UK
{csjhg,m.mirmehdi}@bristol.ac.uk
http://www.bris.ac.uk/vi-lab

Abstract. This paper presents a method for the automatic detection and recognition of text and symbols on the road surface, in the form of painted road markings. Candidates for road markings are detected as maximally stable extremal regions (MSER) in an inverse perspective mapping (IPM) transformed version of the image, which shows the road surface with the effects of perspective distortion removed. Separate recognition stages are then used to recognise words and symbols. Recognition of text-based regions is performed using a third-party optical character recognition (OCR) package, after application of a perspective correction stage. Symbol-based road markings are recognised by extracting histogram of oriented gradient (HOG) features, which are then classified using a support vector machine (SVM) classifier. The proposed method is validated using a data-set of videos, and achieves F-measures of 0.85 for text characters and 0.91 for symbols.

Keywords: Computer vision · Machine learning · Text recognition · Intelligent transportation systems

1 Introduction

Painted text and symbols which appear on the surface of roads in the UK come in three basic forms: text, symbols, and lane division markers. Text provides information to the driver, such as speed limits, warnings, and directions. Symbol-based road markings are displayed in the form of arrows and other ideograms. We propose a method in this paper that detects and recognises such painted text and symbols using a camera mounted inside a car on the driver's rear-view mirror (looking out front). The possible applications for this work are numerous, and include advanced driver assistance systems (ADAS), autonomous vehicles, and surveying of road markings. While information such as the current speed limit or upcoming turnings are also provided in the form of road signs, there is much information, such as 'road merging' warnings or lane specific directional information which may appear exclusively as road markings.

There are several key issues which make the detection and recognition of painted road text and symbols difficult. Road markings suffer badly from wear

© Springer International Publishing Switzerland 2015
A. Fred et al. (Eds.): ICPRAM 2015, LNCS 9493, pp. 124–140, 2015.
DOI: 10.1007/978-3-319-27677-9_8

and deterioration due to the fact that vehicles continuously pass over them, and there is also a large amount of variation between different instances of the same symbol or character, as they are often hand-painted. In addition, common issues such as shadowing, occlusion, and variations in lighting also apply. Despite such difficulties, the problem is constrained in some aspects, with a large amount of a priori knowledge which can be exploited. Much of this a priori knowledge relates to the visual appearance of the text characters and symbols themselves. Text and symbol based road markings appear elongated when viewed from above, so as to improve their readability from the view-point of a driver on the road. This offers a constraint on the size and aspect ratio of candidate regions, when detected in an image that has undergone inverse perspective mapping (IPM).

The text-based aspect of road marking recognition is constrained in comparison with more generalised text recognition [1,2]. The total set of text characters is very limited, consisting only of upper-case characters, numbers, and a small number of punctuation marks. Also, text-based road markings are based on a single typeface, as shown in Fig. 1(top). The total number of symbol-based road markings is also fairly limited, with only 15 different symbols appearing in the data-set, as shown in Fig. 1(bottom).

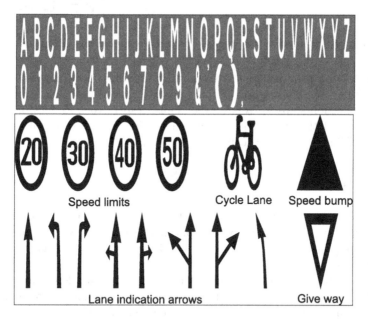

Fig. 1. (Top) full character set for typeface on which text road markings are based (bottom) full set of road marking symbols which appear in the data-set.

The first stage of the algorithm removes perspective distortion from the road surface by applying an IPM, after the vanishing point has been detected. Candidate text characters and symbols are then detected as maximally stable extremal regions (MSER) on the road surface. Road marking candidates are sorted into potential text characters or symbols based on their relative size and proximity, so that they can be classified using separate recognition stages. Symbol-based markings are recognised using histogram of oriented gradients (HOG) and linear support vector machines (SVM). Text words are recognised using an open-source optical character recognition (OCR) engine, Tesseract [3], after a further correction transform has been applied. Recognised words and symbols are matched across consecutive frames so that recognition results can be improved via temporal fusion. The total system pipeline for the algorithm is shown in Fig. 2.

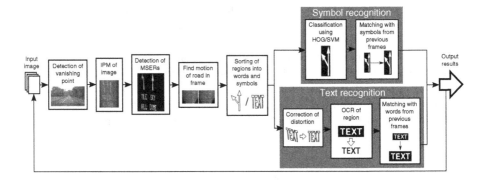

Fig. 2. Pipeline for the proposed painted road markings algorithm.

In Sect. 2, an overview of related work is provided. Section 3 describes the stage for the detection and sorting of candidate regions. Sections 4 and 5 focus on the recognition steps for text and symbols, respectively. In Sect. 6, the temporal aspects of the method are described. In Sect. 7, experimental results are presented. Finally, in Sect. 8, conclusions are drawn.

2   Related Work

Research on road marking detection can be broadly divided into two categories, one of which focusses on lane division markings, such as [4–8], and the other on symbol or text based markings, such as [9–14], which provide semantic information to the driver.

In cases where symbols painted on the road surface are detected and recognised, the total number of symbol types which are classified is generally very limited, often focussing on just arrows or rectangular elements [9–11,13]. For symbol detection, several of these works employ an IPM to remove perspective distortion of the road surface, and hence the markings painted on it, such as [9,11,12,14]. The only papers that deal with the recognition of road surface text are [12,14].

[12] present a method for detecting and recognising both text and symbols on the road surface. An IPM is applied to each frame, after the image vanishing point (VP) has been automatically detected. Regions of interest (ROI) are then

detected in the IPM image by applying an adaptive threshold, and finding CCs in the resulting binary image. After applying some post-processing to the detected shapes, such as orientation normalisation and rejection of complex shapes, the region is classified. The recognition stage involves the extraction of a feature vector from each candidate CC, which includes several shape based features. Each region is then classified using a neural network trained using real road footage. An accumulator of symbols is used to combine results over several frames, and eliminate single frame false positives. The method is limited to recognising only 7 symbols and 16 characters rather than the full alphabet, and is also limited to recognising only 19 unique predefined words. The authors report true positive rates of 85.2 % and 80.7 % for recognition of arrows and text, respectively, with their method taking 60–90 ms to process a single frame.

[14] propose a method for the detection and classification of text and symbols painted on the road surface. ROIs are detected in each frame as MSERs in an IPM transformed version of the image. The FAST feature detector is then used to extract points of interest (POI) from each ROI. A feature vector is then found for each POI using HOG, and the region is classified through comparison with a set of template images. Although Wu and Ranganathan recognise both text and symbols using template matching, entire words are treated as single classes, and as a result only a small subset of words are recognised. In this respect, their proposed method does not provide 'true' text detection, as arbitrary words (such as place names and their abbreviations) are not recognised. The authors report a true positive rate of 90.1 % and a false positive rate of 0.9 % for the combined recognition of arrows and text, at a processing speed of at least 10 frames per second.

The method proposed in this paper improves upon the current state-of-the-art in several ways. Firstly, the proposed algorithm is able to recognise any arbitrary text word which may appear, rather than the limited set of predefined words recognised by the methods presented by [12,14]. This aspect is essential for use with an ADAS or autonomous vehicle, given that important directional information may appear in the form of obscure place names, often with abstract abbreviations, which may not appear in the dictionary. For example, the place name 'Warwick' is sometimes abbreviated to 'W'wick'. Secondly, the range of road symbols recognised is expanded upon compared to existing works which classify up to a maximum of 6 symbols only, e.g. [11–14]. This is an advantage of using a synthetically generated data-set, which allows the system to be trained on any number of template images while retaining the robustness provided by machine learning methods.

3   Detection and Sorting of Candidate Regions

The first stage of the proposed method detects and sorts candidate regions for road markings. Connected components (CC) representing text characters and symbols are detected in an IPM version of the image, and are then sorted into text characters and symbols based on their attributes, and are passed on to the appropriate classification stage.

3.1   Detection of Candidate Regions

Candidate regions are first detected as MSERs in an IPM transformed version of the input frame. Use of IPM eliminates issues caused by perspective distortion, as text and symbols maintain their shape and scale in this transformed image, regardless of their shape and scale in the original frame. Given that road markings appear as high contrast homogeneous regions on the road surface, MSER is considered to be a suitable method for their detection, especially following its successful application to the detection of road signs in [15,16]. Only light-on-dark MSERs are used, given that road markings are always painted in white paint on dark backgrounds. Figure 3 shows an example frame, IPM image, and detected MSERs, which are individually coloured.

Fig. 3. Stages of MSER detection on IPM image, showing original frame (top), IPM image (bottom left), and detected MSERs individually coloured (bottom right) (Color figure online).

A rotated minimum area rectangle (RMAR) is fitted to the CC of each candidate region, the features of which are useful for reducing the total number of candidates. As text and symbols painted on the road are elongated, the angle of their RMAR is expected to be close to $0°$, this also places a constraint on the range of aspect ratios which will appear, as illustrated in Fig. 4.

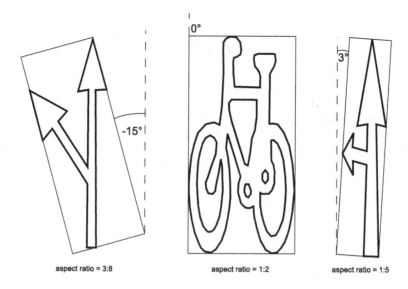

Fig. 4. Dimensions of symbol-based road markings.

The features used for the elimination of false positives are shown in Table 1, with all the values determined empirically through analysis of template images.

3.2   Sorting of Candidates into Words and Symbols

Once road marking candidates have been detected, they can be divided into words and symbols. Text characters contained within words appear in horizontal lines and in close proximity to each other, whereas symbol-based road markings appear in isolation with a greater amount of space around them. Based on this observation, it is possible to sort road marking candidates into text characters and symbols.

First, each candidate region is compared to each other on the basis of size and location. If two candidates are found to be of a similar height, are approximately vertically aligned, and have a small horizontal distance, they are grouped into a single word. To assess the height similarity, a constraint is applied to the ratio of the two heights. To determine whether two characters are vertically aligned, an empirically determined threshold is applied to an overlap measure, ψ, which is calculated as follows:

Table 1. Features for reduction of candidates.

Feature	Min. value	Max. value
Ratio of area of CC to area to RMAR	0.17	0.7
Aspect ratio of RMAR	0.09	0.68
Angle of RMAR	-20°	20°

$$\psi = \frac{min\{E_{y2}, F_{y2}\} - max\{E_{y1}, F_{y1}\}}{max\{E_{y2}, F_{y2}\} - min\{E_{y1}, F_{y1}\}}, \tag{1}$$

where E and F represent the bounding boxes of the two candidate regions. If there is no vertical overlap between the regions, ψ will be equal to 0, and if the regions perfectly overlap it will be equal to 1.

The distance between two characters from the same word is expected to be less than approximately one-third of the width of the widest character, which was determined based on examples from the character set shown in Fig. 1(top). Any candidates grouped together are considered to be text characters. The requirements for the matching of text characters are summarised in Table 2, where all of the described features are invariant to scale.

Table 2. Scale invariant character grouping constraints.

Feature	Min.	Max.
Character height ratio	0.8	1.25
ψ	0.7	1.00
Horizontal character distance	0	(max. character width)*0.35

After the character grouping process, any remaining un-grouped candidate regions are considered to be possible candidates for road symbols. Candidate regions grouped into words are sent to the text recognition stage of the algorithm, and the remaining characters are classified by the symbol recognition stage.

Fig. 5. Algorithm output showing candidates for road markings divided into symbols and words (Color figure online).

An example output for this stage of the algorithm is shown in the right-hand image of Fig. 5, where candidate regions are shown in white, bounding boxes for individual words are shown in orange, and bounding boxes for symbols are shown in green.

4 Recognition of Words

Next, the proposed method attempts to recognise text contained within candidate word regions. An affine transform is applied to minimise the distortion caused by uneven roads, before the region is interpreted using OCR.

4.1 Correction of Perspective Distortion of Words

The IPM stage of the algorithm works on the assumption that the road surface is flat, however, in reality road surfaces often exhibit a slight camber. As a result, words appearing in the IPM road image may sometimes suffer from rotation and shearing, examples of which are shown in the top row of Fig. 6. This perspective distortion reduces the accuracy achieved by the OCR engine, given that it is designed for use with fronto-parallel text.

Fig. 6. Examples of distorted words (top row) with corrected versions (bottom row).

To avoid this distortion, a correction stage is applied to candidate words before recognition. First a RMAR is fitted to all the CCs representing the characters in the word. The angle of the RMAR, ζ, is then used to correct the rotation of the region, using the following transform:

$$\begin{bmatrix} x' \\ y' \end{bmatrix} = \begin{bmatrix} cos(\zeta) & -sin(\zeta) \\ sin(\zeta) & cos(\zeta) \end{bmatrix} \begin{bmatrix} x \\ y \end{bmatrix},$$ (2)

where x and y are pixel coordinates in the original image, and x' and y' are pixel coordinates in the rotated image.

In order to fix the shear of the image, the most common edge orientation is found, and used to calculate the shear mapping. The Sobel filter is used to find the horizontal and vertical derivatives of the image, and from these the magnitude and orientation is found at each pixel. A histogram of orientations is then built, with each pixel weighted by its magnitude. The histogram bin with the highest value is taken to be perpendicular to the shear angle, β, and can be used to calculate the transformation as follows:

$$\begin{bmatrix} x' \\ y' \end{bmatrix} = \begin{bmatrix} 1 & tan(\beta) \\ 0 & 1 \end{bmatrix} \begin{bmatrix} x \\ y \end{bmatrix}. \tag{3}$$

The stages of this process are shown in Fig. 7. Examples of this transformation applied to several images are shown in Fig. 6.

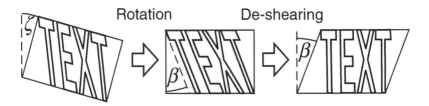

Fig. 7. Diagram illustrating stages of perspective distortion correction.

4.2 OCR of Region

Once the correction stage has been applied to the region, the pixel values are normalised and the textual information is interpreted using the open source OCR engine *Tesseract* [3]. The OCR engine has been retrained based on a road marking typeface, which contains only upper-case characters, numbers, and punctuation marks (see Fig. 1(top) for the full character set). When OCR is applied, a value is returned which represents the confidence of recognition, and if this value is below 50 %, the entire region is rejected.

5 Recognition of Symbols

Once candidate regions have been detected and sorted, the candidates previously selected as possible symbol road markings are classified by a recognition stage. For the classification of symbol-based markings an approach similar to the one described in [15] for the recognition of road signs is used, whereby regions are recognised with HOG descriptors in combination with a classifier. This feature descriptor is deemed to be suitable for encoding the shape of road markings, given that they display clearly defined edges at particular orientations within the image. Each candidate region is resized to 32×192, before a HOG descriptor is calculated. This size is empirically determined to strike a balance between accuracy and efficiency in computation expense. A synthetically generated data-set is used to train the classifier. This data-set is created by applying distortions to a set of template images, to create images which closely resemble the road markings featured in the data-set. This approach eliminates the necessity for a large training set of real images, and allows the system to be trained on any number of symbol classes.

5.1   Selection of Classifier

To select the most efficient classifier for use with the HOG features a comparison between several was made. Classifiers considered included SVM with a linear kernel, SVM with a radial basis function (RBF) kernel, MLP, and Random Forests. Each classifier was trained on a synthetically generated data-set consisting of 1000 examples per symbol class, and a negative class of 5000 examples. The classifiers were then tested on a validation set of real images, the results of which are shown in Table 3. As can be seen from the table, RBF SVM outperforms the other three classification methods, but suffers from a larger classification time. This is likely due to a combination of the mapping process when applied to the data, given its high dimensionality, and the fact that a multi-class SVM classifier is made up of many one-against-one binary classifiers. Although 7.41 ms is not a significant amount of time for a single classification, recognition must be performed on many candidate regions in each frame, therefore making the RBF-kernel SVM potentially unusable for this application. Linear SVM produces the second highest accuracy of all classifiers tested, and also retains a low classification time. Linear SVM is much faster than RBF SVM as the mapping process is removed. Random Forests and MLP have a much lower classification time than RBF SVM, but also suffer from a much lower accuracy. This is largely due to their inability to separate the symbol classes from the background/negative class. The classifier selected was the linear-kernel SVM, due to its high accuracy and fast classification time.

Road markings exhibit a large amount of variation within classes, due to the fact that they are hand painted, and are also subject to greater amounts of deterioration. SVM is able to avoid the problem of over-fitting through careful selection of training parameters. Therefore, for this particular problem SVM proved to be the more suitable classifier.

5.2   Selection of HOG Block Density

For the calculation of HOG features, the optimal block density varies depending on the type of data which is to be classified. This value can be changed by altering the size of each cell, whereby a smaller cell size provides a higher block density. Lower block densities are preferable in cases where the object to be classified is geometrically varied, such as animals or cars. Higher block densities are more applicable to more constrained objects, such as road markings [17]. A comparison is performed to find the most appropriate HOG block density. An SVM classifier is trained on synthetic data with HOG features calculated at

Table 3. Comparison of different classification methods.

Classifier	Accuracy	Average classification time (ms)
SVM (RBF kernel)	93.54 %	7.41
SVM (linear kernel)	90.87 %	0.36
Random forest	75.58 %	0.26
MLP	71.40 %	0.81

number of block density levels, as shown in Fig. 8. The accuracy is then found for each by testing on the same validation test set used in Sect. 5.1. The results of this comparison are shown in the graph in Fig. 8.

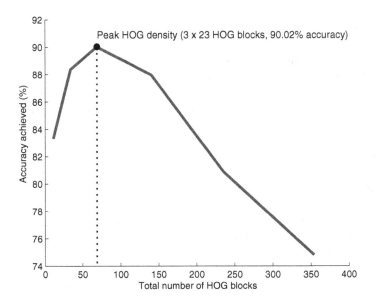

Fig. 8. Graph showing number of HOG feature blocks against accuracy achieved.

It can be seen from the results in Fig. 8 that the peak block density is found at 3 × 23. Once the density increases beyond this value, the accuracy begins to decrease, as a result of intraclass variance.

6   Temporal Information

Temporal information is exploited in order to improve the accuracy of the method. Candidate words and symbols from consecutive frames are matched, and recognition results from several frames are then fused together.

6.1   Calculation of Road Movement

To improve the temporal matching of candidate regions, the motion of the road between consecutive IPM frames is estimated. This movement can be represented by a single 2-D motion vector, which is found by matching MSERs between frames. Use of this 2-D vector allows the current location of previously detected road markings to be estimated, therefore improving the accuracy of matching.

All detected MSERs (symbol, word, or otherwise) from the previous frames are matched to those in the current frame by finding their 'nearest neighbours' based on a descriptor vector consisting of a number of simple features, including width, height, and location. Each MSER in the current frame is compared to each one in the previous frame, and matched to the one with smallest Euclidean distance between feature vectors. A 2-D motion vector for the frame is then calculated by finding the mean of the distances between the centroids of each pair of matched regions. MSERs are well suited to image matching applications such as this due to their high repeatability [18], and conveniently have already been detected in the frame during the road-marking detection process. The top row of Fig. 9 shows two consecutive frames with MSERs marked in blue and matches shown with green lines.

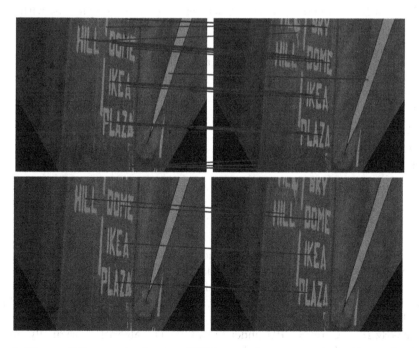

Fig. 9. (Top row) Consecutive frames of IPM image showing matched MSERs, (Bottom row) Consecutive frames of IPM image showing matched words (Color figure online).

6.2   Matching and Temporal Fusion of Words and Symbols

The next stage of the algorithm finds matches between words and symbols in the current frame and previous frame using the road movement vector. Although matching is performed in the previous stage to find the road movement vector, this process does not necessarily match all MSERs correctly. This can be seen in the top-row of Fig. 9 where a text character has been mismatched. Once the road movement vector has been calculated more accurate matching can be performed.

Each current symbol or word is compared to each previously detected symbol or word by computing the Euclidean distance between a set of features. These features include the size and aspect ratio of the RMAR, and the location of the corners of the RMAR once they have been displaced by the calculated road motion vector described above in Sect. 6.1. Each candidate is matched to the previous region with the smallest Euclidean distance between feature vectors. If no match is found, the region is treated as a new candidate. An example of this matching can be seen in the bottom row of Fig. 9 where each text word is matched correctly.

For each series of matched words a histogram is built containing all OCR results, weighted by confidence. The histogram bin with the highest value is taken as the correct result, given that it has been recognised in at least 3 frames. Similarly, a histogram is created for each series of matched symbols, with each bin representing a class of symbol.

Table 4. Results for recognition of text characters.

Method	Precision	Recall	F-measure
OCR with no additional processing	0.66	0.79	0.72
OCR with perspective correction	0.66	0.81	0.73
OCR with temporal fusion only	0.79	0.83	0.81
OCR with perspective correction and temporal fusion	0.86	0.87	0.85

7   Results

We validate both the symbol-based and text-based aspects of the proposed method via comparative analysis on a data-set, consisting of 42,110 frames at a frame rate of 30 fps and a resolution of 1920 × 1088. This test data was shot at a number of different vehicle speeds, featuring different scene types which included urban, suburban, and motorway scenes. The results were obtained on a 3.33 GHz Intel Core i5 CPU under OpenCV. The algorithm originally ran at an average rate of 3 fps on a single core. However, this speed was increased to 7.25 fps exploiting the multi-core capability of the processor and implementing the method as a parallel pipeline.

7.1   Results for Text Detection and Recognition

In order to validate the various stages of the text recognition aspect of the system, recognition results are provided showing the precision, recall, and F-measure, for the detection and recognition of text characters. We present results which compare standard OCR applied to detected regions, OCR with the perspective correction method described in Sect. 4.1, OCR with the temporal fusion method described in Sect. 6, and OCR with both methods. In this manner, the difference in performance which results from the addition of each of these enhancements can be seen. These results are shown in Table 4.

The results demonstrate that the use of perspective correction and temporal fusion provide a significant improvement compared to the raw OCR approach. It can be seen that the recall result for OCR with perspective correction and temporal fusion is higher than precision, likely due to the variation seen in the text characters to be recognised. While individual characters are similar enough to be reliably detected, intraclass variation between characters is large, causing misclassification, and hence, a lower precision. This intraclass variation is due to the fact that the text characters are largely hand painted. Figure 12 contains example output frames of the algorithm, showing road text correctly detected and recognised (Fig. 10).

Fig. 10. Examples of system output showing correctly detected and recognised text.

7.2   Results for Symbol Detection and Recognition

Our proposed algorithm is compared against an existing method proposed by [12] as well as a baseline method in order to validate its performance for the recognition of symbols on the road surface.

The method of [12] was implemented and adapted to detect and recognise symbols on UK roads. An artificial neural network was trained with their proposed feature set, which was extracted from the same synthetic data-set used to train our method.

The baseline method classifies each region using simple template matching. Each detected region is resized to 32 × 192 before its pixels are normalised between 0 and 255. The region is then compared to a number of template images and the template with the smallest difference is selected as the correct class.

From the results shown in Table 5 it can be seen that the proposed method outperforms both the baseline method and the method of [12] (shown in the table as KB2010). The adaptive threshold approach of [12] was unable to detect some of the fainter road markings, causing the recall to be reduced. The recognition stage which used an artificial neural network with shape based features misclassified several non-road marking shapes as road symbols, causing a reduction in precision. The baseline method operates at 1.3 fps, KB2010 at 13.9 fps (average rate), and the proposed method at 7.3 fps.

The baseline method manages a particularly low recall rate, in contrast to the ability of the proposed symbol recognition method in eliminating false positives. Example frames demonstrating the output of the symbol-based stage of the algorithm are shown in Fig. 11.

Table 5. Results for recognition of symbols.

Method	Precision	Recall	F-measure
Baseline method	0.58	0.65	0.61
KB2010	0.74	0.78	0.76
Proposed method	0.91	0.92	0.91

Fig. 11. Examples of algorithm output showing correctly detected and recognised symbols.

7.3   Failure Cases

There were several cases where the detection and recognition of road markings failed. Some examples of these failure cases are shown in Fig. 12. In the example shown on the left-hand side of Fig. 12 it can be seen that the 3 symbol-based road markings have failed to be detected. This is due to the fact that the road markings are faded, and appear very faint. MSER detects high-contrast regions, therefore, the fact that these regions are so faint has resulted in their missed detection. The road marking shown in the right-hand image of Fig. 12 shows a misclassified symbol on the far right hand side, where part of an arrow pointing in two directions has been falsely recognised as a single arrow pointing forward. In this case, the road marking has deteriorated in such a way that the CC representing the region is split, which has caused a section of the symbol to be classified separately.

Example false negative   Example misclassification

Fig. 12. Examples where symbol recognition failed.

8   Conclusion

A method for the automatic detection and recognition of painted text and symbols on the road surface is proposed. Candidate regions are detected in an IPM transformed image, which is then reduced and sorted into words and symbols before being interpreted using separate recognition stages. Text words are recognised using an open-source OCR engine, after a perspective correction stage. Symbols are classified using HOG features and SVM. Temporal fusion is used to combine recognition results across several frames. Results are provided showing precision, recall, and F-measure for a challenging data-set of videos, for which the system produces F-measures of 0.85 and 0.91 for text characters and symbols, respectively.

References

1. Clark, P., Mirmehdi, M.: Recognising text in real scenes. Int. J. Doc. Anal. Recogn. **4**, 243–257 (2002)
2. Merino-Gracia, C., Lenc, K., Mirmehdi, M.: A head-mounted device for recognizing text in natural scenes. In: Iwamura, M., Shafait, F. (eds.) CBDAR 2011. LNCS, vol. 7139, pp. 29–41. Springer, Heidelberg (2012)
3. Google: Tesseract-OCR (2013). http://code.google.com/p/tesseract-ocr/. Accessed 8 October 2013
4. Hanwell, D., Mirmehdi, M.: Detection of lane departure on high-speed roads. In: International Conference on Pattern Recognition Applications and Methods (2009)
5. Chen, Z., Ellis, T.: Automatic lane detection from vehicle motion trajectories. In: 2013 10th IEEE International Conference on Advanced Video and Signal Based Surveillance, pp. 466–471. IEEE (2013)
6. Bottazzi, V.S., Borges, P.V.K., Jo, J.: A vision-based lane detection system combining appearance segmentation and tracking of salient points. In: 2013 IEEE Intelligent Vehicles Symposium (IV), Number IV, pp. 443–448. IEEE (2013)
7. Zhang, F., Stähle, H., Chen, C., Buckl, C., Knoll, A.: A lane marking extraction approach based on random finite set statistics. In: 2013 IEEE Intelligent Vehicles Symposium (IV), pp. 1143–1148 (2013)

8. Huang, J., Liang, H., Wang, Z.: Robust lane marking detection under different road conditions. In: 2013 IEEE International Conference on Robotics and Biomimetics (ROBIO), vol. 1, pp. 1753–1758 (2013)

9. Rebut, J., Bensrhair, A., Toulminet, G.: Image segmentation and pattern recognition for road marking analysis. In: 2004 IEEE International Symposium on Industrial Electronics, vol. 1, pp. 727–732. IEEE (2004)

10. Vacek, S., Schimmel, C., Dillmann, R.: Road-marking analysis for autonomous vehicle guidance. In: European Conference on Mobile Robots, pp. 1–6 (2007)

11. Li, Y., He, K., Jia, P.: Road markers recognition based on shape information. In: 2007 IEEE Intelligent Vehicles Symposium, Number c, pp. 117–122 (2007)

12. Kheyrollahi, A., Breckon, T.P.: Automatic real-time road marking recognition using a feature driven approach. Mach. Vis. Appl. **23**, 123–133 (2010)

13. Danescu, R., Nedevschi, S.: Detection and classification of painted road objects for intersection assistance applications. In: 2010 13th International IEEE Annual Conference on Intelligent Transportation Systems, Number 28, pp. 433–438 (2010)

14. Wu, T., Ranganathan, A.: A practical system for road marking detection and recognition. In: 2012 IEEE Intelligent Vehicles Symposium, pp. 25–30. IEEE (2012)

15. Greenhalgh, J., Mirmehdi, M.: Traffic sign recognition using MSER and random forests. In: European Signal Processing Conference, pp. 1935–1939 (2012)

16. Greenhalgh, J., Mirmehdi, M.: Recognizing text-based traffic signs. IEEE Trans. Intell. Transp. Syst. **16**, 1–10 (2014)

17. Bosch, A., Zisserman, A., Munoz, X.: Representing shape with a spatial pyramid kernel. In: CIVR, pp. 401–408. ACM Press, New York (2007)

18. Mikolajczyk, K., Tuytelaars, T., Schmid, C., Zisserman, A., Matas, J., Schaffalitzky, F., Kadir, T., Gool, L.V.: A comparison of affine region detectors. Int. J. Comput. Vis. **65**, 43–72 (2005)

Applications

Using BLSTM for Spotting Regular Expressions in Handwritten Documents

Gautier Bideault[1(✉)], Luc Mioulet[1], Clément Chatelain[2], and Thierry Paquet[1]

[1] Laboratoire LITIS - EA 4108, Universite de Rouen,
76800 Saint-Etienne-du-Rouvray Cedex, France
gautier.bideault@gmail.com
[2] Laboratoire LITIS - EA 4108, INSA Rouen,
76800 Saint-Etienne-du-Rouvray Cedex, France

Abstract. This article concerns the spotting of regular expressions (REGEX) in handwritten documents using a hybrid model. Spotting REGEX in a document image allow to consider further extraction tasks such as document categorization or named entities extraction. Our model combines state of the art BLSTM recurrent neural network for character recognition and segmentation with a HMM model able to spot the desired sequences. The BLSTM has also been evaluated for spotting without the use of the HMM stage, providing a 100 % precision system. Our experiments on a public handwritten database show interesting results.

1 Introduction

Detecting Regular expression (REGEX) in handwritten documents can be useful for finding sub-string which are relevant for a further higher level information extraction task. It consists in detecting patterns sequence of characters that obey certain rules described using meta models such as lower cases (#[a-z]#), upper cases (#[A-Z]#) or Digits (#[0-9]#). For example, a system of that kind could spot entities. For example, spotting date (#[0-9]{2}/[0-9]{2}/[0-9]{4}#), first name (#[A-Z][a-z]*#), ZIP code and city name of a french postal address (#[0-9]{5} [A-Z]*#). The extraction of these informations allows to consider high level processing stages such as document categorisation, customer identification, Named Entity detection, etc.

The spotting of regular expression is a common task on electronical documents, using Natural Language Processing methods [1,2]. In this case, the REGEX spotting is rather straightforward as it consists in applying exact string matching methods on the ASCII text. When dealing with document images, a recognition step is needed in order to produce the ASCII transcription before processing the input data. The trouble is that this recognition step is subject to errors and uncertainty, making the string matching problematic. Some attempts have been made on printed documents [3,4]. In these works, an OCR is applied on the whole document before applying the regular expression spotting step based on a set of rules that performs an exact matching. In spite of OCR errors,

© Springer International Publishing Switzerland 2015
A. Fred et al. (Eds.): ICPRAM 2015, LNCS 9493, pp. 143–157, 2015.
DOI: 10.1007/978-3-319-27677-9_9

the system provides acceptable performances (Average Precision of 82 % and 72 % of Recall).

However, only a few works concerning regular expression spotting in handwritten documents. The reason is that exact matching methods can not overcome the frequent recognition errors due to the intrinsic difficulties of recognizing handwriting. Therefore, in order to cope with these errors, inexact matching method should be carried out. This can be performed using statistical sequence models such as HMM. Some works have been published within this framework, proposing pattern spotting such as dates [5] and numerical fields [6,7] that involve meta models of characters, namely digits. However, these HMM based approach are limited to very specific fields.

A more generic approach for REGEX spotting in handwritten documents has been addressed using pure HMM approach [8], but led to moderate results (see Sect. 5 for detailed results and comparison). This article presents a REGEX spotting system for handwritten documents. It is based on a combination of HMM statistical sequence model with the state-of-the-art BLSTM neural network. Our alternative hybrid BLSTM/HMM model enables us to benefit from both strong local discrimination, and the generative sequence ability of the HMM. An alternative to the hybrid system is also proposed in this article, using the BLSTM stage without the HMM model. It consists in applying the BLSTM, and searching for the query in the raw recognition results. Surprising results are obtained, namely producing 100 % precision performance.

This paper is organized as follows: first a review of word and REGEX spotting is given in Sect. 2, then we present our REGEX spotting system based on a hybrid BLSTM/HMM in Sect. 3. Section 4 is devoted to the experimental setup and results on both word spotting and regular expression spotting tasks carried out on the RIMES database [9].

2   Related Work

As a REGEX can match sequences with variable length and characters, a REGEX spotting task can be assimilated to a word spotting task where the word belongs to a lexicon which contains all the character string variations admissible by the REGEX. The less constrained the REGEX, the larger the size of the lexicon. Relaxing those constraints makes the REGEX spotting task more complex especially when considering handwritten document images. As regular expression spotting shares many aspects in common with word spotting we now briefly introduce the related works concerning word spotting approaches.

Word spotting in document images has received a lot of attention these last years. Systems proposed in the literature are divided into two main categories: Image based and recognition based systems. The first one, also known as *query-by-example*, operates through the image representation of the keywords [10–14]. Such systems are therefore limited to deal with omni-writer handwriting and require to get an image of the query. The second kind of approaches, also known as *query-by-string* methods, deals with the ASCII representation of the keywords [15–19]. Moving from the image representation to the ASCII representation of the query is performed through a recognition stage. These systems are suitable for omni-writer handwriting and can be used with any string query of any size. In this context, many works have focused on several variants of Hidden Markov Models (HMMs) to process this intrinsically sequential problem [19].

State-of-the-art recognition-based approaches are based on a line of text models [18–20]. The line model generally contains a model of the target word, combined with filler models that describe the out-of-vocabulary words. For example in [20], the authors present an alpha-numerical information extraction system on handwritten unconstrained documents. It relies on a global line modeling allowing a dual representation of the relevant and the irrelevant information. The acceptation or rejection of the matched keyword is controlled by the variation of a hyper-parameter in the HMM line model. A similar approach is presented in [18]. The line model is made of a left and right filler models surrounding the word model. The acceptation or rejection of the matched keyword is controlled by a text line score based on the likelihood ratio of the word text line and the filler text line model. However we know that HMM rely on strong observation independence assumptions and they perform poorly on high dimensional observations. Moreover, they have low discrimination capabilities between character classes due to their inherently generative modelization framework.

Recently a new approach based on recurrent neural networks has overcome these shortcomings. Bilateral Long Short Term Memory (BLSTM) architecture has demonstrated impressive capabilities for omni-writer handwriting recognition [21]. Some primary applications of BLSTM to word spotting have also demonstrated promising results [22,23]. In this system, the BLSTM is combined with the CTC layer which provides character class posterior probabilities. Then a token passing algorithm allows efficient decoding of the spotting line model. Very interesting results have been reported on the IAM Database [22][1].

In this paper, we combine the BLSTM-CTC architecture with a HMM based spotting line model. This two stage architecture is first evaluated for hadnwritten word spotting on the RIMES database. Then we explore some extensions of the system to Regular Expression spotting (REGEX Spotting). This model is described in the following section.

3   BLSTM-CTC/HMM System

In this section, we describe our hybrid model for word and REGEX spotting. We first describe the BLSTM-HMM architecture that has been retained, then we present our word spotting model, based on standard state-of-the-art word spotting framework. And finally, we propose the adaptation of this model for REGEX spotting.

---

[1] Average Precision of 88.15 % and R-precision of 84.34 %.

3.1   Character Recognition and Segmentation

The BLSTM-CTC is a complex Recurrent Neural Network able to manage long term dependencies thanks to its internal buffer structure. Each neuron is specialized to stop a specific character in the input signal. The recurrent architecture allows each neuron to take account of the previous activated neurons (character), possibly at multiple time step earlier in the input signal (thus modeling long term dependencies). This typical architecture allows to take account of character bigrams in addition to the input signal to compute the activation of each neuron. The BLSTM is composed of two recurrent neural networks with Long Short Term Memory neural units. The first one processes the data from left to right whereas the second one proceeds in the reverse order. For each time step, decision is taken combining the two networks output, taking advantage of both left and right context. Such context is essential to have a certain knowledge of the surounding characters, because in most cases sequences of letters are constrained by the properties of the lexicon. The outputs of these two networks are then combined through a Softmax decision layer that provides character posterior probabilities in addition to a non decision class. This decision stage is called the Connectionist Temporal Classification (CTC) [24] that enables the labelling of unsegmented data.

These networks integrate special neural network units: Long Short Term Memory [24] (LSTM). LSTM neurons is composed of a memory cel, an input and three control gates. Each gates control the memory of the cell, i.e. how a given input will affect the memory (input gates), if a new input should reset the memory cell (forget gate) and if the memory of the network should be presented to the following neuron (output gate). This system of control gates allows a very accurate control of the memory cell during the training step. A LSTM layer is fully recurrent, that is to say, the input and the three gates receive at each instant t the input signal at time t and the previous outputs (at time $(t-1)$).

This architecture has shown very impressive results on challenging datasets dedicated to word recognition [9,25] due to its efficient classification and segmentation ability. For these reasons, its efficiency to cope with the low level character identification appears also very promising for handwritten words and REGEX spotting, since such scenario is less contrained by lexicon properties.

The proposed BLSTM/HMM architecture has been chosen in order to take advantage of both generative and discriminative frameworks. As shown on Fig. 2, the input sequence is processed by a BLSTM-CTC network in order to compte character posterior probabilities at every step. Then the probabilities of each labels are fed to the HMM stage (using class posteriors in place of the character likelihood computed by Gaussian Mixtures Models in the traditionnal HMM framework) to perform the alignment of the spotting model. We now describe the HMM line spotting models which enable us to spot either words (cf Sect. 3.2) or REGEX (cf Sect. 3.3).

3.2   Handwritten Word Spotting Model

Our word spotting model describes a line of text that may contain the word to spot. As it is classically proposed in the literature, it is made of the HMM word model surrounded by filler models that represent any other sequence of characters. Figure 1 shows an example of a word spotting model for the word **"sentiments"**. The space model is directly integrated into the filler. By constraining the whole model, we can locate the word at the beginning, in the middle or at the end of the line. The filler model is basically an ergodic model made of every character model. In our problem, we use 99 models corresponding to lower and upper cases, digits, punctuations and space.

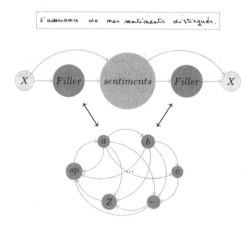

Fig. 1. HMM line model: Detail of every component of the line model.

Decoding a text line is classically achieved using the Viterbi algorithm, the system will outputs the character sequence with the maximum likelihood $P(X|\lambda)$. In order to accept the spotted word or reject it, decoding is generally performed twice: a first pass using the spotting model, and a second pass using a filler model. The likelihood ratio of the two models serves generally as a score for accepting or rejecting the spotted hypothesis. Using the BLSTM-CTC architecture, posterior probabilities are computed that can directly serve as a score for accepting/rejecting the hypothesis, without the need for a filler model. The score of each spotted hypothesis is computed by the average character posteriors over the number of frames spanning the hypothesis. This score is then normalised by the number of characters of the spotted word. Doing this, we choose to rely on the strong discriminative decisions of the BLSTM-CTC and use the HMM only as a sequence model constrained by high level information such as lexicons and/or language models. The graphical representation of the whole word spotting system is shown on Fig. 2.

We now show how this model can be adapted to REGEX spotting.

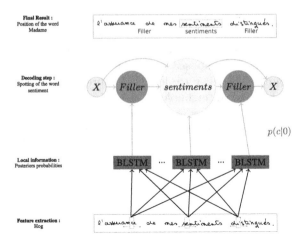

Fig. 2. Hybrid structure BLSTM/HMM: Details of every step of the word spotting task from feature extraction to position of the word **sentiments** in the sentence. The BLSTM/CTC outputs a posteriori probabilities for HMM decoding.

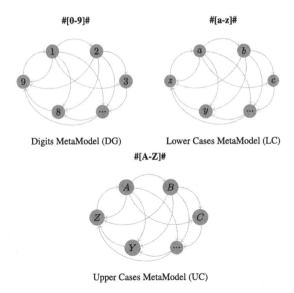

Fig. 3. HMM MetaModels.

3.3   Regular Expression Spotting Model

As previously mentionned, REGEX spotting is a generalisation of the word spotting task, the difference is that the sequences to spot are less constrained and more variable, thus leading to a larger lexicon of admissible expressions.

In order to cope with REGEX queries, we use the HMM stage to model a regular expression with a stochastic model of character sequences. Each meta model is an ergodic model of characters implied in the query, e.g., Lower Cases (#[a-z]#), Upper Cases (#[A-Z]#) or Digits (#[0-9]#), as it is the case for the Filler models. Figure 3 shows examples of meta models for these three examples.

We also need to model the variable length of the queries, which may occurs when using * or + operators (spotting between 0 and ∞ times a character, or spotting between 1 and ∞ times a character) such as in #[0-9]+# which stands for any sequence of at least 1 digit. This is simply modeled by allowing auto transitions over the desired character meta model. Figure 4 shows an example of a model for spotting variable length sequences. The query taken is the sub-string **agr** following by an unconstrained sequence of lower cases (#[a-z]*#), in this example we hope that the system will spot the word **agréer** correctly.

The following models allow searching for a REGEX at the beginning of a line (#[a-z]*ion#), at the end of a line ((#le[a-z]*#)), or both (#[A-Z]o[a-z]#). The line model can also only contain meta models dedicated to spotting sequences of digits of any length, for example (#[0-9]*#) or word beginning by one upper case character and ending with a sequence of lower cases characters of arbitrary length (#[A-Z][a-z]*#). Here, the arbitrary length of the sequence unconstrained (*) is controled by the auto-transition probabilities of the meta model of the HMM.

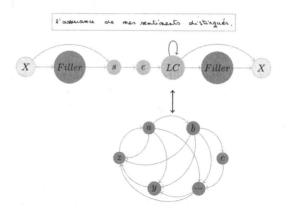

Fig. 4. HMM stage: Spotting of regular expressions #se[a-z]*# (i.e. every word beginning by the sub-string **se** followed by any number of lower case characters).

As the transitions in the HMM meta models are ergodic, the Viterbi alignment will only be driven by the local classification of BLSTM-CTC. The spotting model depends on its discriminant capacity to feed the higher HMM stage with accurate information from the local character recognition stage.

The graphical representation of the whole REGEX spotting system is shown on Fig. 5.

Finally, the integration of meta models and auto transitions into the line model allows spotting of handwritten REGEX. Practically, the line model is

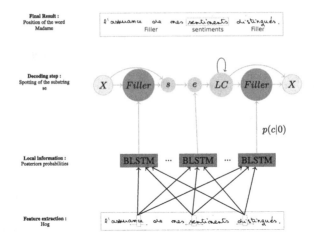

Fig. 5. Hybrid structure BLSTM/HMM: Details of every step of the REGEX spotting task from feature extraction to position of the REGEX **#se[a-z]*#** in the sentence.

build on the fly at the time of querying the data-set, by rewriting the REGEX into a HMM line spotting model. At this time, the "translation" is manually done, but an automatization of this task can be performed for industrial purpose.

4   Experiments

In this section, we give some details about the implementation of the system, starting with a description of the features extraction in Sect. 4.1. The performance of the system are evaluated using the 2011 RIMES database [26], they are summarized in Sect. 4.2.

4.1   Features Set

Our feature vector is based on Histograms of Oriented Gradient (HOG) [27] extracted from windows of 8×64 pixels. During the extraction, the window is dividing into sub-windows of $n \times m$ pixels. For each sub-window a histogram is computed, representing the distribution of the local intensity gradients (edge direction). The histograms of every windows are then merged to obtain the final representation of our feature vector representation. We used 8×8 non-overlapping sub-windows using 8 directions, this produces a 64 dimensional feature vector.

4.2   Results and Discussion

To evaluate the performance of our system, all the experiments have been performed on the RIMES database used for the 2011 ICDAR handwriting recognition competitions [26]. The training database is composed of 1.500 documents,

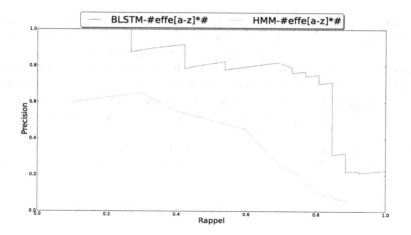

Fig. 6. Regular expression spotting performance with the sub-string **effe** (#effe[a-z]*#).

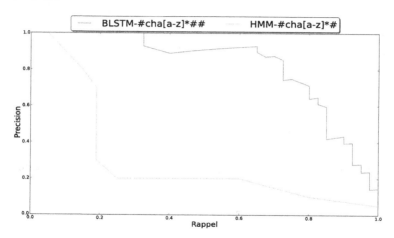

Fig. 7. Regular expression spotting performance with the sub-string **cha** (#cha[a-z]*#).

the validation and test sets are composed respectively of 100 documents. In order to evaluate the spotting system, we compute recall (R) and precision measures (P). To do this, the number of true positives (TP), false positives (FP), and false negatives (FN) are evaluated for all possible threshold values. From these values, a recall-precision curve is computed by accumulating these values over all word queries.

$$R = \frac{TP}{TP + FN} \quad P = \frac{TP}{TP + FP} \tag{1}$$

Regular Expression Results. To evaluate the performance of our system on a regular expression spotting task we performed exactly the same experiments as in [8]. In this study the authors were interested in spotting 4 different REGEX queries corresponding to the the search for the sub-strings "effe", "pa", "com" and "cha" at the beginning of a word (#effe[a-z]*#, #pa[a-z]*#, #com[a-z]*#, #cha[a-z]*#). As for word spotting experiments, results of the HMM system have been added too in order to provide a precise comparison between those systems (cf Figs. 6, 7, 8 and 9).

A first observation is that the system achieves good performance, since most of the REGEX queries lead to a mean-average precision of nearly 75 %, whereas

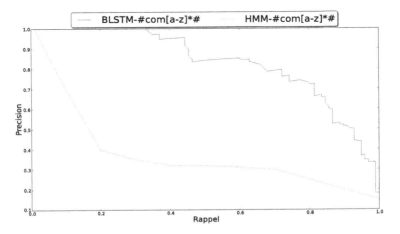

Fig. 8. Regular expression spotting performance with the sub-string **com** (#com[a-z]*#).

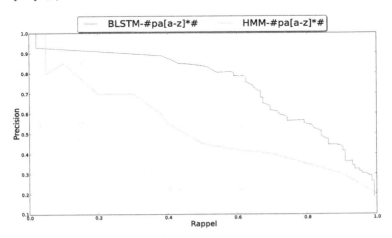

Fig. 9. Regular expression spotting performance with the sub-string **pa** (#pa[a-z]*#).

the queries involve many fewer constraints than for word spotting. Moreover, our results are far beyond the standard HMM approach. We can observe a gap of more than 40 % in the difficult cases (#com[a-z]*#) and (#cha[a-z]*#) and 20 % in easier ones (#effe[a-z]*# and #pa[a-z]*#). We also run more test on other queries such as #[a-z]*er#,#[a-z]*tion#, #[a-z]*tt[a-z]*# and #[a-z]*mm[a-z]*#. The results are still pretty good for both textbf#[a-z]*tion# #[a-z]*mm[a-z]*#. However, the system seems to have trouble dealing with the two other queries. It is certainly due the fact that two consecutive letters such as "t" are really difficult to spot. Concerning the other issue, it is due to the high level of confusion between "r" and other characters such as "n","u", etc.

We have also tested less constrained queries, with the search for REGEX containing any sequence of upper cases characters (#[A-Z]*#), and any sequence of digits (#[0-9]#). This problem is by far more difficult than the previous queries since the corresponding sequences may have variable contents and lengths. For example the digit query should detect the sequence "1" as well as sequence "0123456789". Results are presented in Fig. 11.

Knowing the difficulty of the problem, the performance are still interesting. Note that digit characters are not very frequent in the database. An interesting fact is that the Uppercase query can reach interesting precision scores, whereas the digit query can reach very high recall scores.

In the following section, an additional experiment is proposed in order to maximize the precision.

4.3   Using the BLSTM Without HMM

Recall-Precision curves allow the user to choose the most appropriate threshold for his problem. Indeed, some applications need to maximise the precision whereas some others may privilege the Recall. That is why we perform a REGEX spotting experiment without introducing the higher level HMM stage spotting model. This stage correspond to analysing the raw transcriptions provided by the low level decision stage and then matching the searched keywords on this transcription. Results are shown in Table 1.

The first comment from this experiment is that each query leads to a 100 % precision, which means that the system does not any false alarm. This result was expected since only two cases may occur: the first case is when the BLSTM correctly recognizes every character from the query, leading to a "hit". The second case is when a recognition error occurs within the searched sequence. In this case the query is missed, but no false alarm is produced. Finally, no false alarms are produced using this system, whatever the recognition performance. Despite the lack of constraint of the request, the BLSTM-CTC manage to obtain high value of recall: more than 55 % for 6 of our REGEX and more than 60 % for 4 of them. An interesting fact is that these operating points are significantly beyond the recall-precision curves provided using the HMM stage. The 100 % precision (absence of false alarm) makes this system suitable for keywords spotting dedicated to categorization systems as in [28].

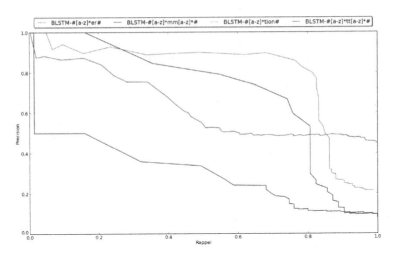

Fig. 10. Regular expression spotting performance for #[a-z]*er#,#[a-z]*mm [a-z]*#, #[a-z]*tion# and #[a-z]*tt[a-z]*#.

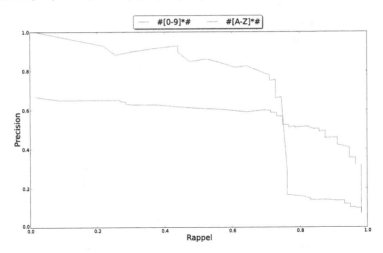

Fig. 11. Regular expression spotting performance with upper case sequence (#[A-Z]*#) and number sequence (#[0-9]#).

Our system seems to have trouble dealing with #[a-z]*tt[a-z]*# and #[a-z]*er#, but as you can see on the Fig. 10 those two REGEX were already badly recognized with the HMM stage. As said before, this is certainly due to the fact that it is difficult to recognize correctly consecutive letters (*tt*) or high confusion level letters (*r*). The BLSTM-CTC seems to be able perform recognition of out vocabulary elements such as named entities for example.

In most of the applications, you can improve your global results thanks to a language model, a lexicon or any kind of high level information. By performing

Table 1. Results of the detection without HMM stage of the following REGEX: #effe[a-z]*#, #pa[a-z]*#, #com[a-z]*#, #cha[a-z]*#, #[a-z]*er#,#[a-z]*tion#, #[a-z]*tt[a-z]*#, #[a-z]*mm[a-z]*#.

REGEX	Recall	Precision
#effe[a-z]*#	63.2 %	100 %
#pa[a-z]*#	57.4 %	100 %
#com[a-z]*#	65.1 %	100 %
#cha[a-z]*#	65.8 %	100 %
#[a-z]*er#	46.5 %	100 %
#[a-z]*tion#	66.4 %	100 %
#[a-z]*tt[a-z]*#	38.3 %	100 %
#[a-z]*mm[a-z]*#	57.1 %	100 %

this experiment, we once again prove the powerful capacity of the BLSTM-CTC to tackle the Sayre paradox as it is able to segment and recognize characters very accurately.

5   Conclusion

In this paper, we have proposed a hybrid system BLSTM-CTC/HMM able to spot any word of REGEX. We have shown that the hybrid system exhibits interesting results, even on weakly constrained queries such as the search for sequences of digits of arbitrary length. We have compared our system for REGEX spotting with some recent work carried out on the same data-set and using the standard HMM framework. Our approach outperforms this system by more than 30 % on the standard word spotting task and by more than 40 % on REGEX spotting. These very promising results allow to envisage the application of higher level spotting systems such as addresses, named entities for which a combination of specific markers (keywords and alpha numerical expressions) is generally used to detect the relevant information. Some additional results show that the BLSTM-CTC provides interesting performance even when no additional constraints are introduced, since it provides some interesting recall values (higher than 60 %). These experiments show that the frontiers of processing handwritten documents are becoming more and more closer to those of processing printed or born digital documents which offers many perspectives for developing applications dealing with handwritten documents in the bear future.

References

1. Hosoya, H., Pierce, B.: Regular expression pattern matching for XML. In: Proceedings of the 28th ACM SIGPLAN-SIGACT Symposium on Principles of Programming Languages, pp. 67–80 (2001)

2. Dengel, A.R., Klein, B.: *smartFIX*: a requirements-driven system for document analysis and understanding. In: Lopresti, D.P., Hu, J., Kashi, R.S. (eds.) DAS 2002. LNCS, vol. 2423, pp. 433–444. Springer, Heidelberg (2002)

3. Spitz, A.: Using character shape codes for word spotting in document images. In: Proceedings of the Symposium on Document Analysis and Information Retrieval, pp. 382–389 (1995)

4. Spitz, A.: Determination of script, language content of document images. IEEE Trans. Pattern Anal. Mach. Intell. **19**, 235–245 (1997)

5. Morita, M.E., Sabourin, R., Bortolozzi, F., Suen, C.Y.: Segmentation and recognition of handwritten dates: an HMM-MLP hybrid approach. Doc. Anal. Recogn. **6**(4), 248–262 (2003)

6. Chatelain, C., Heutte, L., Paquet, T.: A two-stage outlier rejection strategy for numerical field extraction in handwritten documents. In: ICPR, Hong Kong, China, vol. 3, pp. 224–227 (2006)

7. Chatelain, C., Heutte, L., Paquet, T.: Recognition-based vs syntax-directed models for numerical field extraction in handwritten documents. In: ICFHR, Montreal, Canada, 6 p. (2008)

8. Kessentini, Y., Chatelain, C., Paquet, T.: Word spotting and regular expression detection in handwritten documents. In: ICDAR (2013)

9. Grosicki, E., El Abed, H.: ICDAR 2009 handwriting recognition competition. In: 10th International Conference on Document Analysis and Recognition, ICDAR 2009, pp. 1398–1402. IEEE (2009)

10. Rath, T.M., Manmatha, R.: Features for word spotting in historical manuscripts. In: Proceedings of the Seventh International Conference on Document Analysis and Recognition, pp. 218–222. IEEE (2003)

11. Cao, H., Govindaraju, V.: Template-free word spotting in low-quality manuscripts. In: Proceedings of the 6th International Conference on Advances in Pattern Recognition, pp. 135–139 (2007)

12. Adamek, T., O'Connor, N.E., Smeaton, A.F.: Word matching using single closed contours for indexing handwritten historical documents. Int. J. Doc. Anal. Recogn. (IJDAR) **9**, 153–165 (2007)

13. Rusinol, M., Aldavert, D., Toledo, R., Lladós, J.: Browsing heterogeneous document collections by a segmentation-free word spotting method. In: 2011 International Conference on Document Analysis and Recognition (ICDAR), pp. 63–67. IEEE (2011)

14. Rodríguez-Serrano, J.A., Perronnin, F., Lladós, J., Sánchez, G.: A similarity measure between vector sequences with application to handwritten word image retrieval. In: IEEE Conference on Computer Vision and Pattern Recognition, CVPR 2009, pp. 1722–1729. IEEE (2009)

15. Rodríguez-Serrano, J.A., Perronnin, F.: Handwritten word-spotting using hidden markov models and universal vocabularies. Pattern Recogn. **42**, 2106–2116 (2009)

16. Frinken, V., Fischer, A., Manmatha, R., Bunke, H.: A novel word spotting method based on recurrent neural networks. IEEE Trans. Pattern Anal. Mach. Intell. **34**, 211–224 (2012)

17. Thomas, S., Chatelain, C., Heutte, L., Paquet, T., Kessentini, Y.: A deep HMM model for multiple keywords spotting in handwritten documents. Accepted for publication in Pattern Analysis and Applications (2014). doi:10.1007/s10044-014-0433-3

18. Fischer, A., Keller, A., Frinken, V., Bunke, H.: Lexicon-free handwritten word spotting using character HMMs. Pattern Recogn. Lett. **33**, 934–942 (2012)

19. Wshah, S., Kumar, G., Govindaraju, V.: Script independent word spotting in offline handwritten documents based on hidden markov models. In: 2012 International Conference on Frontiers in Handwriting Recognition (ICFHR), pp. 14–19. IEEE (2012)
20. Thomas, S., Chatelain, C., Heutte, L., Paquet, T.: An information extraction model for unconstrained handwritten documents. In: 2010 20th International Conference on Pattern Recognition (ICPR), pp. 3412–3415. IEEE (2010)
21. Graves, A., Liwicki, M., Bunke, H., Schmidhuber, J., Fernández, S.: Unconstrained on-line handwriting recognition with recurrent neural networks. In: Advances in Neural Information Processing Systems, pp. 577–584 (2008)
22. Frinken, V., Fischer, A., Bunke, H.: A novel word spotting algorithm using bidirectional long short-term memory neural networks. In: Schwenker, F., El Gayar, N. (eds.) ANNPR 2010. LNCS, vol. 5998, pp. 185–196. Springer, Heidelberg (2010)
23. Wöllmer, M., Eyben, F., Graves, A., Schuller, B., Rigoll, G.: A tandem BLSTM-DBN architecture for keyword spotting with enhanced context modeling. In: Proceedings of NOLISP (2009)
24. Graves, A., Fernández, S., Gomez, F., Schmidhuber, J.: Connectionist temporal classification: labelling unsegmented sequence data with recurrent neural networks. In: Proceedings of the 23rd International Conference on Machine Learning, pp. 369–376. ACM (2006)
25. Graves, A., Liwicki, M., Fernández, S., Bertolami, R., Bunke, H., Schmidhuber, J.: A novel connectionist system for unconstrained handwriting recognition. IEEE Trans. Pattern Anal. Mach. Intell. **31**, 855–868 (2009)
26. Grosicki, E., El-Abed, H.: ICDAR 2011-french handwriting recognition competition. In: 2011 International Conference on Document Analysis and Recognition (ICDAR), pp. 1459–1463. IEEE (2011)
27. Rodrıguez, J.A., Perronnin, F.: Local gradient histogram features for word spotting in unconstrained handwritten documents. In: International Conference on Frontiers in Handwriting Recognition (2008)
28. Paquet, T., Heutte, L., Koch, G., Chatelain, C.: A categorization system for handwritten documents. Int. J. Doc. Anal. Recogn. **15**, 315–330 (2012)

A Similarity-Based Color Descriptor for Face Detection

Eyal Braunstain[✉] and Isak Gath

Faculty of Biomedical Engineering,
Technion - Israel Institute of Technology, Haifa, Israel
seyalbra@tx.technion.ac.il

Abstract. Most state-of-the-art approaches to object and face detection rely on intensity information and ignore color information, as it usually exhibits variations due to illumination changes and shadows, and due to the lower spatial resolution in color channels than in the intensity image. We propose a new color descriptor, derived from a variant of Local Binary Patterns, designed to achieve invariance to monotonic changes in chroma. The descriptor is produced by histograms of encoded color texture similarity measures of small radially-distributed patches. As it is based on similarities of local patches, we expect the descriptor to exhibit a high degree of invariance to local appearance and pose changes. We demonstrate empirically by simulation the invariance of the descriptor to photometric variations, i.e. illumination changes and image noise, geometric variations, i.e. face pose and camera viewpoint, and discriminative power in a face detection setting. Lastly, we show that the contribution of the presented descriptor to face detection performance is significant and superior to several other color descriptors, which are in use for object detection. This color descriptor can be applied in color-based object detection and recognition tasks.

1 Introduction

Most object and face detection algorithms rely on intensity-based features and ignore color information. This is usually due to its tendency to exhibit variations due to illumination changes and shadows [1], and also to the lower spatial resolution in color channels than in the intensity image (e.g. the works of [2–5]). Face detection performance by a human observer declines when color information is removed from faces [6]. It has been argued that a detector which is based solely on spatial information derived from an intensity image, e.g. histograms of gradients, may fail when the object exhibits changes in spatial structure, e.g. pose, non-rigid motions, occlusions etc. [7]. Specifically, an image color histogram is rotation and scale-invariant.

We hereby review the topic of color representations and descriptors for object detection. Color information has been successfully used for object detection and recognition [1,7–13].

Color can be represented in various color spaces, e.g. RGB, HSV and CIE-Lab, in which uniform changes are perceived uniformly by a human observer [14].

© Springer International Publishing Switzerland 2015
A. Fred et al. (Eds.): ICPRAM 2015, LNCS 9493, pp. 158–171, 2015.
DOI: 10.1007/978-3-319-27677-9_10

Various color descriptors can be designed. The color bins descriptor [7] is composed of multiple 1-D color histograms by projecting colors on a set of 1-D lines in RGB space at 13 different directions. These histograms are concatenated to form the color bins features.

Two color descriptors were examined by [9] for object detection, the Robust Hue descriptor, invariant with respect to the illuminant variations and lighting geometry variations (assuming white illumination), and Opponent Angle (OPP), invariant with respect to illuminant and diffuse lighting (i.e. light coming from all directions).

The trade-off between photometric invariance and discriminative power was examined in [11], where an information theoretic approach to color description for object recognition was proposed. The gains of photometric invariance are weighted against the loss in discriminative power. This is done by formulation of an optimization problem with objective function based on KL-Divergence between visual words and color clusters.

Deformable Part Model (DPM) is used to model objects using spring-like connections between object parts [15,16]. Although DPM achieves very good detection results, in particular through its ability to handle challenging objects (e.g. deformations, view changes and partially occluded objects), the general computational complexity of part-based methods is higher than global feature-based methods [17,18].

The Three-Patch Local Binary Patterns (TPLBP) [19] is a robust variant of the Local Binary Patterns (LBP) descriptor [20], based on histograms of encoded similarity measures of local intensity patches. This descriptor was examined for the face recognition task.

In the present work the focus is not on the design of a new face detection framework, but rather on the design of a novel color descriptor, investigating its possible contribution to face detection. We design a new color descriptor, based on Three-Patch LBP. Our descriptor is computed from histograms of encoded similarities of small local patches of chroma channels in a compact form, utilizing the inter-correlation between image chroma channels. Consequently, the representation of color in an image window is global, i.e. not part-based. We examine the descriptor by ways of its robustness to photometric and geometric variations and discriminative power. We evaluate the contribution of the descriptor in a face detection setting, using the FDDB dataset [21], and show that it exhibits significant contribution to detection rates.

The paper is organized as follows. In Sect. 2 the Three-Patch LBP (TPLBP) descriptor is described briefly, and a multi-scale variant is proposed; in Sect. 3 the new color descriptor is described; in Sect. 4 invariance and discriminative power are evaluated, compared to the Robust Hue and Opponent Angle descriptors [9]; in Sect. 5 we evaluate the color descriptor in a face detection setting, and in Sect. 6 conclusions to this work are provided.

2   Three-Patch LBP Descriptor and a Multi Scale Variant

The Three-Patch LBP [19] descriptor was inspired by the Self-Similarity descriptor [22], which compares a central intensity image patch to surrounding patches from a predefined area, and is invariant to local appearance. For each central pixel, a $w \times w$ patch is considered, centered at that pixel, and S additional patches distributed uniformly in a ring of radius r around that pixel. Given a parameter α (where $\alpha < S$), we take S pairs of patches, α-patches apart, and compare their values to the central patch. A single bit value for the code of the pixel is determined according to which of the two patches is more similar to the central patch. The code has S bits per pixel, and is computed for pixel p by:

$$TPLBP(p) = \sum_{i=1}^{S} f_\tau \left(d\left(C_i, C_p\right) - d\left(C_{i'}, C_p\right)\right) \cdot 2^i$$

$$i' = (i + \alpha) \bmod S$$

(1)

where C_i and $C_{(i+\alpha) \bmod S}$ are two $w \times w$ patches along the patches-ring, α-patches apart, C_p is the central patch, $d\left(\cdot, \cdot\right)$ is a distance measure (metric), e.g. L_2 norm, and the function f_τ is a step threshold function, $f_\tau(x) = 1$ iff $x \geq \tau$. The threshold value τ is chosen slightly larger than zero, to provide stability in uniform regions. The values in the TPLBP code image are in the range $\left[0, 2^S - 1\right]$. Different code words designate different patterns of similarity. Once the image is TPLBP-encoded, the code image is divided into non-overlapping cells, i.e. distinct regions, and a histogram of code words with 2^S bins is constructed for each cell. The histograms of all cells are normalized to unit norm and concatenated to a single vector, which constitutes the TPLBP descriptor.

In the formulation of the LBP descriptor [20], a binary value is assigned according to whether a surrounding pixel is higher or lower than a central pixel. When LBP-encoding a PXP pixels window, Uniform binary patterns are defined by limiting the number of transitions from 0 to 1 or vice versa in the circular binary pattern. Patterns with more than two such transitions are designated non-Uniform, and are assigned to a single label (and therefore a single bin in the LBP histogram). The uniform patterns are considered to provide the majority of micro texture patterns, e.g. edges, corners and spots, while the highly non-uniform patterns, with many 0-1 transitions, can mostly be attributed to image noise.

We stress out that differently from LBP, in TPLBP encoding, surrounding pixels are not thresholded against a central pixel, but surrounding patches are compared by measure of similarity to a central patch. Thus, if a TPLBP pattern exhibits a high number of 0-1 transitions, it may indicate more complicated patterns of similarity of surrounding patches, rather than noise, which is pixel-wise variable.

We propose a Multi-Scale TPLBP descriptor (termed TPLBP-MS), capturing spatial similarities at various scales and resolutions, by concatenating TPLBP descriptors with various parameters r and w. The scale is affected by the radius r and patch resolution by patch size w. Three sets of parameters are used for the encoding operator of Eq. 1, i.e. $(r, S, w) = \{(2, 8, 3), (3, 8, 4), (5, 8, 5)\}$, all with $S = 8$ and $\alpha = 2$, as in [19]. These 3 TPLBP descriptors are concatenated to produce the TPLBP-MS descriptor. Parameters r and w are changed in similar manner in the 3 sets above, thus observing larger scales at lower resolutions.

3   A New Color Descriptor - Coupled-Chroma TPLBP

Many color descriptors are histograms of color values in some color space, e.g. rg-histogram and Opponent Colors histograms [12]. Image color channels contain texture information that is disregarded by color histograms. Our motivation is to formulate a color descriptor that captures the texture information embedded in color channels in a robust manner.

Color descriptors can be evaluated by several main properties: (1) Invariance to photometric changes (e.g. illumination, shadows etc.); (2) Invariance to geometric changes (e.g. camera viewpoint, object pose, scale etc.); (3) Discriminative power, i.e. the ability to distinguish a target object from the rest of the world; (4) Stability, in a sense that the variance of a certain dissimilarity measure between descriptor vectors of samples from a specific distribution (or class) is low. We would like to formulate a color descriptor that adheres to these properties.

We represent color in CIE-Lab space, due to its perceptual uniformity to a human observer. Using Euclidean distance in CIE-Lab space approximates the perceived distance by an observer, hence a detector based on this color space can in some sense approximate the perception of human color vision. In CIE-Lab space, L is the luminance, a and b are the chroma channels. We consider first a color descriptor produced by applying TPLBP to both chroma channels and concatenating the single-channel descriptors to a single descriptor. Images in JPEG format are analyzed, in which the chroma channels are sub-sampled [23], thus spatial resolution in chroma channels is lower than in intensity. Hence, to extract meaningful features from chroma, the appropriate operator should be applied at a coarse resolution, relative to the operator applied to the intensity image. The values of the parameters are chosen accordingly, $(r, S, w) = (5, 8, 4)$, i.e. both the radius and patch dimension are increased. This descriptor is termed Chroma TPLBP (C-TPLBP). It has twice the dimension of TPLBP.

A degree of correlation exists between the chroma channels in CIE-Lab space. This can be observed either from the derived equations of CIE-Lab color space from CIE-XYZ space, or from an experimental perspective, by constructing a 2-D chroma histogram of face images. Elliptically cropped face images from the FDDB dataset [21] with 2500 images are used to fit a 2-D Gaussian density of chroma values a and b by mean and covariance of the data. From the covariance matrix, we have that $\sigma_{ab} = 53.7$, i.e. nonzero correlation between the chroma channels. We presume that coupling the chroma channels information may lead to a robust descriptor, which is also more compact than C-TPLBP, where chroma channels descriptions are computed separately. We propose the following operator:

$$CC - TPLBP\,(p) =$$
$$\sum_{i=1}^{S} f_\tau \left(\sum_{k=a,b} \left(d\left(C_{k,i}, C_{k,p}\right) - d\left(C_{k,i'}, C_{k,p}\right)\right)\right) \cdot 2^i \qquad (2)$$

$$i' = (i + \alpha)\, mod\, S$$

where $C_{k,i}$ is the ith patch of chroma channel k and the inner summation is over chroma channels, a and b. The thresholding function f_τ operates on the sum of differences of patches distance functions, for both chroma channels. Given a parameter α, we take S pairs of patches from each chroma channel, α-patches apart, and for each pair we compare distances to the central patch of the appropriate channel. A single bit value for the code of a pixel is determined as follows - if similarities in both chroma channels correlate, e.g. if in both chroma channels patch C_i is more similar to the central patch C_p than patch $C_{i+\alpha}$, then the appropriate bit will be assigned value 0 (value 1 in the opposite case). Conversely, if dissimilarities of the two channels do not correlate, then by viewing the argument of the function f_τ as $\sum_{k=a,b} d\left(C_{k,i}, C_{k,p}\right) - \sum_{k=a,b} d\left(C_{k,(i+\alpha)\,mod\,S}, C_{k,p}\right)$, the patch with lower sum of distances in both chroma channels is more similar to the center, and the code bit is derived accordingly. The computed code has S bits per pixel, and this descriptor is of the same dimension as TPLBP, i.e. half the dimension of C-TPLBP. This descriptor is termed Coupled-Chroma TPLBP (CC-TPLBP). The parameters are chosen in accordance with those of C-TPLBP,

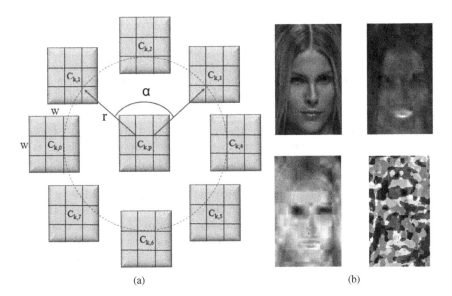

(a)   (b)

Fig. 1. CC-TPLBP code computation. (a) CC-TPLBP operator for a single chroma channel, with parameters $\alpha = 2$, $S = 8$ and $w = 3$. (b) An example of CC-TPLBP code for a color face image. Upper left - face image; upper right - CIE-Lab a chroma; lower left - CIE-Lab b chroma (a and b are presented as gray-level images); lower right -CC-TPLBP code image. The parameters used are $r = 5$, $S = 8$, $w = 4$ (Color figure online).

$(r, S, w) = (5, 8, 4)$ and $\alpha = 2$. We emphasize that different values for the radius (r), number of patches (S), patch dimension (w) and α may be chosen, however, preliminary experiments showed that good discriminative ability was obtained with the parameter values specified above. The histograms are computed on small cells of (20, 20) pixels, thus maintaining the spatial binding of color and shape information in the image by cells delimitation, i.e. late fusion of color and shape [1,24]. CC-TPLBP is invariant to monotonic variations of chroma and luminance. Such variations do not cause any change to the resulting descriptor. In Fig. 1 we present the CC-TPLBP operator, where the index $k = \{a, b\}$ designates the chroma channel, as in Eq. (2), with an example code computation for a color face image. CC-TPLBP can be combined with intensity-based shape features for classification tasks.

4   Evaluation of Color Descriptors

CC-TPLBP is invariant to monotonic changes of both luminance and color channels. Moreover, we expect it to exhibit a high degree of robustness to geometrical changes, e.g. pose, local appearance and camera viewpoint, as it is computed by similarities of radially-distributed image patches. We evaluate CC-TPLBP with respect to properties (1)–(4) described in Sect. 3, compared to the Robust Hue and Opponent Angle (OPP) color descriptors [9]. Opponent Colors are invariant with respect to lighting geometry variations, and are computed from RGB by:

$$O1 = \tfrac{1}{\sqrt{2}} (R - G)$$

$$O2 = \tfrac{1}{\sqrt{6}} (R + G - 2B) \tag{3}$$

The Robust Hue descriptor is computed as histograms on image patches over hue, which is computed from the corresponding RGB values of each pixel, according to:

$$hue = \arctan\left(\frac{O1}{O2}\right) = \arctan\left(\frac{\sqrt{3}\,(R - G)}{R + G - 2B}\right) \tag{4}$$

Hue is invariant with respect to lighting geometry variations when assuming white illumination. Hue is weighted by the saturation, to reduce error. The Opponent Derivative Angle descriptor (OPP) is computed on image patches, by the histogram over the opponent angle:

$$ang_x^O = \arctan\left(\frac{O1_x}{O2_x}\right) \tag{5}$$

where $O1_x$ and $O2_x$ are spatial derivatives of the chromatic opponent channels. OPP is weighted by the chromatic derivative strength, i.e. by $\sqrt{O1_x^2 + O2_x^2}$, and is invariant with respect to diffuse lighting and spatial sharpness. Color histograms are generally considered more invariant to pose and viewpoint changes than shape descriptors [10], but are sensitive to changes of illumination and shading.

We evaluate invariance and discriminative power by the Kullback-Leibler Divergence, a non-symmetric dissimilarity measure between two probability distributions, p and q, expressed as:

$$D_{KL}\left(p\|q\right) = \sum_i p_i \log\left(\frac{p_i}{q_i}\right) \tag{6}$$

where q is considered a model distribution.

We consider descriptors that are constructed from M histograms of M distinct image cells. Referring to CC-TPLBP, each histogram has 2^S bins, producing a descriptor of size $M \times 2^S$. Given two images, each with M cells, we compute M histograms for each image. To compare CC-TPLBP descriptors of these two images, we compute the KL Divergence for each pair of appropriate histograms from both images, i.e. $\{D_{KL}\left(h_{1,m}, h_{2,m}\right)\}_{m=1,..,M}$, where $\{h_{i,m}\}_{i=1,2}^{m=1,...,M}$ is the mth histogram of image i. We define the KL Divergence of image 1 with respect to image 2 by averaging over all image cells, i.e. $D_{KL}^{1,2} = \frac{1}{M}\sum_{m=1}^{M}\left(D_{KL}\left(h_{1,m}, h_{2,m}\right)\right)$. Each single-cell histogram contains $2^8 = 256$ bins.

We evaluate the CC-TPLBP, Hue and OPP descriptors by three experiments, described as follows:

4.1   Invariance to Photometric and Geometric Variations

In the first experiment we evaluate invariance to combined photometric and geometric variations, i.e. illumination and background, face pose and viewpoint. While this does not allow for independent evaluations of invariance to photometric and geometric variations, it simulates a realistic setting for face detection. We use several groups of images of single persons from the LFW Face Recognition dataset [25], each group displays a single person with the above variations. We compute the CC-TPLBP, Hue and OPP histograms for all images in a set, normalized to unit sum, and the KL Divergence between histograms of all image pairs (which is non-symmetric, i.e. $D_{KL}\left(p_i, p_j\right) \neq D_{KL}\left(p_j, p_i\right)$). Table 1 contains statistics of KL Divergence values of all descriptors for several image sets. While the number of images is relatively small, the number of resulting pairing is large and therefore indicative. CC-TPLBP appears to be most robust to these variations, as its mean KL Divergence is by far the lowest from all descriptors on all image sets. CC-TPLBP also exhibits a higher degree of stability than other descriptors, by its lowest variance.

4.2   Invariance to Gaussian Noise

In the second experiment, we test the effects of added noise, using 2500 face images from the FDDB dataset [21], normalized to size 63×39 pixels. According to [10], sensor noise is normally distributed, as additive Gaussian noise is widely used to model thermal noise, and is a limiting behavior of photon counting noise. High Gaussian noise is added to R, G and B channels of all images, i.e. $\{R,G,B\}_D = \{R + n_{xy}^R,\ G + n_{xy}^G,\ B + n_{xy}^B\}$, where $\{n_{xy}^k = n^k\left(x, y\right)\}_{k=R,G,B},$

Table 1. Statistics of KL-Divergence, combined evaluation of photometric and geometric invariance, for several sets of single-person images. KL-Divergence is calculated for all pairs of images in a set. For further explanation, see text.

Person set (No. images / No. pairs)	Descriptor	Mean	Median	STD
Jennifer Aniston (21 / 420)	CC-TPLBP	0.1034	0.096	0.0353
	Hue	0.9666	0.8321	0.5885
	OPP	0.3555	0.2637	0.288
Arnold Schwarzenegger (42 / 1722)	CC-TPLBP	0.1154	0.1	0.0519
	Hue	1.3682	1.2901	0.6468
	OPP	0.6535	0.482	0.551
Vladimir Putin (49 / 2352)	CC-TPLBP	0.1124	0.0988	0.0515
	Hue	1.1799	1.0515	0.6268
	OPP	0.467	0.3582	0.3737

$n(x, y) \sim \mathcal{N}(0, \sigma_n)$, with $\sigma_n = 5$. We calculate KL Divergence between descriptor histograms of original and corrupted images. Statistics of the KL Divergence values are displayed in Table 2. While Hue has an average KL Divergence slightly lower than CC-TPLBP, the latter has significantly lower variance than other descriptors, indicating higher stability under addition of Gaussian noise.

4.3   Discriminative Power

In the third experiment, we examine discriminative power. A descriptor based on color histograms would be effective in distinguishing face patches from distinct objects, e.g. trees or sky patches, but may be less effective in distinguishing a face from skin, e.g. neck, torso. Here a color texture descriptor may be more efficient. We choose randomly 200 face images from the FDDB dataset, and pick 200 background images that give a degree of diversity and challenge for the considered descriptors, i.e. versatility of chroma and texture. Half of the background images do not contain skin at all, and the other half partially contain skin, with variable backgrounds. This image set is constructed to represent the kind of natural setting where the function of the descriptor is to be able to discriminate face patches from non-face skin patches together with versatile non-skin background. Several examples are presented in Fig. 2.

To evaluate discriminative power, we use the KL Divergence similar to [1]. We define a KL-ratio for face sample, considering all face and background samples in the set:

$$KL - ratio_k = \frac{\frac{1}{N_B} \sum_{j \in B} KL(p_j, p_k)}{\frac{1}{N_F - 1} \sum_{i \in F, i \neq k} KL(p_i, p_k)} \quad \forall k \in F \tag{7}$$

where p_k is the descriptor of face patch $k \in F$, p_j is the descriptor of background patch $j \in B$, N_F and N_B are the number of face and background samples, respectively. For a face sample k, Eq. (7) defines the ratio of the average KL Divergence

Fig. 2. Example of face and background images used to examine discriminative power. Top line - sample face images, bottom line - sample background images.

with all non-face patches, divided by the average KL Divergence with all face patches. The higher this ratio for a face patch $k \in F$, the more discriminative the descriptor with respect to this face and data set, as the intra-class KL Divergence is lower than the inter-class KL Divergence. The KL-ratio values of all descriptors on the dataset are displayed in Fig. 3, after low-pass filtering by a uniform averaging filter of size 7. Smoothing is performed in order to reduce

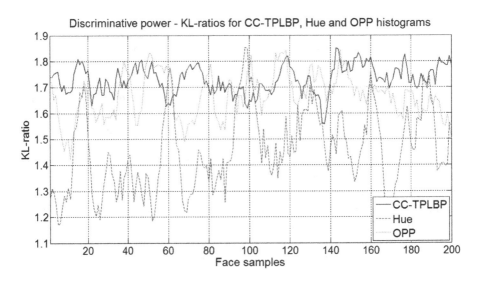

Fig. 3. Discriminative power measure. KL-ratios of 200 face images with 200 background images. Horizontal axis: face sample numbers; vertical axis: KL-ratio values computed by Eq. (7). The displayed KL-ratios are smoothed using a uniform averaging filter of size 7, for further explanation see text. It can be seen that CC-TPLBP (blue curve) has the highest mean KL-ratio and lowest variance, as also seen in Table 3 (Color figure online).

Table 2. Statistics of KL-Divergence, noisy images.

Desc	Mean	Median	STD
CC-TPLBP	0.0553	0.0543	0.0168
Hue	0.0492	0.0397	0.0405
OPP	0.0968	0.0818	0.0579

Table 3. Statistics of KL-ratios; discriminative power. CC-TPLBP is found most discriminative.

Desc	Mean	Median	STD
CC-TPLBP	1.7402	1.7506	0.1678
Hue	1.456	1.3835	0.361
OPP	1.6671	1.6712	0.2185

the noisiness in the original KL-ratio curves. Statistics of the KL-ratios (prior to low-pass filtering) are given in Table 3. We observe that the average KL-ratio for CC-TPLBP is higher than that of Hue and OPP (i.e. higher discriminative power), and that the variance of CC-TPLBP is the lowest, indicating high stability (i.e. low variability of KL-ratios for data samples from a specific class in a dataset).

5 Evaluation of Color Descriptors in a Face Detection Setting

We evaluate the CC-TPLBP color descriptor in a face detection setting.

5.1 Dataset

We use the FDDB benchmark [21], which contains annotations of 5171 faces in 2845 images, divided into 10 folds. five folds are used for training, and five for testing. Training face images are normalized to size 63×39. The background set is constructed from random 63×39 - sized patches from background images of the NICTA dataset [26], i.e. of same size as the face patches.

5.2 Evaluation Protocol

In our face detection system, we use Support Vector Machines [27], a classification method that has been successfully applied for face detection [28,29], as the face classifier. We examine various descriptors combinations, i.e. (1) TPLBP, (2) TPLBP-MS, (3) TPLBP-MS + Hue, (4) TPLBP-MS + OPP, (5) TPLBP-MS + C-TPLBP and (6) TPLBP-MS + CC-TPLBP. For each of (1)–(6) we train a linear-kernel SVM classifier with Soft Margin, where the regularization

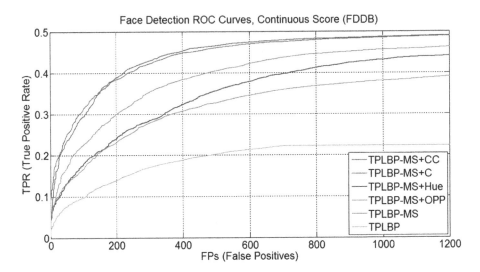

Fig. 4. Face detection ROC curves on FDDB, for various descriptors combinations. It is clearly discerned that both CC-TPLBP and C-TPLBP (red and green lines, respectively) outperform all the other descriptor combinations. In addition, CC-TPLBP is twice more compact than C-TPLBP, making it the more efficient representation (Color figure online).

parameter C is determined by K-fold cross-validation (K = 5). To reduce false alarm rate, we add a confidence measure for an SVM classifier decision, as a probability for a single decision [30]:

$$p\left(\mathbf{w}, \mathbf{x}, y\right) = \frac{1}{1 + \exp\left(-y\left(\mathbf{w} \cdot \mathbf{x} + b\right)\right)} \qquad (8)$$

where $\mathbf{w}$ is the SVM separating hyperplane normal vector, $\mathbf{x}$ is a test sample and y is the classification label. This logistic (sigmoid) function assigns high confidence (i.e. close to 1) to correctly-classified samples which are distant from the hyperplane.

Preprocessing of an image is performed by applying skin detection in CIE-Lab color space, to reduce image area to be scanned by a sliding window method. Various skin detection methods and color spaces can be used [31–35]. We train offline a skin histogram based on chroma (a, b), omitting the luminance L as it is highly dependent on lighting conditions [36]. Skin detection in a test image is performed pixel-wise, by the application of threshold τ_s, i.e. for pixel $p = (x_p, y_p)$ with quantized chroma values $\left(\bar{a}_p, \bar{b}_p\right)$ and histogram value $h\left(\bar{a}_p, \bar{b}_p\right) = h_p$, the pixel is classified as skin if $h_p > \tau_s$. After skin is extracted, we perform a sliding window scan to examine windows at various positions and scales. The confidence measure of Eq. (8) is used by applying a threshold, i.e. if $p\left(\mathbf{w}, \mathbf{x}, y\right) > p_{th}$, the window is classified as a face.

5.3   Results

Face detection performance was evaluated by following the evaluation scheme proposed in [21]. Receiver Operating Characteristic (ROC) were computed, with True Positive rate ($TPR \in [0, 1]$) vs. number of False Positives (FP). In Fig. 4, ROC curves of continuous score [21] are depicted for various descriptor combinations. We observe that each of the descriptor combinations, TPLBP-MS, C-TPLBP and CC-TPLBP produce significant improvements in detection rates, compared to TPLBP. CC-TPLBP leads to similar performance as C-TPLBP, but with a more compact representation.

6   Conclusions

In the present work the focus is not on the design or optimization of a face detection framework, but rather on color representation, or description, for the task of face detection. We proposed a novel color descriptor, CC-TPLBP, which captures the texture information embedded in color channels. CC-TPLBP is by definition invariant to monotonic changes in chroma and luminance channels. A multi-scale variant of TPLBP is designed, termed TPLBP-MS. All experiments were performed in a face detection setting. We examined the invariance of CC-TPLBP, jointly for photometric and geometric variations, i.e. illumination, background, face pose and viewpoint changes, and separately for addition of Gaussian noise, and compared to the Robust Hue and Opponent Angle (OPP) descriptors. Discriminative power was evaluated with respect to the above mentioned descriptors. CC-TPLBP is superior to the other two descriptors. It achieves higher discriminative power and much higher invariance to combined photometric and geometric variations, compared to Hue and OPP, as demonstrated in Sect. 4. The evaluation experiments in a face detection setting demonstrated that (1) TPLBP-MS improves detection rates compared to TPLBP, (2) the addition of CC-TPLBP produces a sharp improvement over TPLBP-MS and (3) CC-TPLBP leads to superior detection rates compared to Hue and OPP.

The CC-TPLBP color-based descriptor can be integrated into face detection frameworks to achieve a substantial improvement in performance using existent color channels information. It can also be used in general color-based object recognition tasks.

References

1. Khan, F. S., Anwer, R. M., van de Weijer, J., Bagdanov, A. D., Vanrell, M., Lopez, A.M.: Color attributes for object detection. In: CVPR, pp. 3306–3313. IEEE (2012)
2. Viola, P., Jones, M.: Robust real-time face detection. Int. J. Comput. Vis. **57**, 137–154 (2004)
3. Mikolajczyk, K., Schmid, C., Zisserman, A.: Human detection based on a probabilistic assembly of robust part detectors. In: Pajdla, T., Matas, J.G. (eds.) ECCV 2004. LNCS, vol. 3021, pp. 69–82. Springer, Heidelberg (2004)

4. Zhang, L., Chu, R.F., Xiang, S., Liao, S.C., Li, S.Z.: Face detection based on multi-block LBP representation. In: Lee, S.-W., Li, S.Z. (eds.) ICB 2007. LNCS, vol. 4642, pp. 11–18. Springer, Heidelberg (2007)
5. Li, H., Hua, G., Lin, Z., Brandt, J., Yang, J.: Probabilistic elastic part model for unsupervised face detector adaptation. In: The IEEE International Conference on Computer Vision (ICCV) (2013)
6. Bindemann, M., Burton, A.M.: The role of color in human face detection. Cogn. Sci. **33**, 1144–1156 (2009)
7. Wei, Y., Sun, J., Tang, X., Shum, H. Y.: Interactive offline tracking for color objects. In: ICCV, pp. 1–8 (2007)
8. Gevers, T., Smeulders, A.: Color based object recognition. Pattern Recogn. **32**, 453–464 (1997)
9. van de Weijer, J., Schmid, C.: Coloring local feature extraction. In: Leonardis, A., Bischof, H., Pinz, A. (eds.) ECCV 2006. LNCS, vol. 3952, pp. 334–348. Springer, Heidelberg (2006)
10. Diplaros, A., Gevers, T., Patras, I.: Combining color and shape information for illumination-viewpoint invariant object recognition. IEEE Trans. Image Process. **15**, 1–11 (2006)
11. Khan, R., van de Weijer, J., Khan, F. S., Muselet, D., Ducottet, C., Barat, C.: Discriminative color descriptors. In: CVPR, pp. 2866–2873. IEEE (2013)
12. Van de Sande, K.E.A., Gevers, T., Snoek, C.G.M.: Evaluating color descriptors for object and scene recognition. IEEE Trans. Pattern Anal. Mach. Intell. **32**, 1582–1596 (2010)
13. Khan, F.S., van de Weijer, J., Vanrell, M.: Modulating shape features by color attention for object recognition. Int. J. Comput. Vis. **98**, 49–64 (2012)
14. Jain, A.K.: Fundamentals of Digital Image Processing. Prentice-Hall Inc, Upper Saddle River (1989)
15. Felzenszwalb, P.F., Girshick, R.B., McAllester, D., Ramanan, D.: Object detection with discriminatively trained part-based models. IEEE Trans. Pattern Anal. Mach. Intell. **32**, 1627–1645 (2010)
16. Zhu, X., Ramanan, D.: Face detection, pose estimation, and landmark localization in the wild. In: CVPR, pp. 2879–2886 (2012)
17. Bergholdt, M., Kappes, J., Schmidt, S., Schnörr, C.: A study of parts-based object class detection using complete graphs. Int. J. Comput. Vis. **87**, 93–117 (2010)
18. Heisele, B., Ho, P., Wu, J., Poggio, T.: Face recognition: component-based versus global approaches. J. Comput. Vis. Image Underst. - Spec. Issue Face Recogn. **91**(1–2), 6–21 (2003)
19. Wolf, L., Hassner, T., Taigman, Y.: Descriptor based methods in the wild. In: Real-Life Images Workshop at the European Conference on Computer Vision (ECCV) (2008)
20. Ojala, T., Pietikäinen, M., Mäenpää, T.: Multiresolution gray-scale and rotation invariant texture classification with local binary patterns. IEEE Trans. Pattern Anal. Mach. Intell. **24**, 971–987 (2002)
21. Jain, V., Learned-Miller, E.: Fddb: a benchmark for face detection in unconstrained settings. Technical report UM-CS-2010-009. University of Massachusetts, Amherst (2010)
22. Shechtman, E., Irani, M.: Matching local self-similarities across images and videos. In: IEEE Conference on Computer Vision and Pattern Recognition (CVPR 2007) (2007)
23. Guo, L., Meng, Y.: Psnr-based optimization of jpeg baseline compression on color images. In: ICIP, pp. 1145–1148. IEEE (2006)

24. Snoek, C. G. M.: Early versus late fusion in semantic video analysis. In: In ACM Multimedia, pp. 399–402 (2005)
25. Huang, G. B., Ramesh, M., Berg, T., Learned-Miller, E.: Labeled faces in the wild: a database for studying face recognition in unconstrained environments. Technical report 07–49, University of Massachusetts, Amherst (2007)
26. Overett, G., Petersson, L., Brewer, N., Pettersson, N., Andersson, L.: A new pedestrian dataset for supervised learning. In: IEEE Intelligent Vehivles Symposium, Eindhoven, The Netherlands (2008)
27. Cortes, C., Vapnik, V.: Support-vector networks. Mach. Learn **20**, 273–297 (1995)
28. Romdhani, S., Torr, P., Schölkopf, B.: Efficient face detection by a cascaded support-vector machine expansion. R. Soc. Lond Proc. Ser. A **460**, 3283–3297 (2004)
29. Osuna, E., Freund, R., Girosi, F.: Training support vector machines: an application to face detection, pp. 130–136 (1997)
30. Platt, J. C.: Probabilistic outputs for support vector machines and comparisons to regularized likelihood methods. In: Advances in Large Margin Classifiers, pp. 61–74. MIT Press (1999)
31. hsuan Yang, M., Ahuja, N.: Gaussian mixture model for human skin color and its applications in image and video databases. In: Proceedings of SPIE 1999 and its Application in Image and Video Databases, San Jose, CA, pp. 458–466 (1999)
32. Jones, M.J., Rehg, J.M.: Statistical color models with application to skin detection. Int. J. Comput. Vis. **46**, 81–96 (2002)
33. Zarit, B.D., Super, B.J., Quek, F.K.H.: Comparison of five color models in skin pixel classification. In: International Workshop on ICCV 1999, pp. 58–63 (1999)
34. Terrillon, J. C., Fukamachi, H., Akamatsu, S., Shirazi, M. N.: Comparative performance of different skin chrominance models and chrominance spaces for the automatic detection of human faces in color images. In: FG, pp. 54–63 (2000)
35. Braunstain, E., Gath, I.: Combined supervised / unsupervised algorithm for skin detection: a preliminary phase for face detection. In: Petrosino, A. (ed.) ICIAP 2013, Part I. LNCS, vol. 8156, pp. 351–360. Springer, Heidelberg (2013)
36. Cai, J., Goshtasby, A.A.: Detecting human faces in color images. Image Vis. Comput. **18**, 63–75 (1999)

Pose Estimation and Movement Detection for Mobility Assessment of Elderly People in an Ambient Assisted Living Application

Julia Richter(✉), Christian Wiede, and Gangolf Hirtz

Department of Electrical Engineering and Information Technology,
Technische Universität Chemnitz, Reichenhainer Str. 70, 09126 Chemnitz, Germany
{julia.richter,christian.wiede,g.hirtz}@etit.tu-chemnitz.de

Abstract. In European countries, the increasing number of elderly with dementia causes serious problems for the society, especially with regard to the caring sector. As technical support systems can be of assistance to caregivers and patients, a mobility assessment system for demented people is presented. The grade of mobility is measured by means of the person's pose and movements in a monitored area. For this purpose, pose estimation and movement detection algorithms have been developed. These algorithms process 3-D data, which are provided by an optical stereo sensor installed in a living environment. The experiments demonstrated that the algorithms work robustly. In connection with a human machine interface, the system facilitates a mobilisation as well as a more valid assessment of the patient's medical condition than it is presently the case. Moreover, recent advances with regard to action recognition as well as an outlook about necessary developments are presented.

Keywords: Pose estimation · Stereo vision · Image understanding · Video analysis · 3-D image processing · Machine learning · Support vector machine · Ambient assisted living

1 Introduction

The increasing life expectancy is an important achievement of modern medicine. Over the coming years, the number of elderly people will continually rise and with it the number of demented people [3]. Due to this development, care facilities will encounter challenges in maintaining the quality of human care.

People in an early state of dementia should remain in their familiar household as long as possible in order to mitigate these problems. The encouragement of their cognitive, social and physical functions will also help to keep their quality of life at high level. Next to activation, assessing the need of care in regular intervals is another task medical experts are facing. Since the health status of a person is examined only irregularly at present, the result is highly dependent on the form on the inspection day and might be further influenced by the fact that patients can prepare for the inspection. Additionally, many patients put particular concern

© Springer International Publishing Switzerland 2015
A. Fred et al. (Eds.): ICPRAM 2015, LNCS 9493, pp. 172–184, 2015.
DOI: 10.1007/978-3-319-27677-9_11

on personal hygiene on that day and when questioned about their physical and psychological comfort, they usually feel embarrassed and avoid talking about their problems. The medical findings are therefore not always reliable.

In this paper, only persons living alone at home without the care of a partner are considered. The focus lies on the physical capabilities of the demented person – and particularly his or her mobility. This parameter was measured by the detection of the general pose (i. e. standing, sitting and lying) and of the person's movements in the living environment. To this end, a single, wide angle stereo camera was mounted at the ceiling. The information gathered about the general pose and the movements were recorded over a certain period of time. If long periods of inactivity were detected, the demented person was encouraged to do some exercises or to go for a walk. The communication was realized via a human machine interface, i. e. a tablet or a monitor, on which the messages appeared, optionally in combination with an acoustic signal. Furthermore, statistics were calculated from the recorded data. At a later time, such statistics could be analysed by medical personnel to notice considerable changes in a patient's mobility and to draw reliable conclusions about the need of care.

2   Related Work

Various works address the subject of supporting elderly people in their home environment. The assistance concepts are closely related to the topic of AAL (Ambient Assisted Living). Their unobtrusive integration into the living environment is one of the most important requirement for AAL systems.

Clement et al. detected ADLs (Activities of Daily Living) with the help of 'Smartmeters', which measure the energy consumption of household devices [5]. A Semi-Markov model was trained in order to construct behaviour profiles of persons and to draw conclusions about their state of health. Kalfhues et al. analysed a person's behaviour by means of several sensors integrated in a flat, e. g. motion detectors, contact sensors and pressure sensors [8]. Link et al. employed optical stereo sensors to discern emergencies, i. e. falls and predefined emergency gestures [10]. Chronological sequences of the height of the body centre and the angle between the main body axis and the floor were analysed. Belbachier et al., who also applied stereo sensors to detect falls [2], used a neural network-based approach to classify the fall event. The major advantage of optical sensors is their easy integration into a flat. A considerable amount of additional information can be obtained by applying image processing algorithms, especially in connection with RGB-D sensors, which deliver red, blue and green channel images as well as depth information. Therefore, we decided to use a stereo camera in our study. Although other sensors that provide RGB-D data, such as the Kinect, could also be installed in a flat, they show features that have proved to be disadvantageous with regard to the application field of AAL: Firstly, if the Kinect is mounted at the ceiling, the range and the field of view do not cover the complete room. It would be necessary to integrate several Kinect sensors at different places in a flat, which is hardly applicable. Secondly, the resolution is not sufficient enough for the recognition of objects that are far away from the sensor. When, thirdly, several Kinects are installed for better coverage of the room, they are apt to influence each other, due to their active technique for determining depth information. Consequently, although the Kinect is highly performant for a variety of applications, we considered this sensor as unsuitable for AAL purposes.

The approaches listed above either address ADL detection or emergency scenarios. In the context of assessing the health status of persons, several former projects have focused especially on the analysis of mobility. Scanaill et al. employed body-worn sensors for mobility telemonitoring [13]. However, this type of sensor unsuitable for demented persons, as this group tends to forget to put them on or puts them off intentionally. In the work of Steen et al., another way of measuring mobility was presented [14]. In first field tests, several participants' flats were equipped with laser scanners, motion detectors and contact sensors. By means of these sensors, the persons could be localised within their flats. Apart from this, the traversing time between the sensors as well as walking speeds were computed. These field tests gave evidence that the evaluation of sensor data allows conclusions about mobility.

In addition to a person's location and the movements, we think that the pose, i. e. standing, sitting and lying, provides also an indication of a person's mobility. We therefore introduce a pose estimation algorithm, which detects the pose of a person within the area observed by a single stereo camera.

There is a variety of pose estimation algorithms that use optical sensors. They differ, for example, with respect to such parameters as camera type (mono, stereo), inclusion of temporal information and utilisation of explicit human models. Ning et al. discerned the human pose using a single monocular image [11]. By modifying a bag-of-words approach, they were able to increase the discriminative power of features. They also introduced a selective and invariant local descriptor, which does not require background subtraction. The poses walking, boxing and jogging could be classified after supervised learning. Agarwal et al. determined the pose from monocular silhouettes by regression [1] and thus needed neither a body model nor labelled body parts. Along with spatial configurations of body parts, Ferrari et al. additionally considered the temporal information in their study [6]. Haritaoglu et al. employed an overhead stereo camera in order to recognize the 'pick' movement of customers while shopping [7]. In this study, a three dimensional silhouette was computed by back-projecting image points to their corresponding world points by the use of depth information and calibration parameters. The persons' localizations were found at regions with significant peaks in the occupancy map. The pose is determined by calculating shape features instead of using an explicit model. Other approaches applied the Kinect sensor. Their results proved that the Kinect, when suitable for the particular application, leads to results of high quality. Ye et al. estimated the pose from a single depth map of the Kinect [15]. They then compared this map with mesh models from a database. In a first step, a similar pose was searched by point cloud alignment using principal component analysis and nearest neighbour search. In a second step, the found pose was refined. Missing information of occluded parts could be replaced by data from the corresponding mesh model. As a result, skeleton joints comparable to the Kinect skeleton output could be determined.

Another study addressed the design of a scale and viewing angle robust feature vector, which describes a person's head-to-shoulder signature [9]: Points between head and shoulder are first assigned to vertical slices. The points within each slice are then projected to a virtual overhead view and the feature vector is eventually composed of the slices' spans. The authors aim at detecting persons in a 3-D point cloud. However, this approach can also be adapted and utilized for pose estimation.

3 Mobility Assessment

This section describes the algorithms for movement detection and pose estimation. First of all, the person has to be detected and localized within the monitored area. Therefore, the stereo camera is extrinsically calibrated with respect to a defined world coordinate system. The 2-D position is measured in relation to the origin of this coordinate system. On the basis of this position, the person is classified as 'moving' if the position changes considerably between two successive frames in a video sequence. The pose estimation requires three steps. Firstly, 3-D points belonging to the person are extracted from the back-projected point cloud. Secondly, discriminative feature vectors, which allow a reliable classification, are designed. Finally, a suitable machine learning technique is selected and a model is trained with feature vectors generated from training examples.

3.1 Person Localisation

The person localisation is performed on the back-projected 3-D point cloud obtained from the stereo camera [12]. Hypotheses of possible foreground regions are generated in a first step, so a mixture of Gaussian algorithm is applied to the world z-map, which represents the z component, i.e. the height, of the corresponding world point for every pixel.

The mixture model is calculated for every pixel in the map and updated for every new frame according to the new pixel value. The model was described by [16] and is expressed as follows:

$$p(x^{(t)}|\chi_T; BG + FG) \sim \sum_{m=1}^{M} \hat{\pi}_m^{(t)} \cdot N(x^{(t)}; \hat{\mu}_m^{(t)}, \hat{\sigma}_m^{2(t)}) \tag{1}$$

$p(x^{(t)}|\chi_T; BG + FG)$ is the probability density function for the value x of a pixel in the z-map for frame t with the history χ_T. This density function models both the background BG and the foreground FG. M denotes the number of Gaussian distributions N. Each distribution is characterised by its mean value $\hat{\mu}_m^{(t)}$ and its variance $\hat{\sigma}_m^{2(t)}$. $\hat{\pi}_m^{(t)}$ denotes the influence of every single distribution on the mixture model.

In a second step, the points within the foreground mask are projected on a virtual overhead plan view. The final determination of the persons' positions is executed on this view. The detected person is characterised by a centre point $\overrightarrow{p} = (x, y, z)$, the expansion in each direction – $expansion_x$ and $expansion_y$ – and an orientation α related to the world coordinate system. An example of detected persons is illustrated in Fig. 1.

Fig. 1. Example point cloud with detected persons [12]. Detected persons are visualised via red cuboids defined by a 3-D centre point and expansions in each direction. White areas indicate regions, where 3-D world points cannot be calculated due to the lack of depth information (Color figure online).

3.2   Movement Detection

For movement detection, only vectors $\vec{p}_{xy}$ containing the x and y component of the 3-D centre point $\vec{p}$ are processed.

The distance $distance_{frame}$ that a person moves between two frames depends on the $frame\,rate$ and can be estimated with:

$$distance_{frame} = v_{movement} \cdot t_{frame} = \frac{v_{movement}}{frame\,rate}. \tag{2}$$

Provided a person is walking with a speed $v_{movement}$ of at least $0.5\,\text{m/s}$ and the frame rate is about 5 FPS, the $distance_{frame}$ is estimated at 100 mm. We consider a person to be moving when a threshold distance of more than X m is covered. Therefore, we utilize a sliding window containing the vectors $\vec{p}_{xy}^{(t-i)}$ with $i = \{0, ..., 4\}$. Each $distance_j$ crossed between two successive frames is calculated according to Eq. 3 with $j = \{0, ..., 3\}$. It is the Euclidean norm between the person's position in the frame $t - j$ and the position in the previous frame $(t - j - 1)$.

$$distance_j = \left\| \vec{p}_{xy}^{(t-j)} - \vec{p}_{xy}^{(t-j-1)} \right\|. \tag{3}$$

Afterwards, the distances are summed up to the final $distance$ between the five frames of the sliding window:

$$distance = \sum_{j=0}^{3} distance_j. \tag{4}$$

The distance between two frames is only added to the sum if its value exceeds $distance_{frame}$. Furthermore, the threshold X mentioned above for this sum is estimated according to the product of the estimated distance between two frames $distance_{frame}$ and the number of distances $nDist$ within the window:

$$X = distance_{\text{frame}} \cdot nDist$$
$$= 100\,\frac{\text{mm}}{\text{frame}} \cdot 4\,\text{frames} = 400\,\text{mm}. \tag{5}$$

Moreover, the decision about movement or non-movement is realised via a finite state machine consisting of the two states 'movement' and 'non-movement'. At the transitions, the distance is compared with two different thresholds T_{high} and T_{low} that are slightly lower/higher than the estimated threshold distance X (hysteresis):

$$T_{\text{high}} = 500\,\text{mm},$$
$$T_{\text{low}} = 300\,\text{mm}. \tag{6}$$

The hysteresis suppresses oscillations near the estimated threshold value. Finally, the value of $movement^{(t)}$ is recorded over time, so that it can be analysed later. Generally, these threshold values can be adjusted when conditions in terms of velocity and frame rate are altering.

3.3 Pose Estimation

Point Cloud Extraction. The presented pose estimation algorithm processes 3-D world points belonging to the person. Every point of the point cloud has therefore to be classified as person or non-person. For that purpose, both the previously calculated cuboid and the foreground mask are used for classification. The algorithm is outlined in the following pseudo code, which is performed for every detected person. The geometric context is illustrated in Fig. 2.

```
R = sqrt(expansion.x^2 + expansion.y^2);
for all points:
if (foregound
      && z < 2*expansion.z
      && expansion.x < R
      && expansion.y < R )
{
      (xT,yT) = CoordinateTransformation(x, y);
      if (!( xT < expansion.x
          && yT < expansion.y ))
      {
          deletePoint(x,y);
      }
}
else
{
      deletePoint(x,y);
}
```

Points are removed from the cloud if they belong neither to the foreground nor to the interior of the cuboid. In order to reduce processing power, it is first checked whether a point (x_{pc}, y_{pc}) is within the person's radius R. If this is the case, the point is transformed from the world coordinate system (x_w, y_w) to the person's coordinate system (x_p, y_p), which enables a direct comparison of the point coordinate with the corresponding expansion $expansion_x$ and $expansion_y$. The person's coordinate system is defined by its origin, namely the 2-D centre point $\vec{p}_{xy}$, and its rotation angle α.

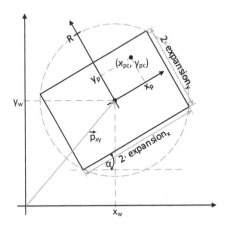

Fig. 2. Classifying points from the point cloud as person or non-person by means of coordinate transformation. If a point in the point cloud lies inside the circle defined by radius R, this very point is transformed from the world coordinate system to the person's coordinate system. Provided that the point has been classified as foreground, it belongs to the person if its x and y component fall below the corresponding expansion.

The remaining points are denoted as the person's point cloud $points_{person}$. Figure 3 shows the extracted point clouds of three persons.

Fig. 3. Point clouds of three persons.

Feature Vector Generation. The determination of a person's pose is based on the points extracted in the previous step. In order to train a machine, a discriminative feature vector has to be designed first. For that purpose, the point cloud is divided into 20 vertical bins of 110 mm height each, which start at a z value of −100 mm. During the extrinsic calibration, the origin of the world coordinate system is set on the floor plane of the room. The plane formed by the x and the y axis runs parallel to the floor while the z axis is directed at the ceiling. Therefore, the floor is defined by a z value around zero. According to their z component, all points are assigned to one of these bins. Consequently, each bin contains the number of points that fall within a certain z range. All bins together form a feature vector. In a final step, the feature vector is normalized by dividing every item by the total number of points n. The process of feature vector generation is visualised in Fig. 4.

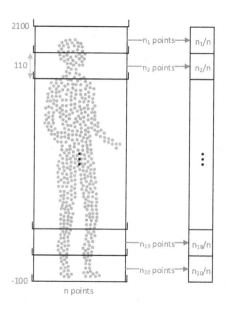

Fig. 4. Feature vector generation from point cloud. All numbers in mm.

Training. After the feature vector generation, a machine was trained in a supervised manner, i. e. with labelled training samples . Video sequences with three different persons (P3, P4 and P7) were recorded for this purpose in a laboratory flat and manually labelled (about 3000 images). Furthermore, a linear Support Vector Machine (SVM) was chosen. The SVM is a *discriminative, maximum margin* classifier. The term '*discriminative*' means that the variable to be predicted, i. e. the posterior probability, is modelled whereas '*maximum margin*' refers to the fact that an optimization problem is formulated: A separating hyperplane has to be determined, so that the margin between two adjacent classes is maximized. The outer vectors of the classes form the support vectors. These are the vectors with the minimum distance to the separating hyperplane. We decided to

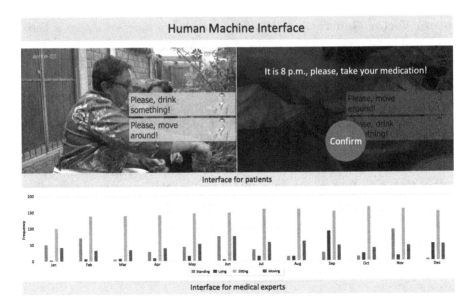

Fig. 5. Design for all: Interface for both patients and for medical personnel. The images at the top show the interface for the patient. Reminders appear time-controlled. The patient can remove the messages either by touching the display (touch screen) or by performing the action, e.g. when movement is detected by the sensor. The image at the bottom shows a graph that presents pose and movements over a month. Other intervals can also be selected.

use this type of classifier, because it ranks among the classifiers with the best performance if the amount of training data is limited [4].

3.4   Human Machine Interface

The medical staff can view the statistically prepared mobility data via a web interface. Additionally, if no movement is detected over a certain period of time, which can be specified beforehand, a reminder appears on a tablet as well as on a touch display. This touch display might be a TV set, so that the person is activated while watching TV, for example. In that way, the person can be immediately addressed in an unobtrusive way. Examples of such scenarios are illustrated in Fig. 5.

4   Experimental Results

In order to determine the performance of the trained pose classifier, we recorded several test sequences. A total number of 2958 samples was classified during the test.

The first test case consisted of realistic scenarios in the laboratory flat with two elderly volunteers (P1 and P2). In the second test case, we attached high importance to the fact that the test sequences had been recorded in a completely different environment compared to the scene where the training sequences have been recorded. We installed, therefore, a test set-up with a stereo camera similar to the one in the laboratory flat. The sequences were recorded with four persons (P3 - P6), of whom two had already participated in the training sequences (P3 and P4).

Table 1 shows the results for the elderly persons in the laboratory flat while Tables 2 and 3 indicate the classification results for both types of test persons in the special test set-up. The letters L and C in the table headings stand for classified pose and labelled pose respectively. All numbers are percentages.

The experiments show that the classification results are of high quality. These first tests also revealed that the algorithm does work reliably in different surroundings and with different persons. The misclassification rate for 'Lying' in Table 1 is obviously very high compared to the other scenarios. This is, however, caused by the sparse and noisy point cloud at the place, where the person was

Table 1. Classification results for persons P1 and P2.

C \ L	Standing	Sitting	Lying
Standing	100	0	6.5
Sitting	0	100	0
Lying	0	0	93.5

Table 2. Classification results for persons P3 and P4.

C \ L	Standing	Sitting	Lying
Standing	97.6	0	0
Sitting	0	100	0
Lying	2.4	0	100

Table 3. Classification results for persons P5 and P6.

C \ L	Standing	Sitting	Lying
Standing	100	0	0
Sitting	0	100	1
Lying	0	0	99

lying at this time. The location was relatively far away from the stereo sensor, so that the stereo matching algorithm reached its limits.

For the purpose of movement evaluation, we recorded and labelled a video sequence, in which persons were either walking through the room or standing somewhere at the spot. We could thus compare the labels with the output of the algorithm (moving/non-moving) and calculate the true-positive rate TPR and the false-positive rate FPR were calculated. $mov_{\text{detected}|\text{neg}}$ denotes the number of frames where movement was detected although the label was non-movement, $mov_{\text{detected}|\text{pos}}$ the number of frames where movement was detected and the label was movement, $mov_{\text{neg,labelled}}$ the number of frames labelled as non-movement, $mov_{\text{pos,labelled}}$ the number of frames labelled as movement.

$$TPR = \frac{mov_{\text{detected}|\text{pos}}}{mov_{\text{pos,labelled}}} = \frac{288}{298} \approx 96.6\,\% \tag{7}$$

$$FPR = \frac{mov_{\text{detected}|\text{neg}}}{mov_{\text{neg,labelled}}} = \frac{5}{193} \approx 2.6\,\% \tag{8}$$

These values show that significant movements between different positions in the monitored area are detected by the algorithm.

5   Action Recognition

Latest developments of our system aim at monitoring and analysing activities important for the need of care of demented persons. Such activities are related to nourishment, social contacts and personal hygiene. On the basis of the registered activities, assistance, such as reminders or activation messages, can be provided for patients. Caring personnel could benefit from the generated information by involving it in their care planning. New advances in our project show that activities, such as drinking, can be reasonably well detected by means of machine learning techniques. Figure 6 shows a feature vector example that has been used for training a machine in order to recognise drinking from a bottle.

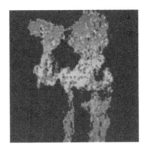

Fig. 6. Feature vector for recognising drinking.

Further activities can be recognised by combining the person's position and pose with context information, for example the knowledge about the location of furniture or of different utensils in the flat. When a person is localised in the bed and at the same time is detected to be lying for a longer time, then it is assumed that the person is sleeping. In addition to sleeping in the bed, actions such as resting in an armchair, taking a shower and washing hands while standing in front of a basin are several examples of activities the system is capable to detect at the moment.

6   Conclusions

In this paper, we presented an approach to measure significant indicators for mobility, i.e. a person's pose and movement. The most significant finding to emerge from this study is that the proposed machine learning technique works reliably in different environments and with different persons. In combination with movement detection (e.g. crossing a room), conclusions about a person's mobility can be drawn. In that way, long-term diagnostics involving mobility observations can lead to more reliable diagnoses of the health status, which will result in a better assessment of the need of care. Moreover, activation and mobilization by means of a HMI can support the demented persons in preserving their functional abilities.

7   Future Work

Further work needs to be done to increase the accuracy of the action registration and to extend the range of detectable activities.

An essential aspect of our future studies will be the conduction of field tests in cooperation with our medical partners. The application of the system in the field over a longer period of time will provide data for a long-term statistical data analysis and for system validation. Since the focus of the presented approach lies on the patient, the HMI has to be attuned to the special needs of demented people, which shall result in a patient-oriented assistance and assessment system.

With regard to the demographic developments, the quality of care for demented people has to be ensured. The proposed approach can contribute to a more valid assessment and to the preservation of the patient's quality of life. Not only would this be of high benefit for our caring sector, but it could also increase the quality of life of demented persons and their relatives.

Acknowledgements. This project was funded by the European Fund for Regional Development (EFRE).

References

1. Agarwal, A., Triggs, B.: Recovering 3D human pose from monocular images. IEEE Trans. Pattern Anal. Mach. Intell. **28**(1), 44–58 (2006)

2. Belbachir, A. N., Litzenberger, M., Schraml, S., Hofstatter, M., Bauer, D., Schon, P., Humenberger, M., Sulzbachner, C., Lunden, T., Merne, M.: CARE: a dynamic stereo vision sensor system for fall detection. In: 2012 IEEE International Symposium on Circuits and Systems (ISCAS), pp. 731–734. IEEE (2012)
3. Berlin Institut für Bevölkerung und Entwicklung: Demenz-Report. http://www.berlin-institut.org/fileadmin/user_upload/Demenz/Demenz_online.pdf   (2011). Accessed 07 July 2014
4. Bradski, G., Kaehler, A.: Learning OpenCV: Computer Vision with the OpenCV Library. O'Reilly Media, Sebastopol (2008)
5. Clement, J., Ploennigs, J., Kabitzsch, K.: Erkennung verschachtelter ADLs durch Smartmeter. Lebensqualität im Wandel von Demografie und Technik (2013)
6. Ferrari, V., Marin-Jimenez, M., Zisserman, A.: Progressive search space reduction for human pose estimation. In: IEEE Conference on Computer Vision and Pattern Recognition, CVPR 2008, pp. 1–8. IEEE (2008)
7. Haritaoglu, I., Beymer, D., Flickner, M.: Ghost 3d: detecting body posture and parts using stereo. In: Proceedings Workshop on Motion and Video Computing, pp. 175–180. IEEE (2002)
8. Kalfhues, A.J., Hübschen, M., Löhrke, E., Nunner, G., Perszewski, H., Schulze, J.E., Stevens, T.: JUTTA–JUsT-in-Time Assistance: Betreuung und Pflege nach Bedarf. In: Shire, K.A., Leimeister, J.M. (eds.) Technologiegestützte Dienstleistungsinnovation in der Gesundheitswirtschaft, pp. 325–349. Springer, Heidelberg (2012)
9. Kirchner, N., Alempijevic, A., Virgona, A.: Head-to-shoulder signature for person recognition. In: 2012 IEEE International Conference on Robotics and Automation (ICRA), pp. 1226–1231. IEEE (2012)
10. Link, N., Steiner, B., Pflüger, M., Kroll, J., Egeler, R.: safe@ home-Erste Erfahrungen aus dem Praxiseinsatz zur Notfallerkennung mit optischen Sensoren. Lebensqualität im Wandel von Demografie und Technik (2013)
11. Ning, H., Xu, W., Gong, Y., Huang, T.: Discriminative learning of visual words for 3D human pose estimation. In: IEEE Conference on Computer Vision and Pattern Recognition, CVPR 2008, pp. 1–8. IEEE (2008)
12. Richter, J., Findeisen, M., Hirtz, G.: Assessment and care system based on people detection for elderly suffering from dementia. In: IEEE Fourth International Conference on Consumer Electronics, ICCEBerlin 2014, pp. 59–63. IEEE (2014)
13. Scanaill, C.N., Carew, S., Barralon, P., Noury, N., Lyons, D., Lyons, G.M.: A review of approaches to mobility telemonitoring of the elderly in their living environment. Anna. Biomed. Eng. **34**(4), 547–563 (2006). http://link.springer.com/article/10.1007/s10439-005-9068-2
14. Steen, E. E., Frenken, T., Frenken, M., Hein, A.: Functional Assessment in Elderlies Homes: Early Results from a Field Trial. Lebensqualität im Wandel von Demografie und Technik (2013)
15. Ye, M., Wang, X., Yang, R., Ren, L., Pollefeys, M.: Accurate 3d pose estimation from a single depth image. In: 2011 IEEE International Conference on Computer Vision (ICCV), pp. 731–738. IEEE (2011)
16. Zivkovic, Z.: Improved adaptive gaussian mixture model for background subtraction. In: Proceedings of the 17th International Conference on Pattern Recognition, ICPR 2004, vol. 2, pp. 28–31. IEEE (2004)

A Non-rigid Face Tracking Method for Wide Rotation Using Synthetic Data

Ngoc-Trung Tran[1,2(✉)], Fakhreddine Ababsa[2], and Maurice Charbit[1]

[1] LTCI-CNRS, Telecom ParisTECH, 37-39, Rue Dareau, 75014 Paris, France
[2] IBISC, University of Evry, 40, Rue du Pelvoux, 91020 Evry, France
`trung-ngoc.tran@telecom-paristech.fr`

Abstract. This paper propose a new method for wide-rotation non-rigid face tracking that is still a challenging problem in computer vision community. Our method consists of training and tracking phases. In training, we propose to use a large off-line synthetic database to overcome the problem of data collection. The local appearance models are then trained using linear Support Vector Machine (SVM). In tracking, we propose a two-step approach: (i) The first step uses baseline matching for a good initialization. The matching strategy between the current frame and a set of adaptive keyframes is also involved to be recoverable in terms of failed tracking. (ii) The second step estimates the model parameters using a heuristic method via pose-wise SVMs. The combination makes our approach work robustly with wide rotation, up to 90° of vertical axis. In addition, our method appears to be robust even in the presence of fast movements thanks to baseline matching. Compared to state-of-the-art methods, our method shows a good compromise of rigid and non-rigid parameter accuracies. This study gives a promising perspective because of the good results in terms of pose estimation (average error is less than 4° on BUFT dataset) and landmark tracking precision (5.8 pixel error compared to 6.8 of one state-of-the-art method on Talking Face video. These results highlight the potential of using synthetic data to track non-rigid face in unconstrained poses.

Keywords: 3D face tracking · Out-of-plane tracking · Rigid tracking · Non-rigid tracking · Face matching · Synthesized face · Face matching

1 Introduction

Non-rigid face tracking is an important topic, which has been having a great attention since last decades. It is useful in many domains such as: video monitoring, human computer interface, biometric. The problem gets much more challenging if occurring out-of-plane rotation, the illumination changes, the presence of many people, or occlusions. In our study, we propose an approach to track non-rigid face at out-of-plane rotation, even the profile face. In other words, our method gets involved in the estimation of six rigid face parameters, namely

© Springer International Publishing Switzerland 2015
A. Fred et al. (Eds.): ICPRAM 2015, LNCS 9493, pp. 185–198, 2015.
DOI: 10.1007/978-3-319-27677-9_12

the 3D translation and the three axial rotations[1]), and non-rigid parameters at the same time.

For non-rigid face tracking, a set of landmarks are considered as the face shape model. Since the pioneer work of [1], it is well-known that Active Appearance Model (AAM) provides an efficient way to represent and track frontal faces. Many works [2–4] have suggested improvements in terms of fitting accuracy or profile-view tracking. Constrained Local Model (CLM) has been proposed by [5] that consists of an exhaustive local search around landmarks constrained by a shape model. [6,7] both improved this method in terms of accuracy and speed; more specifically, [7] can track single face with vertical rotation up to 90° in well-controlled environment. The Cascaded Pose Regression (CPR), which is firstly proposed by [8], has recently shown remarkable performance [9,10]. This method shows the high accuracy and real-time speed, merely it is restricted at the near-frontal face tracking. Most of the methods work at constrained views because of two reasons: (i) The acquisition of ground-truth for unconstrained views is really expensive in practice and (ii) how to handle the hidden landmarks on invisible side is hard.

The literature also mentioned other face models such as: cylinder [11–13], ellipsoid [14] or mesh [15]. Most of these methods can estimate the three large rotations even on the profile-view, but it is worth noting that they handle with rigid rather than non-rigid facial expression. On the other hand, the popular 3D Candide-3 model has been defined to manage rigid and non-rigid parameters. [16] used Kalman Filter to the interest points in a video sequence based on the adaptive rendered keyframe, and this work is semi-automatic and is insufficient to work in quick movement. [17] used Mahalanobis distances of local features with the constraint of the face model, to capture both rigid and non-rigid head motions. [18] learned a linear model between model parameters and the face's appearance. These methods poorly works on profile-view. [19] extended Candide face to work with the profile, but their objective function, combining structure and appearance features with dynamic modeling, appears to slowly converge due to the high dimensionality. [20] proposed an adaptive Bayesian approach to track principal components of landmark appearance. Their algorithm appears to be robust for tracking landmarks, but unable to recover when tracking is lost. Let us notice that these methods use the synthetic database to train tracking models. For pose estimation, The pose estimation performance of mentioned methods can be improved more if integrating Kalman Filtering [21] or Particle Filtering [22].

A face tracking framework is robust if it can operate with a wide range of pose views, face expression, environmental changes and occlusions, and also have recovering capability. In [11,12], the authors utilized dynamic templates based on cylinder model in order to handle with lighting and self-occlusion. Local features can be considered [7,10], since local descriptors are not much affected by facial expressions and self-occlusion. In order to have recovering capability,

---

[1] In the literature, the terms Yaw (or Pan), Pitch (or Tilt), and Roll are adopted for the three axial rotations.

tracking-by-detection or wide baseline matching [15,23,24] have been applied. The primary idea is to match the current frame with preceding-stored keyframes. The matching is sufficient to fast movements, illumination, and able to recover the lost tracking. However, the matching is only suitable to work with rigid parameters; moreover, these methods degrade when the number of keypoints detected on the face is not enough. Recently, [25] propose the combination of traditional tracking techniques and deep learning to have a proficient performance of pose tracking. Many commercial products also exist, i.e. [26], which shows effect results in pose and face animation tracking, but this product needs to work in controlled environments of illumination and movements. In addition, it has to wait for the frontal view to re-initialize the model when the face is lost.

In this paper, our contribution is two-folds: (i) using a large offline synthetic database to train tracking models, (ii) proposing a two-step tracking approach to track non-rigid face. These points are immediately introduced in more detail. Firstly, a large synthesized database is built to avoid the expensive and time-consuming manual annotation. To the best of our knowledge, although there were some papers worked with the synthetic data [18,27,28], our paper is the first study that investigates the large offline synthetic dataset for the free-pose tracking of non-rigid face. Secondly, the tracking approach consists of two steps: a) The first step benefits 2D SIFT matching between the current frame and some preceding-stored keyframes to estimate only rigid parameters. By this way, our method is sustainable to fast movement and recoverable in terms of lost tracking. b) The second step obtains the whole set of parameters (rigid and non-rigid) by a heuristic method using pose-wise SVMs. This way can efficiently align a 3D model into the profile face in similar manner of the frontal face fitting. The combination of three descriptors is also considered to have better local representation.

The remaining of this paper is organized as follows: Sect. 2 describes the face model and the used descriptors. Section 3 discusses the pipeline of the proposed framework. Experimental results and analysis are presented in Sect. 4. Finally, we provide in Sect. 5 some conclusions and further perspectives.

2   Face Representation

2.1   Shape Representation

Candide-3, initially proposed by [29], is a popular face model managing both facial shape and animation. It consists of $N = 113$ vertices representing 168 surfaces. If g $\in R^{3N}$ denotes the vector of dimension $3N$, obtained by concatenation of the three components of the N vertices, the model writes:

$$g(\sigma, \alpha) = \bar{g} + S\sigma + A\alpha \tag{1}$$

where $\bar{g}$ denotes the mean value of g. The known matrices S) $\in R^{3N \times 14}$ and A) $\in R^{3N \times 65}$ are Shape and Animation Units that control respectively shape and animation through σ and α parameters. Among the 65 components of animation control α, 11 ones are associated to track eyebrows, eyes and lips. Rotation and translation also need to be estimated during tracking. Therefore, the full model parameter, denoted Θ, has 17 dimensions: 3 dimensions of rotation (r_x, r_y, r_z), 3 dimensions of translation (t_x, t_y, t_z) and 11 dimensions of animation r_a: $\Theta = [r_x\, r_y\, r_z\, t_x\, t_y\, t_z\, r_a]^T$. Notice that both σ and Θ are estimated at first frame, but only Θ is estimated at next frames because we assume that the shape parameter does not change. In next section, $\Theta(t)$ indicates the model parameters at time.

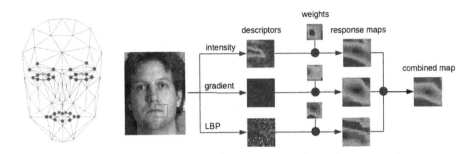

Fig. 1. (a) The Candide-3 model with facial points in our method. (b) The way to compute the response map at the mouth corner using three descriptors via SVM weights.

2.2 Projection

We assume the perspective projection, in which the camera calibration has been obtained from empirical experiments. In our case, the intrinsic camera matrix is written as follows:

$$\begin{bmatrix} f_x & 0 & c_x \\ 0 & f_y & c_y \\ 0 & 0 & 1 \end{bmatrix} \tag{2}$$

where the focal length of camera $f_x = f_y = 1000$ pixels and the coordinates of a camera's principal point (c_x, c_y) as a center of the 2D video frame. The such focal length is defined because it is shown in [30] that the focal length does not require to be accurately known if the distance between the 3D object and camera is much larger than the 3D object depth. Notice that because of the perspective projection assumption, the depth t_z is directly related to scale parameter.

2.3 Appearance Representation

The facial appearance is represented by a set of $N_p = 30$ landmarks (Fig. 1). The local patch of a landmark is described by three local descriptors: intensity, gradient and Local Binary Patterns (LBP) [31] because the combination of multiple descriptors are more discriminative and robust. This combination is fast enough if using linear SVM like [6]. The patch size is 15×15 in our study.

3 Our Method

We present the framework into three sub-sections: (i) the model training from synthesized dataset, (ii) the robust initialization using wide baseline matching, and (iii) the fitting strategy using pose-wise SVMs.

3.1 Model Training from Synthesized Data

We consider the synthesized data for the training because of some reasons: (i) Most of available datasets were built for frontal face alignment [32]. The others contain profile information such as ALFW [32], Multi-PIE [33], but the range of *Pitch* or *Roll* are restricted. In addition, the number of landmarks of frontal and profile faces is different. That makes a gap, how to track from frontal to profile faces (ii) The campaign of ground-truth for building a new dataset is very expensive and (iii) The hidden landmarks could be localized in synthesized dataset, so the gap between frontal and profile tracking could be bridged.

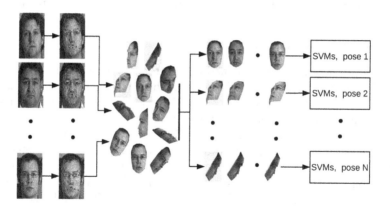

Fig. 2. From left to right of training process: 143 frontal images, landmark annotation and 3D model alignment, synthesized images rendering, and pose-wise SVMs training.

The training process is shown in Fig. 2. At first, we select 143 frontal images (143 different subjects) from Multi-PIE. We then align 3D face model into the known landmarks of each image by POSIT [34] and warp the image texture to the model. Afterwards, rendering is deployed to generate a set of synthesized images of different poses. Finally, all synthesized images are clustered into pose-wise groups before extracting local features and training landmark models by linear SVM classifiers. In terms of rendering, we only consider the three rotations to generate synthesize data. Indeed, we can assume that the translation parameters do not considerably affect the facial appearance. Because of storage and computational problems, the data are rendered in following ranges: 15 of $Yaw \in [-70:10:70]$, 11 of $Pitch \in [-50:10:50]$, 7 of $Roll \in [-30:10:30]$.

The empirical experiments show that the mentioned ranges are sufficient for a robust tracking.

Linear SVMs are used for landmark model training because of its computational efficiency [6], and the combination of three descriptors (Sect. 2.3) makes robust response maps more robust. Because of large pose variation of the dataset, the pose-wise SVMs are trained as follows. The total rendered images are splitted into 1155 $(15(Yaw) \times 11(Pitch) \times 7(Roll))$ pose-wise groups (143 images/group). Each group is used to train 90 pose-wise linear SVMs (30 landmarks $\times$ 3 descriptors) in similar manner of [6]. So, the total of 103950 classifiers (namely ζ) needs to be trained. In the other words, let us denote $C_{x,y,z}) \in \zeta$ is one classifier that is trained on the specific pose x ($\in$ 1155 poses), the landmark id y ($\in [1,..,30]$) and the descriptor type z ($\in$ [intensity, gradient, LBP]). With the given descriptor of local region ϕ_z, $C_{x,y,z}(\phi_z)$ returns the map of confidence levels, called *response map*. This map is the confidence matrix of how correct the landmark may be localized. See Fig. 1. The number of classifiers seems too great, but it is applicable in practice because this training is once offline and just a few of classifier is employed at each time in tracking. In order to train a such big number of linear SVM classifiers, a very fast linear SVM, libLinear [35], is one suitable tool.

3.2   Robust Initialization

In non-rigid face tracking, the aligned model from the previous frame was usually used as the initialization for the current frame. This initialization is hard to robustly work with fast motions. Some others, e.g. [7], (in the implementation) adaptively localized the current face position via the maximum response of template matching [36]. However, the false positive detection can happen, and the recovery in terms of lost tracking is impossible if the face detection is not involved. In fact, the information from some previous frames could provide a more robust initialization. [24] showed impressive results of pose tracking by matching via keypoint learning. Yet, we propose to use the simpler strategy for initialization. Our method uses the SIFT matching like [23] and estimate the rigid parameters closely to [15]. It is sustainable enough to fast motions and provides the accurate recovery before fitting the face model by pose-wise SVMs in the next step.

First of all, 2D SIFT points are detected. We base on the projections from the 3D model of the keyframe k onto the 2D current frame t to estimate the rotation and translation (rigid parameters). Let us denote n_k and n_t are the numbers of SIFT points detected respectively on a keyframe k and a frame $t > k$, and

$$l_k = \{l_k^0, l_k^1, ..., l_k^{n_k}\} \quad \text{and} \quad l_t = \{l_t^0, l_t^1, ..., l_t^{n_t}\} \tag{3}$$

are their respective locations. Let define the 3D points L_k, which associated with the 2D points l_k, are the intersections between the 3D model and the straight lines passing through the projection center of the camera and the 2D locations l_k. Because some points can be invisible (that are ignored), the number of L_k could be different to m. If $R_{k,t}$ and $T_{k,t}$ denote respectively the rotation and translation from frame k to frame t, we can write that, for $i = 1$ to m, the predicted i-th point at frame t could be written:

$$\widehat{l_t^i} = K\Phi(L_k^i) \quad \text{where} \quad \Phi(L_k^i) = (R_{k,t} \circ T_{k,t})L_k^i \tag{4}$$

where $\circ$ denotes the composition operator and K is the intrinsic camera matrix. To determine $R_{k,t}$ and $T_{k,t}$, we use the following least squares algorithm:

$$\{\hat{R}_{k,t}, \hat{T}_{k,t}\} = \arg \min_{R_{k,t}), T_{k,t}} \sum_{l_k^j \leftrightarrows l_t^i} \left(l_t^i - K\Phi(L_k^j)\right)^2 \tag{5}$$

where $\sum_{l_k^j \leftrightarrows l_t^i}$ denotes the sum over the couples (i, j) obtained by matching RANSAC algorithm of [37] between the keyframe k and the current frame t. This transformation is denoted $l_k^j \leftrightarrows l_t^i$. Before using RANSAC, we use the Flann matcher in both directions (from the keyframe k to the current frame t and vice versa) and return their intersection as a result. Finally, the optimization of the expression (5)) is effected numerically via the Levenberg-Marquardt algorithm.

3.3 Matching Strategy by Keyframes

Wide baseline matching via SIFT is deployed to estimate rigid parameters as the initialization for the next step. After detecting the landmarks at first frame using the landmark detector, we align 3D face models into landmarks and estimate the rigid parameters $\check{\Theta}(1)$ of $\Theta(1)$ and then using the pose-wise models (Sect. 3.4 to estimate the rigid and non-rigid parameters simultaneously. The information of this first frame such as the 2D face region and its rigid parameters, 2D and 3D corresponding points, and the value of objective function in (Sect. 3.4 are saved as a *keyframe*. To find the rigid parameters $\check{\Theta}(2)$ at the second frame, the first

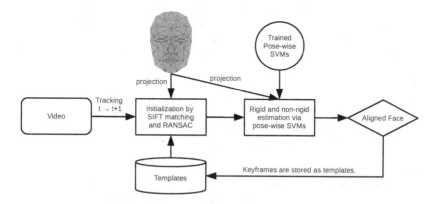

Fig. 3. Our two-step approach from the frame t to $t+1$. First step uses SIFT matching to estimate the rigid parameters. Second step uses pose-wise SVMs to re-estimate rigid and non-rigid paramters. The aligned current frames are stored as keyframes if they satisfy some given conditions.

keyframe is used to match with the second one via the method as reported in Sect. 3.4. If the number of matching points is less than a given threshold $T_p = 25$, Haris detector [38] and KLT [39] is considered instead exactly the same like the SIFT matching for rigid estimation. The method reported in (Sect. 3.4 then is applied to estimate again parameters. The same strategy is applied to coming frames. To estimate rigid parameters of frame t, it is matched to all preceding-stored keyframes $\mathcal{K}_t$ to select a *candidate keyframe* k. The *candidate keyframe* is the *keyframe* that have the maximum number of matching points with the current frame (after removing the ouliers by RANSAC). This number should be larger than T_p; otherwise, we estimate parameters using Harris points tracked by KLT from the previous frame. After model alignment, the current frame is registered as a keyframe into the set of keyframes if its three residuals (*Yaw*, *Pitch* or *Roll*) is out of value ranges of the whole set of preceding keyframes. The first keyframe is fixed and other keyframes can be updated. The updating happens if the *keyframe* is *candidate keyframe* and its value of (Sect. 3.4 is bigger than current frame's. To make sure unless bad keyframes were registered, we detect the face position parallel by the face detector and compute the distance between this position and where is detected by matching. The keyframe used for matching (*candidate keyframe*) is withdrawn from the set of keyframes if this distance is too large. Our method is fully automatic, and no manual keyframe is selected before tracking the video sequence. The strategy of tracking is reported in {Algorithm 1}.

Algorithm 1. The proposed matching strategy in our method.

INPUT: Given a set of keyframes $\mathcal{K}_t$ the $\check{\Theta}(t-1)$ of aligned model at frame t-1.
OUTPUT: $\check{\Theta}(t)$ at frame t.

1: Detecting the number of SIFT points n_t on frame t.
2: Matching between frame t and all of keyframes from $\mathcal{K}_t$ with RANSAC.
3: Determining the keyframe candidate that has the maximum matching keypoints.
4: **if** The number of matching points $> T_p$ **then**
5:    Estimating $\check{\Theta}(t)$ at frame t using keyframe candidate via the method in (Sect. 3.2).
6:    Estimating $\Theta(t)$ using method in (Sect. 3.4).
7:    **if** frame t satisfied as a new keyframe **then**
8:       Extracting the keyframe information from frame t to store into $\mathcal{K}_t \rightarrow \mathcal{K}_{t+1}$.
9:    **else if** the frame t updates the *keyframe* candidate **then**
10:      Extracting the keyframe to replace keyframe candidate in $\mathcal{K}_t$.
11:   **end if**
12: **else**
13:   Detecting Harris points and using KLT tracker do exactly like steps from 5 to 11 except the keyframe is now replaced by the previous frame.
14: **end if**

3.4   Fitting via Pose-Wise Classifiers

The previous step provides precisely the initial pose of face model. This pose permits to determine which pose-wise SVMs among the set of SVMs (ζ) should be chosen for fitting. For simplicity, $\check{\Theta}(t)$ and $\Theta(t)$ are represented as $\check{\Theta}$ and Θ. As above that $\check{\Theta}$ is the rotation components of the current model parameter Θ that are estimated after the initialization step. m groups of SVMs ($C_{\check{\Theta}_{i},y,z}$, $i = 1, ..., m$), where $\check{\Theta}_i$ is m nearest values of $\check{\Theta}$, are chosen for fitting. $v = 4$ obtains the best performance in our empirical experiments. Given the Θ parameter of 3D face model, $x_k(\Theta)$ denotes the projection of the k-th landmark on the current frame. The response map of $x_k(\Theta)$ is computed independently by each group as follows: Three local descriptors $\phi_z), z \in$ [intensity ($gray$), gradient ($grad$), LBP (lbp)], are extracted around the landmark k-th. The combined response map is the element-by-element multiplication of response maps (normalized into $[0, 1]$) that are computed independently by descriptors: $w = C_{\Theta_i,k,gray}(\phi_{gray}). * C_{\Theta_i,k,grad}(\phi_{grad}). * C_{\Theta_i,k,lbp}(\phi_{lbp})$, see Fig. 1. This final combined response map is applied to detect candidates of landmark location. The same procedure is applied for all landmarks. It is worth noting that the face is normalized to one reference face before extracting feature descriptors.

If picking up the highest score position as the candidate of k-th landmark, m candidates have to be considered (from v pose-wise SVMs $C_{\Theta_i,k,z}, i = 1, ..., v$). However, the highest score is not always the best one through observations and other peaks are probably the candidates as well. By this investigation, we keep more than one candidate (if it is local peak and its score is bigger than 70 % of the highest one) before determining the best by shape constraints. The set of candidates of k-th landmark detected by v classifiers $C_{\Theta_i,k,z}$ are merged together. Let us denote Ω_k is this merged set of candidates. The rigid and non-rigid parameters can be estimated via the objective function:

$$\hat{\Theta} = \arg \min_{p_k \in \Omega_k, \Theta} \sum_{k=1}^{n} w_k \|x_k(\Theta) - p_k\|_2^2 \tag{6}$$

where $x_k(\Theta)$ is the projection of kth landmark corresponding to Θ. Meanwhile, the position $p_k \in \Omega_k$ with the confidence w_k (from its response map) to be the candidate of the kth landmark. The optimization problem in Eq. 6 is combinatorial. In our work, we propose a heuristic method, which is based on ICP (Iterative Closest Points) [40] algorithm, to find the solution. The proposed approach consists of two iterative sub-steps: (i) looking for the closest candidate p_k from $x_k(\Theta)$, and (ii) estimating the update $\Delta\Theta$ using gradient method. As represented in Algorithm 2. The update $\Delta\Theta$ can be computed via the approximation of Taylor expansion that was mentioned similarly in [7], where J_k is the Jacobian matrix of the kth landmark.

$$\Delta\Theta = \left(\sum_{k=1}^{n} w_k J_k^T J_k \right)^{-1} \left(\sum_{k=1}^{n} w_k J_k^T (p_k - x_k(\Theta)) \right) \tag{7}$$

Algorithm 2. The fitting algorithm.
INPUT: n sets of Ω_k and the Θ at previous frame.
OUTPUT: Θ at current frame.

1: **repeat**
2:    Localizing 2D coordinate projection of landmarks $x_k(\Theta), k = 1, .., N_p$.
3:    Looking $v = 4$ nearest candidates p_k from $x_k(\Theta)$ in Ω_k.
4:    Selecting the candidate p_k from v ones that has highest SVM score w_k.
5:    Computing Jacobian matrices J_k at Θ.
6:    Computing updates $\triangle\Theta$ using {Eqn. (7)}.
7:    $\Theta \leftarrow \Theta + \triangle\Theta$
8: **until** Θ converged.

---

4   Experimental Results

Boston University Face Tracking (BUFT) database of [11] and Talking Face video[2]) are adopted to evaluate the precision of pose estimation and landmark tracking respectively. VidTimid videos of [41], and Honda/UCSD of [42] are also used to investigate profile-face tracking capability.

BUFT: The pose ground-truth is captured by magnetic sensors *"Flock and Birds"* with an accuracy of less than $1°$. The uniform-light set, which is used to evaluate, has a total of 45 video sequences (320×240 resolution) for 5 subjects (9 videos per subject) with available ground-truth of pose Yaw (or Pan), $Pitch$ (or $Tilt$), $Roll$. The precision is measured by Mean Absolute Error (MAE) of three directions between the estimation and ground-truth over tracked frames: $E_{yaw}, E_{pitch}, E_{roll}$ and $F_m = \frac{1}{3}(E_{yaw} + E_{pitch} + E_{roll})$ where $E_{yaw} = \frac{1}{N_s}\sum |\Theta^i_{yaw} - \hat{\Theta}^i_{yaw}|$ (similarly for the $Pitch$ and $Roll$). N_s is the number of frames and $\Theta^i_{yaw}, \hat{\Theta}^i_{yaw}$ are the estimated value and ground-truth of Yaw respectively.

Since BUFT videos have low resolution and the number of SIFT points is often not enough to apply the matching, our result (Table 1) is still comparable to state-of-the-art methods. Our method achieves the same mean error E_m as [7,13, 20], but worse than [12,19,23–25]. With the use of offline training of synthesized data, the result is promising. The algorithm is better than [20] at Yaw and $Roll$ precision and [7] at $Pitch$ and $Roll$ precision. The fully automatic method is marked (*) in Table 1; otherwise, it is the manual method. In addtion of rigid tracking, our method is able to track non-rigid parameters (+) in Table 1. The other methods having better results than us, is able to estimate only the rigid parameter or is a manual method. Otherwise, our method can estimate both rigid and non-rigid parameters, recover the lost-tracking while the training data is synthetic.

The Talking Face Video: is a freely 5000-frames video sequence of a talking face with available ground-truth of 68 facial points on the whole video.

---

[2] http://www-prima.inrialpes.fr/FGnet/data/01-TalkingFace/talking_face.html.

Table 1. The pose precision of our method and state-of-the-art methods on uniform-light set of BUFT dataset.

Approach	E_{yaw}	E_{pitch}	E_{roll}	E_m
[24])	3.8	2.7	1.9	2.8
[12])	3.8	3.2	1.4	2.8
[19]) (+)	4.4	3.3	2.0	3.2
[23]) (*)	4.6	3.7	2.1	3.5
[25]) (*)	4.3	3.8	2.6	3.5
[13]) (*)	5.0	3.7	2.9	3.9
[7]) (*,+)	4.3	4.8	2.6	3.9
[20]) (+)	5.4	3.9	2.4	3.9
Our method (*,+)	**5.0**	**4.5**	**2.2**	3.9

The Root-Mean-Squared (RMS) error is used to measure the landmark tracking (non-rigid) precision. Although the number of landmarks of methods is different, the same evaluation scheme could be still applied on the same number of selected landmarks. Twelve landmarks at corners of eyes, nose and mouth are chosen. The Fig. 4) shows the RMS of our method (red curve), and FaceTracker (blue curve) [7] on the Talking Face video. The vertical axis is the RMS error (in pixel) and the horizontal axis is the frame number. The result shows that even though our method just learned from the synthesized data, what we obtain is comparable to the state-of-the-art method, even more robust. The average precision of the entire video of our method is 5.8 pixels and FaceTracker is 6.8 pixels.

VidTimid and Honda/UCSD: The VidTimid is captured in resolution 512x384 pixels at the good office environment. Honda/UCSD dataset at resolution 640x480, is much more challenging than VidTimid that provides a wide range of different poses at different conditions such as face partly occlusion, scale changes, illumination, etc. The ground-truth of pose or landmarks is unavailable in these databases;

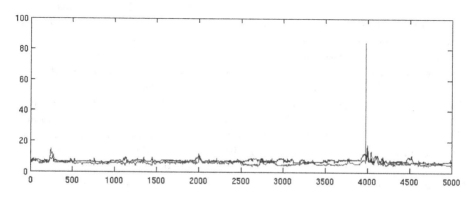

Fig. 4. The RMS of our framework (red curve) and FaceTracker [7] (blue curve). The vertical axis is RMS error (in pixel) and the horizontal axis is the frame number (Color figure online).

Fig. 5. Our tracking method on some sample videos of VidTimid and Honda/UCSD.

hence, they are used for visualizing purpose of the profile tracking. Our framework again demonstrates its capability even in more complex movements of the head. In fact, our method is more robust than FaceTracker in terms of keeping track unloosing and it can recover face quickly without waiting for frontal face reset as FaceTracker. See Fig. 5. Some full videos in paper can be found at here[3], in which one our own video is also recorded for evaluation. Our method is again more robust on FaceTracker on this video.

Although real-time computation is unsustainable (about 5s/frame on Desktop 3.1 GHz, 8G RAM) due to Matlab implementation. In which, the first step is about 3s/frame because of SIFT matching. The C/C++ implementation and the replacement of SIFT by another faster descriptor is a possible future work. In addition, our method is not robust with complex background because no background is included in our synthetic training data. The aware of background in training process may be a possible solution.

5   Conclusions

We presented a robust framework for wide rotation tracking of non-rigid face. Our method used the large synthesized dataset rendering from a small set of annotated frontal views. This dataset was divided into groups to train pose-wise linear SVM classifiers. The response map of one landmark is the combination of response maps from three descriptors: intensity, gradient and LBP. Through keeping some candidates from one combined response map, we apply an heuristic method to choose the best one via the constraint of 3D shape model. In addition, the SIFT matching makes our method robust to fast movements and provides a good initial rigid parameters. Through keyframes, our method can do recover the lost tracking quickly without waiting for frontal-view reset. Our method is more robust than one state-of-the-art method in terms of the profile tracking and comparable in landmark tracking. However, our method is still limited to work with complex background because of that no complex background is included in training synthesized images. It can be more efficient if the backgrounds of synthesized images are more complex. In addition, the usage of SIFT matching is slow and it needs to be improved for a real-time performance as future direction.

---

[3] http://www.youtube.com/watch?v=yqAh1_2uaPA.

References

1. Cootes, T.F., Edwards, G.J., Taylor, C.J.: Active appearance models. TPAMI **23**, 681–685 (2001)
2. Xiao, J., Baker, S., Matthews, I., Kanade, T.: Real-time combined 2d+3d active appearance models. CVPR. **2**, 535–542 (2004)
3. Gross, R., Matthews, I., Baker, S.: Active appearance models with occlusion. IVC **24**, 593–604 (2006)
4. Matthews, I., Baker, S.: Active appearance models revisited. IJCV **60**, 135–164 (2004)
5. Cristinacce, D., Cootes, T. F.: Feature detection and tracking with constrained local models. In: BMVC. (2006)
6. Wang, Y., Lucey, S., Cohn, J.: Enforcing convexity for improved alignment with constrained local models. In: CVPR (2008)
7. Saragih, J.M., Lucey, S., Cohn, J.F.: Deformable model fitting by regularized landmark mean-shift. IJCV **91**, 200–215 (2011)
8. Dollar, P., Welinder, P., Perona, P.: Cascaded pose regression. In: CVPR (2010)
9. Cao, X., Wei, Y., Wen, F., Sun, J.: Face alignment by explicit shape regression. In: CVPR (2012)
10. Xiong, X., la Torre Frade, F.D.: Supervised descent method and its applications to face alignment. In: CVPR (2013)
11. Cascia, M.L., Sclaroff, S., Athitsos, V.: Fast, reliable head tracking under varying illumination: an approach based on registration of texture-mapped 3d models. TPAMI **22**, 322–336 (2000)
12. Xiao, J., Moriyama, T., Kanade, T., Cohn, J.: Robust full-motion recovery of head by dynamic templates and re-registration techniques. Int. J. Imaging Syst. Technol. **13**, 85–94 (2003)
13. Morency, L. P., Whitehill, J., Movellan, J. R.: Generalized adaptive view-based appearance model: Integrated framework for monocular head pose estimation. In: FG. (2008)
14. An, K. H., Chung, M. J.: 3d head tracking and pose-robust 2d texture map-based face recognition using a simple ellipsoid model. In: IROS, pp. 307–312 (2008)
15. Vacchetti, L., Lepetit, V., Fua, P.: Stable real-time 3d tracking using online and offline information. TPAMI **26**, 1385–1391 (2004)
16. Ström, J.: Model-based real-time head tracking. EURASIP **2002**, 1039–1052 (2002)
17. Chen, Y., Davoine, F.: Simultaneous tracking of rigid head motion and non-rigid facial animation by analyzing local features statistically. In: BMVC. (2006)
18. Alonso, J., Davoine, F., Charbit, M.: A linear estimation method for 3d pose and facial animation tracking. In: CVPR (2007)
19. Lefevre, S., Odobez, J. M.: Structure and appearance features for robust 3d facial actions tracking. In: ICME (2009)
20. Tran, N.-T., Ababsa, F.-E., Charbit, M., Feldmar, J., Petrovska-Delacrétaz, D., Chollet, G.: 3D face pose and animation tracking via eigen-decomposition based Bayesian approach. In: Bebis, G., Boyle, R., Parvin, B., Koracin, D., Li, B., Porikli, F., Zordan, V., Klosowski, J., Coquillart, S., Luo, X., Chen, M., Gotz, D. (eds.) ISVC 2013, Part I. LNCS, vol. 8033, pp. 562–571. Springer, Heidelberg (2013)
21. Ababsa, F.: Robust extended kalman filtering for camera pose tracking using 2d to 3d lines correspondences. In: IEEE/ASME International Conference on Advanced Intelligent Mechatronics, pp. 1834–1838 (2009)

22. Ababsa, F., Mallem, M.: Robust line tracking using a particle filter for camera pose estimation. In: Proceedings of the ACM Symposium on Virtual Reality Software and Technology. (2006)
23. Jang, J. S., Kanade, T.: Robust 3d head tracking by online feature registration. In: FG (2008)
24. Wang, H., Davoine, F., Lepetit, V., Chaillou, C., Pan, C.: 3-d head tracking via invariant keypoint learning. IEEE Trans. Circ. Syst. Video Technol. **22**, 1113–1126 (2012)
25. Asteriadis, S., Karpouzis, K., Kollias, S.: Visual focus of attention in non-calibrated environments using gaze estimation. IJCV **107**, 293–316 (2014)
26. FaceAPI. (http://www.seeingmachines.com)
27. Gu, L., Kanade, T.: 3d alignment of face in a single image. In: CVPR (2006)
28. Su, Y., Ai, H., Lao, S.: Multi-view face alignment using 3d shape model for view estimation. In: Proceedings of the Third International Conference on Advances in Biometrics (2009)
29. Ahlberg, J.: Candide-3 - an updated parameterised face. Technical report, Department of Electrical Engineering, Linkoping University, Sweden (2001)
30. Aggarwal, G., Veeraraghavan, A., Chellappa, R.: 3D facial pose tracking in uncalibrated videos. In: Pal, S.K., Bandyopadhyay, S., Biswas, S. (eds.) PReMI 2005. LNCS, vol. 3776, pp. 515–520. Springer, Heidelberg (2005)
31. Ojala, T., Pietikäinen, M., Harwood, D.: A comparative study of texture measures with classification based on featured distributions. PR **29**, 51–59 (1996)
32. Koestinger, M., Wohlhart, P., Roth, P. M., Bischof, H.: Annotated facial landmarks in the wild: a large-scale, real-world database for facial landmark localization. In: First IEEE International Workshop on Benchmarking Facial Image Analysis Technologies (2011)
33. Gross, R., Matthews, I., Cohn, J.F., Kanade, T., Baker, S.: Multi-pie. IVC **28**, 807–813 (2010)
34. Dementhon, D.F., Davis, L.S.: Model-based object pose in 25 lines of code. IJCV **15**, 123–141 (1995)
35. Fan, R.E., Chang, K.W., Hsieh, C.J., Wang, X.R., Lin, C.J.: LIBLINEAR: A library for large linear classification. JMLR **9**, 1871–1874 (2008)
36. Lewis, J.P.: Fast normalized cross-correlation. In: Proceedings of Vision Interface, vol. 1995, pp. 120–123 (1995)
37. Fischler, M.A., Bolles, R.C.: Random sample consensus: a paradigm for model fitting with applications to image analysis and automated cartography. Commun. ACM **24**, 381–395 (1981)
38. Harris, C., Stephens, M.: A combined corner and edge detector. In: Fourth Alvey Vision Conference, pp. 147–151 (1988)
39. Tomasi, C., Kanade, T.: Detection and tracking of point features. Technical report, International Journal of Computer Vision (1991)
40. Besl, P.J., McKay, N.D.: A method for registration of 3-d shapes. TPAMI **14**, 239–256 (1992)
41. Sanderson, C.: The VidTIMIT Database. Technical report, IDIAP (2002)
42. Lee, K., Ho, J., Yang, M., Kriegman, D.: Video-based face recognition using probabilistic appearance manifolds. In: Proceedings of the IEEE Computer Society Conference on Computer Vision and Pattern Recognition, vol. 1, pp. 313–320 (2003)

3-D Face Recognition Using Geodesic-Map Representation and Statistical Shape Modelling

Wei Quan[(✉)], Bogdan J. Matuszewski, and Lik-Kwan Shark

Applied Digital Signal and Image Processing (ADSIP) Research Centre,
University of Central Lancashire, Preston PR1 2HE, UK
{WQuan,BMatuszewski1,LShark}@uclan.ac.uk
http://www.springer.com/lncs

Abstract. 3-D face recognition research has received significant attention in the past two decades because of the rapid development in imaging technology and ever increasing security demand of modern society. One of its challenges is to cope with non-rigid deformation among faces, which is often caused by the changes of appearance and facial expression. Popular solutions to deal with this problem are to detect the deformable parts of the face and exclude them, or to represent a face in terms of sparse signature points, curves or patterns that are invariant to deformation. Such approaches, however, may lead to loss of information which is important for classification. In this paper, we propose a new geodesic-map representation with statistical shape modelling for handling the non-rigid deformation challenge in face recognition. The proposed representation captures all geometrical information from the entire 3-D face and provides a compact and expression-free map that preserves intrinsic geometrical information. As a result, the search for dense points correspondence in the face recognition task can be speeded up by using a simple image-based method instead of time-consuming, recursive closest distance search in 3-D space. An experimental investigation was conducted on 3-D face scans using publicly available databases and compared with the benchmark approaches. The experimental results demonstrate that the proposed scheme provides a highly competitive new solution for 3-D face recognition.

Keywords: 3-D face recognition · Non-rigid deformation · Shape modelling · Geodesic-map representation

1 Introduction

Face recognition is one of the most common biometrics with unique advantages, such as naturalness, non-contact and non-intrusiveness. Its related research has been for many years of great interest to computer vision and pattern recognition communities, which has been exploited for applications such as public security [1], fraud prevention [2] and crime prevention and detection [3]. A fair amount of efforts have been made on the development of 2-D face recognition systems using intensity images as input data in the past. Despite 2-D face recognition systems being able to perform well under constrained conditions, they are

© Springer International Publishing Switzerland 2015
A. Fred et al. (Eds.): ICPRAM 2015, LNCS 9493, pp. 199–212, 2015.
DOI: 10.1007/978-3-319-27677-9_13

still facing great difficulties as facial appearances can vary significantly even for the same individual due to differences in pose, lighting conditions and expressions [4]. Using the 3-D geometry of the face instead of its 2-D appearance is expected to alleviate the difficulties since the human face is a natural 3-D entity [5]. According to the type of features used, the relevant work for 3-D face recognition can be roughly classified into three major categories, which are geometrical feature-based, shape descriptor-based and prominent region-based approaches.

Geometrical feature-based methods achieve the face recognition task using structural information extracted from 3-D faces, such as landmarks, salient curves and geodesic-like patterns. Landmarks are representative key facial points often associated with in order to construct a feature space. Shi et al. [6] introduced a method based on the so called 'soft' landmarks, i.e., the landmarks that are easily located on actual skin surfaces, such as eye corners, mouth corners, nose edge, etc. It showed that these landmarks vary significantly if different subjects are used to generate them. The use of anthropometric facial fiducial landmarks for the face recognition was presented in [7]. Salient curves are a kind of the discriminative surface curves extracted from 3-D faces. The symmetric profile curve from the intersection between the symmetry plane and the 3-D facial scans was described in [8]. Three facial curves which intersect the facial scan using horizontal and vertical planes as well as a cylinder were proposed in [9]. A geodesic is a locally length-minimizing curve along the surface and it contains information related to the intrinsic geometry of an object. Mpiperis et al. [10] proposed a geodesic polar representation of the facial surface. With this representation, the intrinsic surface attributes do not change under isometric deformations and therefore it can be used for expression-invariant face recognition. Based on the similar concept, a method using the similarity measurement of local geodesic patch has been proposed by Hajati et al. [11].

Shape descriptor-based methods look into attributes of local surfaces and encode the 3-D face into special designed patterns, which are often invariant to the orientation of faces. A representation of free-form surfaces based on the point signature was proposed for 3-D face recognition by Chua et al. [12]. The approach uses the point signature extracted from the rigid parts of a face to overcome the challenge of facial expressions. A similar representation, named local shape map, was proposed in [13]. Tanaka et al. [14] introduced a special shape descriptor based on the extended Gaussian image (EGI) and local surface curvature. The EGI was used as a mediate feature after curvature-based segmentation on which principal directions are mapped as local features. Huang et al. [15] adopted a multi-scale extended local binary pattern (eLBP) as an accurate descriptor of local shape changes for 3-D face identification. A hybrid matching approach based on scale-invariant feature transform (SIFT) was designed to measure similarities between control and test face scans once they were represented by the multi-scale eLBP.

Prominent region-based methods use dense point clouds detected from specific regions to form feature vectors. The similarity between two faces is determined based on their relationship in the feature space. Queirolo et al. [16] proposed a union of the segmented regions from faces for the purpose of face identification. These regions include the circular nose area, elliptical nose area and upper head. Regions segmented by median and Gaussian curvature were utilised for the feature construction in [17]. Gupta et al. [7] manually placed anthropometric points on faces and used a feature vector based on the anthropometric distance between points for face recognition. Xu et al. [18] proposed a method which converted a 3-D face into a regular mesh and created a feature vector to encode the 3-D shape of face based on this regular mesh.

In this paper, we propose a new geodesic-map representation for 3-D faces, which is an extension of original work proposed by Quan et al. [19]. The proposed method preserves the intrinsic geometrical information related to the identity of the face. It can be considered as the expression-free representation for faces from the same person and is able to reduce the non-rigid deformation effect in the face recognition task. The method first creates the geodesic strip for each extracted landmarks on a single face based on the geodesic distance measurement between surface points and the landmark. Then it combines the calculated geodesic strips for all extracted landmark in order to form a map. This map is the new representation of the face. In the subsequent stage of the statistical shape modelling, the search for the dense points correspondence is therefore simplified to an image-based method using the calculated geodesic-map instead of the iterative search in 3-D space. This helps to improve the efficiency of the whole face recognition task, including training and testing.

The rest of the paper is organized as follows: Sect. 2 introduces the sparse facial landmark detection. Section 3 describes the processing steps for generating the geodesic-map representation for 3-D faces. Section 4 explains the mechanism for dense point correspondence search across all training datasets. Section 5 presents the statistical shape models used in this work. Section 6 illustrates the model matching process. The experimental results using the statistical shape models for face recognition are presented in Sect. 7. Finally, concluding remarks and possible future work are given in Sect. 8.

2 Sparse Facial Landmark Detection

Landmarks are often used to assist the process of data registration in order to determine the coefficients of a transformation function. A minimum of three pairs of corresponding landmarks are needed if the transformation is considered as rigid and more is required when the transformation has more degrees of freedom. In this work, a small number of landmarks are extracted from 3-D faces, which are used for generating the geodesic-map representation and dense point correspondence search at the later stage. Using a combination of the shape index [20] and the intersecting profiles of facial symmetry plane [21], a set of 12 key landmarks can be extracted, which are two upper nose base, two nose corners, upper and lower lip tips, two inner eye corners, two outside eye corners and two mouth corners.

The general strategy of this landmark detection process is to use the Gaussian curvature and mean curvature to locate a set of candidates for each landmark along the intersecting profiles of facial symmetry plane, and then select the candidate with the shape index as the key landmark. The shape index $S(p)$ at point p is calculated as:

$$S(p) = \frac{1}{2} - \frac{1}{\pi}\tan^{-1}\frac{K_1(p) + K_2(p)}{K_1(p) - K_2(p)} \tag{1}$$

where $K_1(p)$ and $K_2(p)$ are the maximum and minimum local curvature at point p, respectively. According to the value of shape index, between zero and one, each point can be classified into six types of shape, such as cup, rut, saddle, ridge and cap. Figure 1 demonstrates the locations of all 12 key landmarks extracted. Since the extracted landmarks are sparse and around the facial areas that are anatomically stable, as they are well defined for all faces, and invariant to facial expressions, they are more likely to be robustly detected than landmarks located in other parts of the face.

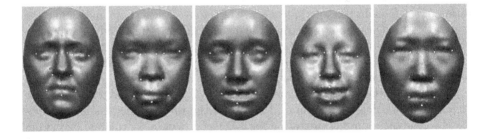

Fig. 1. Examples of sparse landmarks detected on 3-D faces.

3   Geodesic-Map Representation

A geodesic is a generalization of an Euclidean distance and is defined as the length of the shortest path between two points along a continuous surface [22]. Bronstein et al. [23] proposed a face recognition method based on transformation, ψ, mapping an original face $\mathbb{S}$ with the given geodesic distance $d_{\mathbb{S}}(\xi_1, \xi_2)$ onto another space $\mathbb{S}'$ with the Euclidean distance $d_{\mathbb{S}'}(\psi(\xi_1), \psi(\xi_2))$ in such a way that corresponding distances are preserved:

$$d_{\mathbb{S}}(\xi_1, \xi_2) = d_{\mathbb{S}'}(\psi(\xi_1), \psi(\xi_2)) \tag{2}$$

This means that the surface information represented by the geodesic distance between different points on the surface is preserved. Such mapping is invariant to rigid transformation as well as any non-rigid deformation which does not change the distance between the points on the surface. Based on the assumption that for the same subject facial expressions do not change geodesic distance, the dense point correspondence between two faces of the same subject can be estimated

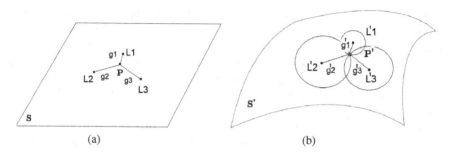

Fig. 2. Geodesic distances for searching point correspondence: (a) original surface; (b) deformed surface from (a).

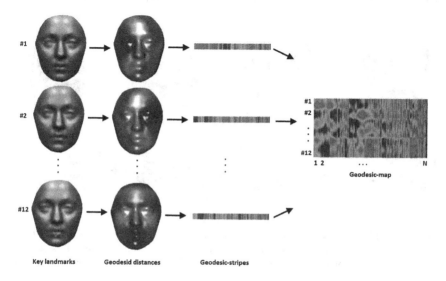

Fig. 3. Geodesic-map representation for a 3-D face (N is the number of surface points on the face).

using geodesic distances between surface points to a number of fixed points on both surfaces. Figure 2 illustrates the search for point $\mathbf{P}'$ corresponding to the given point $\mathbf{P}$ on another surface, where $\mathbf{L1}$, $\mathbf{L2}$, $\mathbf{L3}$ are three landmark points on one surface with the corresponding geodesic distances to $\mathbf{P}$ denoted by $\mathbf{g_1}$, $\mathbf{g2}$, $\mathbf{g3}$; $\mathbf{L'1}$, $\mathbf{L'2}$, $\mathbf{L'3}$ are three landmark points on the other surface with the corresponding geodesic distances to $\mathbf{P}'$ denoted by $\mathbf{g'1}$, $\mathbf{g'2}$ and $\mathbf{g'3}$. $\mathbf{P}'$ is said to correspond to $\mathbf{P}$ if it is found that $\mathbf{g1} = \mathbf{g'1}$, $\mathbf{g2} = \mathbf{g'2}$ and $\mathbf{g3} = \mathbf{g'3}$. For a unique solution, a minimum of three fixed landmark points are needed.

The geodesic distance is used in this paper to assist the dense point correspondence search across 3-D faces. 12 key landmarks extracted using the method described in Sect. 2 are considered as the fixed surface point landmarks on 3-D faces. A geodesic-map representation is proposed to simplify overall correspondence search. The geodesic map is built using the following three steps. The first step is to compute the geodesic distances between 12 key landmarks and all surface points on a 3-D face. The second step is to re-arrange the related geodesic distances of each key landmark to an geodesic-stripe. The final step is to combine all the geodesic-stripes in order to form the geodesic-map. An example of generating the geodesic-map representation for a 3-D face is illustrated in Fig. 3, where the colour of the surface represents geodesic-distances to a specific landmark. In the geodesic-map, the row index corresponds to the order of landmarks and the column index matches the order of the surface points. From the figure, it can be seen that the 3-D faces are transformed from $\mathbb{R}^3$ space to a $\mathbb{R}^2$ image space and this enables an efficient dense point correspondence search. Figure 4 shows examples of 3-D faces and their corresponding geodesic-maps.

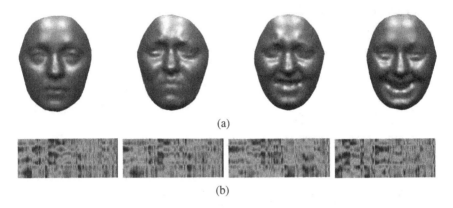

(a)

(b)

Fig. 4. Examples of 3-D face scans with the corresponding geodesic-maps: (a) faces from the same person with four expressions (from left to right): neutral, anger, fear and happiness; (b) corresponding geodesic-maps calculated using 12 key landmarks.

4   Geodesic-Map Matching

Having the geodesic-maps created, the pair-wise dense point correspondences among faces can be estimated using standard image-based matching techniques. For the faces from the same person, this can be achieved by using cross-correlation [22] in which geodesic-map's column of the given face is cross-correlated with the target face geodesic-map. The geodesic-map's column of the target face with the highest cross-correlation value is considered as being in correspondence with the point in questions from the given face as shown in Fig. 5.

For the faces from different persons, the geodesic-map cannot be directly used for the correspondence search simply because the characteristics of geodesic distance. Computation of the point correspondence between faces of different subjects is required to construct the statistical shape model for the face recognition task. To tackle this problem, a data warping process is introduced prior to the geodesic-map matching process when it applying to faces from different subjects. The data warping is based on the Thin-plate Splines (TPS)

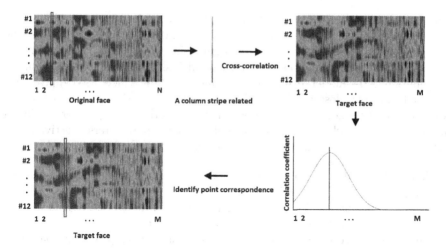

Fig. 5. Pair-wise point correspondence search procedure using geodesic-map representation (N and M are the number of surface points on the original face and target face, respectively).

warping technique [24] and applies to the target face. 12 pairs of extracted key landmarks from both original and target faces are used as the control points for the calculation the warping function. It is then used to warp the whole target face to match the one from the original face so that the standard geodesic-map matching described above can be carried out. This process is able to minimise non-rigid deformation caused by changes of person.

5   Dimensionality Reduction

Statistical models have been successfully used for face analysis and recognition for many years. The core of the models is the dimensionality reduction, which often serves the purpose of feature vector extraction. PCA is often the popular choices, which produces a compact representation based on low dimensional linear manifolds [25]. However, the models fail to discover the underlying non-linear structure of facial data especially for faces containing facial expressions. Another choice is Locality Preserving Projection (LPP) and it is able to handle a wider range of data variability while preserving local structure linked to the nonlinear structure of facial data. In this work the statistical model, LPP, was used to evaluate their performance for the task of face recognition. The detail of the method can be found in [26].

6   New Dataset Fitting

Given the eigenvectors of statistical models extracted from the training dataset, the estimation of feature vectors in order to synthesise shape for faces from a new dataset, using the constructed statistical model is the next processing stage. This is usually achieved by a recursive data registration in which the shape and pose parameters are iteratively estimated in turn. While pose parameters control the orientation and position of the model, shape parameters encapsulate deformation of the model. Instead of applying one of the widely used approaches, modified Iterative Closest Point (ICP) registration, a hybrid fitting based on the combination of geodesic-map representation and feature sub-space projection is proposed in this work. In order to solve all unknown parameters effectively, the following standard optimization scheme is used:

1. Create the geodesic-map representation for both the model and a new face using the method described in Sect. 3.
2. Estimate the dense point correspondence between model and new face using the geodesic-map matching process explained in Sect. 4.
3. Calculate feature vector, α, for the new face using back-projection based on the created feature sub-space, described as

$$\alpha = \mathbf{W}_{opt}^{T}\widehat{\mathbf{x}} \tag{3}$$

where $\widehat{\mathbf{x}}$ is the surface points related to estimated dense correspondences from the new face and $\mathbf{W}_{opt}$ is the matrix containing feature vectors.
4. Generate a new instance of the statistical model, $\mathbf{Q}$, using the feature vector α, as

$$\mathbf{Q} = \mathbf{W}_{opt}\widehat{\alpha} \tag{4}$$

and repeat steps 2 to 4 until the preset convergence condition is reached.

In this optimization scheme, the geodesic-map representation and map matching serve the similar purpose as applied to the training dataset in which it estimates correspondence between both the models and new face. Since the models have learnt non-rigid deformation from faces across different identities in the training set and can be adapted to match the deformation in the new dataset, the TPS warping techniques described in Sect. 4 is no longer needed. Furthermore the use of the proposed method speeds up the whole fitting process for the new dataset and saves up to 70 % computation time on average compared with the widely used modified ICP registration [19,27]. A few examples of the fitting results generated using the LPP-based method are shown in Fig. 6. From the figure it can be seen that the shape of synthesised faces are very close to new faces.

It is worth noticing that the feature vector α controls shape of the models in order to match it to the new face. Therefore it contains geometrical information of the face and is used as the feature vector for the classification of face identity in this work. A variety of classification methods can be applied, including, Nearest-Neighbour, Naive Bayesian, Support Vector Machine, etc. For the sake of simplicity and to demonstrate the discriminative nature of the shape parameters α for the proposed feature vector, the Nearest-Neighbour classifier is chosen for the face classification in this work.

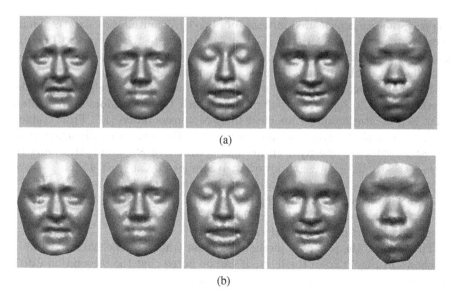

(a)

(b)

Fig. 6. Example of new dataset fitting: (a) new faces; (b) synthesised faces after the fitting.

7   Experimental Results

To show the effectiveness of the proposed method for the purpose of face recognition tasks, two publicly available 3-D facial databases, BU-3DFE and Gavab, were exploited for the evaluation in this work. The BU-3DFE database consists of 2,500 3-D faces from 100 people, with age ranging from 18 to 70 years old, with a variety of ethnic origins including White, Black, East-Asian, Middle East Asian, Indian and Hispanic Latino [28]. Each person has seven basic expressions. The Gavab database contains 549 face scans from 61 different subjects [29]. Each subject was scanned 9 times for different poses and expressions, giving six neutral scans and three scans with an expression. The scan with missing data contains one scan while looking up ($+35°$), one while looking down ($-35°$), one for the left profile ($-90°$), one for the right profile ($+90°$) as well as one with random poses.

7.1   Facial Expression Changes

The robustness to facial expression variation is an important aspect in face recognition. To test the face recognition invariance with respect to face articulation, a series of tests were run and the performance of the proposed method is compared with that of the state-of-the-art methods, including Patch Geodesic Moments [10], Geodesic Polar Representation [11] and Canonical Image Representation [23]. In order to make a direct comparison with the results reported in [11], the same experimental protocol used in [11] is adopted here.

The performance is measured in terms of rank-1 recognition rate and the Cumulative Matching Characteristics (CMC) [30]. In the test, all faces with neutral expression from BU-3DFE database are used to form the statistical models, while the rest of the database is used as the testing faces.

The rank-1 recognition rates of the proposed approaches are given in Table 1 together with the reported results of the Patch Geodesic Moments, Geodesic Polar Representation and Canonical Image Representation [11]. From Table 1, it can be seen that among the four methods LPP-based approach achieved the highest recognition rate with an average accuracy of 89%, outperforming the state-of-the-art 3D expression-invariant techniques by at least 4%. It is worth noticing that the recognition rates for different expressions range from 87% to 94%. This shows that the proposed statistical shape modelling scheme can handle facial expression changes well but still introduces uncertainty into face recognition task caused by facial expressions.

The CMC of the proposed methods together with those benchmark methods are shown in Fig. 7. From the figure it can be noted that the recognition rate of the proposed LPP-based method is always the highest.

Table 1. Performance comparison under facial expression changes.

Expression	LPP-based	Patch geodesic moments	Geodesic polar representation	Canonical image representation
Anger	93%	93%	-	-
Disgust	80%	79%	-	-
Fear	90%	82%	-	-
Happiness	92%	86%	-	-
Sadness	94%	85%	-	-
Surprise	87%	84%	-	-
Overall	**89.3%**	**84.8%**	**80.3%**	**77.2%**

7.2   Data Resolution Variation

In many practical applications, the data resolution usually varies because of the specification of data acquisition system, the need of data storage or the use of preprocessing. It often requires face recognition system to cope with low-resolution data. In order to evaluate the capability of the proposed methods in terms of handling low-resolution data in the face recognition task, a set of experiments were conducted using data with 75%, 50% and 25% of the original resolution as the test dataset. The original resolution is approximately 5,000 surface points for each face. In terms of experimental strategy, all 2,500 faces from BU-3DFE database were used in the experiments. The faces were divided into ten subsets with each subset containing all 100 subjects and all seven expressions. One subset is selected for testing while the remaining subsets were used for training. Such experiments are repeated ten times with a different subset selected for testing each time.

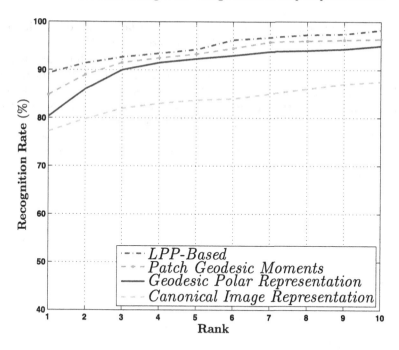

Fig. 7. Recognition rate obtained under facial expression changes.

The faces in the training set are not used for the testing. Figure 8 reports the CMC of the proposed method. From the figures it can been seen that the method is able to achieve reasonable recognition rates with data resolution of 75 % and 50 % compared to the resolution of 25 %.

Table 2. Recognition comparison on missing data using Gavab database.

	LPP-based	Sparse representation	Ridge images	Concave and convex region	Elastic radial curves
Frontal	93 %	92 %	82 %	86 %	97 %
Looking up	94 %	-	85 %	80 %	100 %
Looking down	93 %	-	87 %	79 %	98 %

7.3 Missing Data

In order to evaluate the missing data challenge of the proposed method, and compare with the results achieved by the existing benchmark methods reported in [31], the same experimental protocol introduced in [31] was used here. The benchmark methods include sparse representation [32], 3-D ridge images [33], concave and convex regions [34] and elastic radial curves [31]. In the experiment, the frontal scans with neutral expression of each person was taken as the training set. The rest of the scans were used for testing. Since the proposed approach is not designed for working on facial scans with a large part of missing data, the scans for the left and right profiles were not included in testing. Table 2 illustrates the results of the rank-1 recognition accuracy for different categories of testing faces. From the table, it can be seen that the proposed approach provides a high recognition accuracy on both expression and pose variations and outperforms majority of the existing methods and its performance is close to the best recognition accuracy achieved by the elastic radial curves [31].

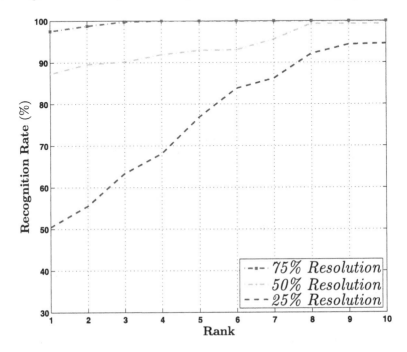

Fig. 8. Recognition rate obtained with various resolution.

8   Conclusions

This paper presents an effective representation with the statistical shape modelling scheme for 3-D face recognition. Given a set of training faces with a variety of facial expressions, the proposed scheme is able to effectively estimate the accurate dense point correspondence among the faces using the geodesic-map, construct the statistical shape models and synthesise appropriate shapes for a new face. The face recognition experiments show that the proposed method is handling well non-rigid deformations caused by the changes in the appearance (e.g. due to weight gain was not tested here) as well as certain level of missing data. It provides evidence that the proposed method can cope with the face recognition task with lower data resolution. The use of geodesic-map also helps improve the efficiency of the entire face recognition task. The research will be extended further by taking into consideration other practical factors, such as independent database, lack of training samples and occlusion.

References

1. Chellppa, R., Wilson, C., Sirohey, S.: Human and machine recognition of faces: a survey. Proc. IEEE **83**(5), 705–740 (1995)
2. Jafri, R., Ardabilian, H.R.: A survery of face recognition techniques. J. Inf. Process. Syst. **53**(2), 41–66 (2009)
3. Kong, S.G., Heo, J., Abidi, B.R., Paik, J., Abidi, M.A.: Recent advances in visual and infraraed face recognition - a review. Comput. Vis. Image Underst. **97**, 103–135 (2005)
4. Lu, Y., Zhou, J., Yu, S.: A survey of face detection, extraction and recognition. Comput. Inf. Pattern Recogn. **22**(2), 163–195 (2003)
5. Lu, X., Jain, A.K.: Deformation modeling for robust 3D face matching. IEEE Trans. Pattern Anal. Mach. Intell. **30**(8), 1346–1356 (2005)
6. Shi, J., Samal, A., Marx, D.: How effective are landmarks and their geometry for face recognition? Comput. Vis. Image Underst. **102**(2006), 117–133 (2006)
7. Gupta, S., Aggatwal, J.K., Markey, M.K., Bovik, A.C.: 3D face recognition founded on the structural diversity of human faces. In: IEEE Conference on Computer Vision and Pattern Recognition, Minnesota (2007)
8. Zhang, L., Razdan, A., Farin, G., Femiani, J., Bae, M., Lockwood, C.: 3D face authentication and recognition based on bilateral symmetry analysis. Vis. Comput. **22**(1), 43–55 (2006)
9. Nagamin, T., Uemura, T., Masuda, I.: 3D facial image analysis for human identification. In: IEEE Conference on Computer Vision and Pattern Recognition, Champaign (1992)
10. Mpiperis, I., Malassiotis, S., Strintzis, M.G.: 3-D face recognition with the geodesic polar representation. IEEE Trans. Inf. Forensics Secur. **2**(3), 537–547 (2007)
11. Hajati, F., Raie, A.A., Gao, Y.: 2.D face recognition using patch geodesci moments. Pattern Recogn. **45**(2012), 969–982 (2012)
12. Chua, C., Han, F., Ho, Y.: 3D human face recognition using point signature. In: IEEE International Conference on Automatic Face and Gesture Recognition, Washington D.C. (2000)
13. Wu, Z., Wang, Y., Pan, G.: 3D face recognition using local shape map. In: IEEE International Conference on Image Processing, Singapore (2004)
14. Tanaka, H.T., Ikeda, M., Chiaki, H.: Curvature-based face surface recognition using spherical correlation - principal directions for curved object recognition. In: IEEE International Conference on Automatic Face and Gesture Recognition, Nara (1998)
15. Huang, D., Ardabilian, M., Wang, Y., Chen, Y.: 3-D face recognition using eLBP-based facial description and local feature hybrid matching. IEEE Trans. Inf. Forensics Secur. **7**(5), 1551–1565 (2012)
16. Queirolo, C.C., Silva, L., Bellon, O.R.P., Segundo, M.P.: 3D face recognition using simulated annealing and the surface interpenetration measure. IEEE Trans. Pattern Anal. Mach. Intell. **32**(2), 206–219 (2010)
17. Moreno, A. B., Sanchez, A., Velez, J.F., Diaz, F.J.: Face recognition using 3D surface-extracted descriptors. In: International Conference on Irish Machine Vision and Image Processing, Coleraine (2003)
18. Xu, C., Tan, T., Li, S., Wang, Y., Zhong, C.: Learning effective intrinsic features to boost 3D-based face recognition. In: Leonardis, A., Bischof, H., Pinz, A. (eds.) ECCV 2006. LNCS, vol. 3952, pp. 416–427. Springer, Heidelberg (2006)
19. Quan, W., Matuszewski, B.J., Shark, L.-K.: 3-D shape matching for face analysis and recognition. In: International Conference on Pattern Recognition Applications and Methods, Lisbon (2015)

20. Lu, X., Jain, A.K., Colbry, D.: Matching 2.5D face scans to 3D models. IEEE Trans. Pattern Anal. Mach. Intell. **28**(1), 31–43 (2006)

21. Quan, W., Matuszewski, B.J., Shark, L.-K.: Facial asymmetry analysis based on 3-D dynamic scans. In: IEEE International Conference on System, Man and Cybernetics. Seoul (2012)

22. Bouttier, J., Francesco, P.D., Guitter, E.: Geodesic distance in planar graphs. Nucl. Phys. **663**(3), 535–567 (2003)

23. Bronstein, A.M., Bronstein, M.M., Kimmel, R.: Three-dimensional face recognition. Int. J. Comput. Vision **64**(1), 5–30 (2005)

24. Bookstein, F.L.: Principal warps: thin-plate splines and decomposition of deformations. IEEE Trans. Pattern Anal. Mach. Intell. **11**(6), 567–585 (1989)

25. Belhumeur, P.N., Hespanha, J.P., Kriegman, D.J.: Eigenfaes vs. fisherfaces: recognition using class specific linear projection. IEEE Trans. Pattern Anal. Mach. Intell. **19**(7), 711–720 (1997)

26. He, X., Yan, S., Hu, Y., Niyogi, P., Zhang, H.-J.: Face recognition using laplacianfaces. IEEE Trans. Pattern Anal. Mach. Intell. **27**(3), 328–340 (2005)

27. Quan, W., Matuszewski, B.J., Shark, L.-K.: Facial expression biometrics using statistical shape models. EURASIP J. Adv. Signal Process. **2009**(1), 1–17 (2009)

28. Yin, L., Wei. X., Sun, Y., Wang, J., Rosato, M.J.: A 3D facial expression database for facial behavior research. In: IEEE International Conference on Automatic Face and Gesture Recognition, Dublin (2006)

29. Moreno, A.B., Sanchez, A.: GavabDB: a 3D face database. In: COST Workshop on Biometrics on the Internet: Fundamentals, Advances and Applications, Nara (2004)

30. Rizvi, S.A., Philips, P.J., Moon H.: The FERET verification testing protocol for face recognition algorithms. In: IEEE International Conference on Automatic Face and Gesture Recognition, Nara (1998)

31. Drira, H., Amor, B.B., Mohamed, D., Srivastava, A.: Pose and expression-invariant 3-D face recognition using elastic radial curves. In: British Machine Vision Conference, Aberystwyth (2010)

32. Li, X., Jia, T., Zhang, H.: Expression-insensitive 3D face recognition using sparse representation. In: Conference on Computer Vision and Pattern Recognition, Kyoto (2009)

33. Mahoor, M.H., Abdel-Mottaleb, M.: Face recognition based on 3D ridge images obtained from range data. Pattern Recogn. **42**(3), 445–451 (2009)

34. Berretti, S., Del Bimbo, A., Pala, P.: 3D face recognition by modeling the arrangement of concave and convex regions. In: Marchand-Maillet, S., Bruno, E., Nürnberger, A., Detyniecki, M. (eds.) AMR 2006. LNCS, vol. 4398, pp. 108–118. Springer, Heidelberg (2007)

Learning Discriminative Mid-Level Patches for Fast Scene Classification

Angran Lin[✉], Xuhui Jia, and Kwok Ping Chan

Department of Computer Science,
The University of Hong Kong, Hong Kong, China
{arlin,xhjia,kpchan}@cs.hku.hk
http://www.cs.hku.hk

Abstract. Discriminative mid-level patch based approaches have become increasingly popular in the past few years. The reason of their popularity can be attributed to the fact that discriminative patches have the ability to accumulate low level features to form high level descriptors for objects and images. Unfortunately, state-of-the-art algorithms to discover those patches heavily rely on SVM related techniques, which consume a lot of computation resources in training. To overcome this shortage and apply discriminative part based techniques to more complicated computer vision problems with larger datasets, we proposed a fast, simple yet powerful way to mine part classifiers automatically with only class labels provided. Our experiments showed that our method, the Fast Exemplar Clustering, is 20 times faster than the commonly used SVM based methods while at the same time attaining competitive accuracy on scene classification.

Keywords: Discriminative mid-level patches · Fast scene classification · Fast exemplar clustering

1 Introduction

Scene classification is not an easy task due to the various visual appearances of different scenes and the complexity in their compositions. Recently, new approaches using discriminatively trained part classifiers are applied to this problem and achieved better performance than conventional methods [1,2]. This is not surprising since part classifiers have the ability to accumulate low level features to generate high level descriptors for each image, which carry information of the visual elements that appear frequently to better describe our real world.

Scene classification is not the only computer vision topic that benefits from part based models. As a matter of fact, in the last few years part based models have been applied to topics like object detection [3,4], motion detection [5] and video classification [6,7]. The reason why part based methods become so popular can be attributed to two reasons. Firstly, they focus on a key problem in computer vision. The relationship between discriminative patches and images can be described as an analogy to the relationship between words and articles.

© Springer International Publishing Switzerland 2015
A. Fred et al. (Eds.): ICPRAM 2015, LNCS 9493, pp. 213–228, 2015.
DOI: 10.1007/978-3-319-27677-9_14

The Bag of Words (BoW) models, including Locality-constrained Linear Coding (LLC) BoW [8] or Improved Fisher Vectors (IFV) [9] succeeded in answering this question to some extent, but the idea of training part classifiers that are visually discriminative may have pushed us one step further. In particular, [10,11] have shown the benefits of using desired patches as visual words [12]. Secondly, most techniques in the framework are shared among different computer vision tasks, which suggests a great potential for this technique.

Given all these advantages, there are still several issues remain unsolved. The most important one is the computation consumption in the training stage. Most discriminative mid-level discovery algorithm rely on a max-margin framework which uses variants of SVMs like exemplar SVM [1] and miSVM [3]. To achieve broad coverage and better purity, thousands of training rounds are required. Moreover, in each round the classifiers/detectors are learned in an iterative manner. Thus, the complexity of using a standard procedure that involves hard negative mining for a huge amount of classifiers would be surprisingly high. It leaves us a major challenge: a simple, efficient and effective method is yet to be found.

In this paper, we proposed a fast algorithm to discover discriminative mid-level patches. It is named Fast Exemplar Clustering (FEC), which works extremely fast while at the same time, attaining competitive accuracy. As a comparison, the MIT 67 indoor scene classification problem in Sect. 5 spent only one day in training on an ordinary Core-i5 computer, while the commonly used methods today would take several days on a cluster [13].

The vastly improved efficiency of FEC method benefits from two factors. The first one is that FEC only requires spatial information of feature vectors and classifiers are trained using their distance measure rather than iteratively solving a time consuming optimization problem. The second one is that FEC uses only local information instead of global information to train classifiers. When the number of patches increases, the training time of SVM based methods for each round may increase sharply while for FEC the time consumption will rise slowly in an $O(logN)$ manner with the help of data structures like R-tree.

The biggest challenge of FEC is the risk of over-fitting. However, we managed to solve it by using a properly designed evaluation function described in Sect. 3.3 together with a large validation set. Our experiments showed that the patches discovered by FEC were both discriminative and representative. In summary, the contributions of this paper are:

1. A novel algorithm for efficiently and effectively detecting discriminative image parts is developed, which demonstrated promising performance in the task of part-based scene classification. Besides, our approach can be seamlessly integrated into bag of visual words models to improve the results of many computer vision problems.
2. A rich training dataset for outdoor scene detection and classification (Outdoor Sight 20) is built. To our best knowledge, this is the first dataset designed for discovering meaningful mid-level patches of outdoor scenes with good in-class consistency. Our dataset consists of images covering 20 famous tourist attractions around the world.

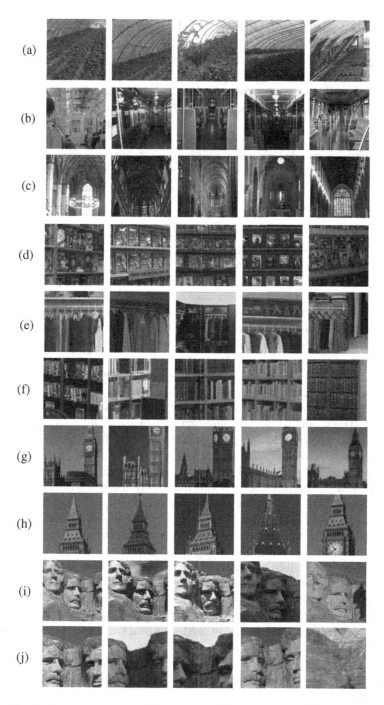

Fig. 1. Visual elements extracted from classes (a) greenhouse (b) inside subway (c) church inside (d) video store (e) closet (f) library of MIT Indoor 67 dataset and (g) (h) Big Ben (i) (j) Mount Rushmore of our Outdoor Sight 20 dataset.

In the experiments, we evaluated our novel FEC method on the public benchmark: MIT Indoor 67 dataset, and the newly created Outdoor Sight 20 dataset, achieving extremely efficient performance (about 20x faster) while maintaining close to state-of-the-art accuracy.

Some of our results are shown in Fig. 1. (a)–(f) are discriminative visual elements extracted from MIT Indoor 67, while (g)–(j) come from our Outdoor Sight 20. As shown in the figure, our method not only captures discriminative and representative visual elements from training data with only class labels provided, but also discovers and distinguishes different visual elements of the same concept, like (g) and (h), which is naturally capable of recognizing different scenes.

2   Related Work

The practice of using parts to represent images has been adopted for quite a long time [14]. Since parts are considered more semantically meaningful compared to some low level features, the introduction of image descriptor generated by algorithms like ScSPM [15], LLC [8] and IFV [9] presented the promising future of parts. The idea of training classifiers discriminatively improved the performance of object detection [11]. However, the discovery of parts are still heavily relied on the training data. Some used the bounding box information on which several assumptions between the parts and the ground truth were based [16], while others relied on partial correspondence [17] to generate meaningful patches.

It was not until recent years that the issue of discovering discriminative mid-level patches automatically with little or no supervision was raised. Patch discovery using geometric information showed that such method has the ability to learn and extract semantically meaningful visual elements for image classification [10,18,19]. Unsupervised learning of patches which are frequent and discriminative in an iterative manner boosted the performance of object detection [13]. [1] summarized a simple and general framework to mine discriminative patches using exemplar SVM [20] and showed that this framework was efficient in scene classification in combination with the use of bag-of-parts and bag-of-visual-words models.

Recent works on discriminative mid-level patches can be categorized into two groups. One is to apply this method to other computer vision problems like video representation [6], 2D-3D alignment [21], movement prediction [22,23] or learning image attributes [19,24]. The other is to collect Internet images to enrich the visual database of discriminative mid-level patches [3,25]. In these works the most widely used types of classifiers are mainly variants of SVM. They can achieve satisfactory accuracy but the huge time consumption really becomes a factor that must be considered if we want to apply this technique to large scale computer vision problems [26,27].

3   Discovering Discriminative Patches: Designed for Speed

Since our purpose is to speed up the training procedure of the model, we designed it to run very fast from the very beginning. We followed the idea that discriminative patches would be learned and discovered in a framework which had three stages: seeding, expansion and selection [1]. Generally speaking, to discover discriminative patches and the corresponding classifiers that to be able to recognize them, we first need to get a bunch of seed patches from the given images. Since the number of patches is enormous, a selection procedure is carried out. They will then be used to train classifiers using our FEC method. Subsequently the classifiers will be ranked using an evaluation function to test whether they are discriminative and representative enough. Those who have top rankings will be kept and used to represent images in the way described in Sect. 4.

3.1 Patch Selection and Feature Extraction

Commonly used ways in patch selection can be divided into two categories. One is to include all possible patches in an image or randomly select some [13], the other is to use some techniques like saliency detection [3] or superpixels [1] to reasonably remove the patches that are unlikely to contain meaningful information to reduce the problem scale and speed up the training procedure. Patch selection is an essential and indispensable part for a method which aims to run very fast as the training time can be reduced significantly with little impact on the results.

In our method, we introduced a very light-weight way by detecting the number of edges in a patch. The rationale is that we believe the most important feature that human uses to identify different objects and scenes is shape. Edge detection is able to discover the shapes of objects in patches while the number of edges inside a patch somehow suggests the importance of the patch. Intuitively, a patch containing few edges may be a part of the background which lacks discriminativeness, while a patch containing a lot of edges may involve too much details which lacks representativeness. As a result, to ensure our patch selection procedure are able to choose patches that are meaningful, we shall select those with neither too many edges nor too few edges. Figure 2 shows how this works. (a) presents the initial training image from MIT Indoor 67 and its edges detected using Canny method [28]. (b) and (c) are some patches with too few or too many edges. (d) shows the patches with modest number of edges, which contain only one or two objects and their spatial relationship. Even though edge detection is rather simple, it is very efficient and effective to find the patches that we need.

In our experiment, we selected patches with sizes of 80 * 80, 120 * 120, 180 * 180, 270 * 270. To avoid duplicates, very similar patches with close feature vectors (i.e. the city-block distance is smaller than a threshold $\delta = 0.01$) from the same image were removed. Then the percentage of the area covered by edges in each patch was calculated and a number of patches with medium number of edges among all the patches were kept for each image. We used the HOG feature [29] to represent each patch.

3.2 Classifier Training

The training procedure of the classifiers is the most time consuming section in discriminative part based techniques. Traditional approaches use SVM variants like exemplar SVM [20] and miSVM [30]. For example, Juneja introduced the exemplar SVM and the outcome was satisfactory in terms of classification results [1]. In each round, one patch from a certain class is treated as the positive input, while all patches from other classes [10] are used as negative inputs. After the SVM is trained, it is used to find the top best patches whose scores are highest among the current class. These patches are added into the positive input and the SVM is trained iteratively for several times. It is undeniable that these methods are able to mine discriminative part classifiers eventually. However, the total number of trained SVMs during the training procedure can reach millions and will take lots of time.

We managed to solve this problem by using an efficient type of classifier instead. We call it fast exemplar clustering (FEC). It follows the idea that each patch will be given a chance to see whether it is able to become a cluster [1].

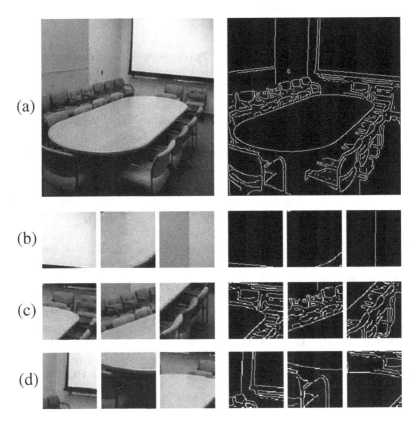

Fig. 2. Images and their 'canny' edges, (a) original image (b) patches with few edges (c) patches with too many edges (d) patches with modest number of edges.

Each cluster will then be tested to see if it is discriminative and representative among all the clusters.

The training procedure is shown in Fig. 3(b) and Algorithm 1. Each exemplar cluster will be trained only twice. For a specific patch, it is treated to be a cluster center at first. Then the 10 closest patches whose class labels are the same as the initial patch will be added to the cluster. The cluster center is recalculated using the mean value of these points, followed by adding the next 10 closest and non-duplicate patches with the same class label into the cluster. Each cluster C_i is represented by a clustering center P_i and a radius r_i which is equal to the largest distance between the cluster center and the patches inside the cluster. A classifier can then be built from the resultant cluster. The center and the

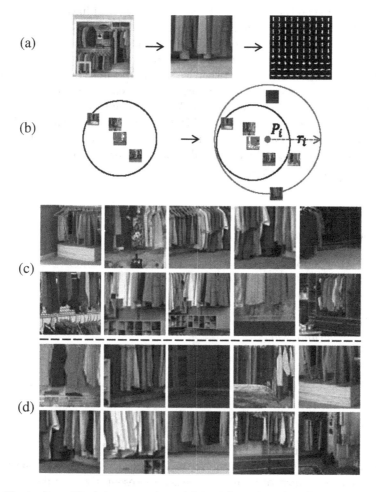

Fig. 3. Illustration of training procedure: (a) initial patch and its HOG representation (b) illustration of cluster expansion using FEC (c) example of patches added in first round of training (d) example of patches added in second round of training.

radius form an Euclidean ball which naturally divide the feature space into two parts, the inner part of the cluster and the outer part. The purpose of training is to transform the initial patch which is specific and particular to a visual concept which is generalized and meaningful.

Since we use the distance measure of feature vectors to form a cluster, the biggest challenge is the risk of over-fitting. The reason why we train only twice in clustering is that we want the clusters to be both **generalized** and **diverse** at the same time to help get rid of over-fitting. **Generalization** means that the cluster can represent not only the initial patch itself but also the patches that are visually similar to the initial patch. Generalization ensures that the patch chosen is representative and common. We want the clusters to be **diverse** since we still do not know which cluster can really represent a discriminative visual concept. If we are able to keep the diversity of the clusters, we will have more chances of obtaining the best classifier when ranking and filtering them in Sect. 3.3.

We did several experiments to decide the optimal number of training rounds, in which two results are really revealing. In one experiment we clustered until the center converges, while in the other we simply did not cluster at all, i.e., we used the initial patch as the center directly with fixed radius for all clusters. It turned out both of them worked poorly. We looked into the results and found that the first way resulted in a lot of identical classifiers which lack diversity, while the latter way resulted in serious over-fitting since one classifier is built merely on one data point. Good generalization and broad coverage are the key to find high quality classifiers.

Algorithm 1. Build exemplar cluster from a patch.

---

    function BUILDCLUSTER($patch$)
        $cluster \leftarrow [\ ]$
        for $i = 1 \rightarrow 2$ **do**
            $euclidean(patch, patchesInSameClass)$
            $add\ 10\ closestpatches \rightarrow cluster$
            $patch \leftarrow mean(cluster)$
        end for
        $center \leftarrow mean(cluster)$
        $radius \leftarrow max(euclidean(center, cluster))$
        return $< center, radius >$
    end function

3.3 Classifier Selection

Though we have obtained a bunch of classifiers $C = \{C_i\}$ centered at $P = \{P_i\}$ with radius of $r = \{r_i\}$ in the training procedure, the number of classifiers is still enormous and most of them are neither representative nor discriminative. To test whether a classifier C_i is good enough, we try to find all the patches inside the Euclidean ball centered at P_i with radius r_i, and compare the class labels of these patches with the class label of the classifier. Denote n_i to be the number of patches inside the ball and p_i to be the number of patches inside the ball with the same class label as the classifier's. Then the accuracy of each classifier is p_i/n_i.

However, if we use accuracy as the only evaluation criteria, it is very likely that the classifiers will only recognize features from very few images. It may lead to the absence of representativeness. To overcome this, we count the number of true positive patches that each image contributed and calculate the variance σ_i^2 of these numbers. A smaller σ_i^2 indicates that the true positive patches come from more training images, which suggests that the classifier is more representative than other classifiers with higher σ_i^2 values. The scoring function is then formulated as

$$F(C_i) = \frac{p_i}{n_i} log(\frac{M}{\sigma_i^2 + N} + 1). \tag{1}$$

M, N are scaling constants to normalize the contribution of the two parts. The argument M, N are calculated by $\arg\min_{M,N} \sum_{\forall j, C_j \in C} (\frac{(p_i)_j}{(n_i)_j} - log(\frac{M}{(\sigma_i^2)_j + N} + 1))^2$. Actually according to our experiment results, the actual value of M, N doesn't have much impact on the results as long as it roughly balances the two parts.

Figure 4 shows the best classifier selected using different evaluation criteria. (a) shows the result of evaluating with accuracy only. The five nearest patches come from 3 different images. Even though they are visually consistent, they did not reveal the nature that really makes 'computer room' different from other classes. (b) shows the top classifier evaluated using our evaluation function. The five nearest patches come from 5 different images. The resultant classifier is more representative.

In addition to evaluating the classifiers on the training set, we introduced a large validation set to be used in the same fashion described above. A number of classifiers with top rankings will be chosen as discriminative classifiers. Figure 1 shows the results.

Fig. 4. Evaluation comparison of classifier trained on class 'computer room': (a) evaluate using only accuracy (b) evaluate using function (1) in Sect. 3.3.

4   Image Representation and Scene Classification

Since it is very hard to judge whether a patch classifier is good or not, we need to test our classifiers using a traditional computer vision task. In our experiment we introduced scene classification to compare our results with others to show that patch classifiers discovered in our method are both meaningful and useful.

For the task of scene classification, we need to first represent each image as a vector. We followed the idea of 'bag-of-parts' (BoP) [1] and used the discriminative classifiers learned in Sect. 3 to generate the mid-level descriptor for each image in a spatial pyramid manner [31] using 1×1 and 2×2 grids. In practice, patches are extracted using a sliding window and each patch together with its flipped mirror is evaluated using the part classifiers. As a result, each image is represented by a $5\,mn$ dimensional vector, in which m represents the number of classifiers kept for each class in Sect. 3.3 and n is the total number of classes.

Scene classification accuracy can be further improved if BoP representation is used in combinition with Bag of Words (BoW) models like Locality-constrained Linear Coding (LLC) BoW [8] or Improved Fisher Vectors (IFV) [9]. However, to make sure our comparison is on an even base, we presented our results using only the BoP representation. We tested the union representations though in Sect. 5 as a reference.

One-vs-rest classifiers are trained to classify the scenes. Linear SVM is used for BoP representation and linear encoding. For the IFV encoding, Hellinger kernel is used.

5   Experiments and Results

The framework of FEC is simple and runs extremely fast. It is not surprising that people will question the effectiveness and correctness of these classifiers and the corresponding image descriptor generated in Sect. 4. In order to test the classifiers we obtained, we focused on the task of scene classification using two datasets. One is the MIT Indoor 67 dataset [32], the other is the Outdoor Sight 20 that we created.

MIT Indoor 67 consists of 5 main scene categories, including store, home, public places, leisure and working place. Each category contains several specific classes, making a total of 67 classes. This dataset is quite challenging thus widely used in scene classification problems.

Outdoor Sight 20 is a dataset we created which consists of outdoor views of 20 famous tourist attractions around the world such as Big Ben, The Eiffel Tower and The Great Wall of China. To test the ability of distinguishing different scenes, a 21st class which contains images of non-tourist attractions is introduced. Part of the sample images are shown in Fig. 5. We built this dataset since we wanted to test our models on both indoor and outdoor scenes. As a complementary of the MIT Indoor 67 dataset, it is specifically designed to include only outdoor images, most of which are photos taken from different angles with various lighting conditions while some are sketches or drawings. Among all the

Fig. 5. Sample images of Outdoor Sight 20 dataset of classes (a) Big Ben (b) Buckingham Palace (c) Mount Rushmore (d) Notre Dame (e) Parthenon (f) St. Paul's Cathedral (g) St. Peter's Basilica (h) Sydney Opera House (i) The Eiffel Tower (j) The Great Wall of China. The rest are: The Brandenburg Gate, The Colosseum, The Golden Gate Bridge, The Kremlin, The Leaning Tower of Pisa, The Pyramids of Giza, The Statue of Liberty, The Taj Mahal, The White House, Tower Bridge with an additional class of none attraction images.

images, the majority have a good within-class consistency since they are portrayals of the same object while some are even difficult for human to classify due to a lot of shared characteristics, like (g) and (f) of Fig. 5.

In our experiment on MIT Indoor 67 dataset, we draw 100 random images from each class. They are partitioned into training set containing 80 images and test set from the remaining 20 images. The training set is further split equally into two parts to be used as training part and validation part, each with 40 images. 50 classifiers for each class are kept to recognize the visual words (Fig. 6).

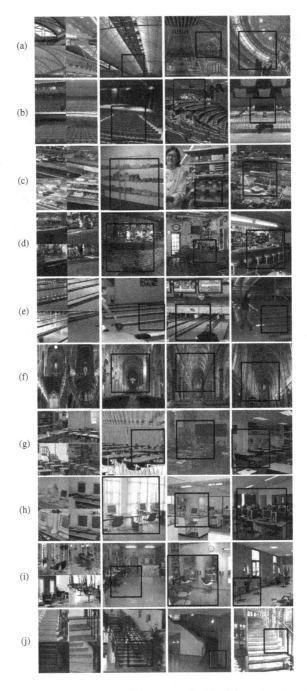

Fig. 6. Classifiers trained on classes: (a) airport inside (b) auditorium (c) bakery (d) bar (e) bowling (f) church inside (g) classroom (h) computer room (i) hair salon (j) staircase of MIT Indoor 67 dataset. The left four patches of each part show how this classifier is trained and the three images on the right show their detections on the testing image.

To test the discriminatively trained mid-level patches, we compared our results (FEC + BoP) with ROI [32], MM-scene [33], DPM [34], CENTRIST [35], Object Bank [36], RBoW [37], Patches [13], Hybrid-Parts [38], LPR [39], exemplar SVM + BoP [1] and IVC [3]. The results are shown in Table 1. Even though our method did not achieve the highest accuracy, it should be clarified that we did not mean to produce best scene classification result. We presented these numbers to show that the patches we obtained in the way described in Sect. 3 are indeed meaningful and could be used as discriminative classifiers in various computer vision problems.

Table 1. Test results on MIT Indoor 67 dataset.

Method	Accuracy (%)
ROI	26.05
MM-scene	28.00
DPM	30.04
CENTRIST	36.90
Object Bank	37.60
RBoW	37.93
Patches	38.10
Hybrid-Parts	39.80
LPR	44.84
IVC (miSVM)	47.60
Exemplar SVM + BoP	46.10
FEC + BoP (Ours)	40.30

We compared the training time required to obtain discriminative mid-level patches with exemplar SVM [20] and ours. On an ordinary Quad-core i5-3570 computer with 16 GB RAM installed using Matlab 2013b, the exemplar SVM took around 3 weeks to train while ours took only 1 day (20x faster). This is an impressive result as the accuracy did not show an enormous drop compared to the exemplar SVM + BoP method.

As is mentioned in Sect. 4, the accuracy can be further improved if BoP representation is used in combination with BoW features. In our experiment,

Table 2. Test results on Outdoor Sight 20 dataset. Comparison between accuracy and training time for part classifier is presented.

Method	Acc. (%)	Time ($\approx$)
Exemplar SVM + BoP	85.75	5 days
FEC + BoP (Ours)	79.25	7 h

the FEC + BoP + LLC and FEC + BoP + IFV achieved the accuracy of 49.55 % and 53.81 % respectively using parameters suggested in [40].

For the Outdoor Sight 20 dataset, we followed the exact same procedure as MIT Indoor 67 dataset on the same computers with the same number of images used in training, testing and validation for each class. We compared our results with exemplar SVM + BoP [1] in Table 2 to show that our FEC could train discriminative mid-level patches as well as the exemplar SVM with much less time.

6   Conclusion

In this paper a novel approach to learn discriminative mid-level patches from training data with only class labels provided is presented. The motivation is that current discriminative patch learning methods are too time-consuming and can hardly be applied to complicated computer vision problems with large dataset. To begin with, we trained part classifiers using the FEC algorithm. Under proper validation settings and appropriately designed evaluation function, we obtained classifiers whose accuracy could compete with state-of-the-art SVM based classifiers. We tested our classifiers on scene classification using MIT Indoor 67 and our Outdoor Sight 20. Both results revealed they were as good as classifiers generated by the contemporary methods. Our classifiers could be further applied to other computer vision problems like scene classification, video classification, object detection, 2D-3D matching.

Acknowledgement. This work is supported by the Hong Kong RGC General Research Fund GRF HKU/710412E.

References

1. Juneja, M., Vedaldi, A., Jawahar, C., Zisserman, A.: Blocks that shout: distinctive parts for scene classification. In: 2013 IEEE Conference on CVPR, pp. 923–930. IEEE (2013)
2. Sun, J., Ponce, J.: Learning discriminative part detectors for image classification and cosegmentation. In: 2013 IEEE International Conference on Computer Vision (ICCV), pp. 3400–3407. IEEE (2013)
3. Li, Q., Wu, J., Tu, Z.: Harvesting mid-level visual concepts from large-scale internet images. In: 2013 IEEE Conference on CVPR, pp. 851–858. IEEE (2013)
4. Rios-Cabrera, R., Tuytelaars, T.: Discriminatively trained templates for 3D object detection: a real time scalable approach. In: 2013 IEEE International Conference on Computer Vision (ICCV), pp. 2048–2055. IEEE (2013)
5. Wang, L., Qiao, Y., Tang, X.: Motionlets: mid-level 3D parts for human motion recognition. In: 2013 IEEE Conference on Computer Vision and Pattern Recognition (CVPR), pp. 2674–2681. IEEE (2013)
6. Jain, A., Gupta, A., Rodriguez, M., Davis, L.S.: Representing videos using mid-level discriminative patches. In: 2013 IEEE Conference on CVPR, pp. 2571–2578. IEEE (2013)

7. Tang, K., Sukthankar, R., Yagnik, J., Fei-Fei, L.: Discriminative segment annotation in weakly labeled video. In: 2013 IEEE Conference on Computer Vision and Pattern Recognition (CVPR), pp. 2483–2490. IEEE (2013)

8. Shabou, A., LeBorgne, H.: Locality-constrained and spatially regularized coding for scene categorization. In: 2012 IEEE Conference on CVPR, pp. 3618–3625. IEEE (2012)

9. Perronnin, F., Liu, Y., Sánchez, J., Poirier, H.: Large-scale image retrieval with compressed fisher vectors. In: 2010 IEEE Conference on CVPR, pp. 3384–3391. IEEE (2010)

10. Doersch, C., Singh, S., Gupta, A., Sivic, J., Efros, A.A.: What makes paris look like paris? ACM Trans. Graph. (TOG) **31**, 101 (2012)

11. Felzenszwalb, P.F., Girshick, R.B., McAllester, D., Ramanan, D.: Object detection with discriminatively trained part-based models. IEEE Trans. Pattern Anal. Mach. Intell. (PAMI) **32**, 1627–1645 (2010)

12. Mittelman, R., Lee, H., Kuipers, B., Savarese, S.: Weakly supervised learning of mid-level features with beta-bernoulli process restricted boltzmann machines. In: IEEE Conference on Computer Vision and Pattern Recognition, pp. 476–483 (2013)

13. Singh, S., Gupta, A., Efros, A.A.: Unsupervised discovery of mid-level discriminative patches. In: Fitzgibbon, A., Lazebnik, S., Perona, P., Sato, Y., Schmid, C. (eds.) ECCV 2012, Part II. LNCS, vol. 7573, pp. 73–86. Springer, Heidelberg (2012)

14. Agarwal, S., Awan, A., Roth, D.: Learning to detect objects in images via a sparse, part-based representation. IEEE Trans. Pattern Anal. Mach. Intell. (PAMI) **26**, 1475–1490 (2004)

15. Yang, J., Yu, K., Gong, Y., Huang, T.: Linear spatial pyramid matching using sparse coding for image classification. In: 2009 IEEE Conference on CVPR, pp. 1794–1801. IEEE (2009)

16. Felzenszwalb, P., McAllester, D., Ramanan, D.: A discriminatively trained, multi-scale, deformable part model. In: 2008 IEEE Conference on CVPR, pp. 1–8. IEEE (2008)

17. Maji, S., Shakhnarovich, G.: Part discovery from partial correspondence. In: 2013 IEEE Conference on CVPR, pp. 931–938. IEEE (2013)

18. Shen, L., Wang, S., Sun, G., Jiang, S., Huang, Q.: Multi-level discriminative dictionary learning towards hierarchical visual categorization. In: 2013 IEEE Conference on Computer Vision and Pattern Recognition (CVPR), pp. 383–390. IEEE (2013)

19. Lee, Y.J., Efros, A.A., Hebert, M.: Style-aware mid-level representation for discovering visual connections in space and time. In: 2013 IEEE International Conference on ICCV, pp. 1857–1864. IEEE (2013)

20. Malisiewicz, T., Gupta, A., Efros, A. A.: Ensemble of exemplar-svms for object detection and beyond. In: 2011 IEEE International Conference on ECCV, pp. 89–96. IEEE (2011)

21. Aubry, M., Maturana, D., Efros, A. A., Russell, B.C., Sivic, J.: Seeing 3D chairs: exemplar part-based 2D–3D alignment using a large dataset of cad models. In: 2014 IEEE Conference on CVPR. IEEE (2014)

22. Walker, J., Gupta, A., Hebert, M.: Patch to the future: unsupervised visual prediction. In: 2014 IEEE Conference on CVPR. IEEE (2014)

23. Lim, J. J., Zitnick, C. L., Dollár, P.: Sketch tokens: a learned mid-level representation for contour and object detection. In: 2013 IEEE Conference on Computer Vision and Pattern Recognition (CVPR), pp. 3158–3165. IEEE (2013)

24. Sandeep, R.N., Verma, Y., Jawahar, C.: Relative parts: distinctive parts for learning relative attributes. In: 2014 IEEE Conference on CVPR. IEEE (2014)

25. Chen, X., Shrivastava, A., Gupta, A.: Neil: extracting visual knowledge from web data. In: 2013 IEEE International Conference on ICCV, pp. 1409–1416. IEEE (2013)

26. Jia, X., Zhu, X., Lin, A., Chan, K.P.: Face alignment using structured random regressors combined with statistical shape model fitting. In: 28th International Conference on Image and Vision Computing New Zealand, IVCNZ 2013, Wellington, New Zealand, 27–29 November 2013, pp. 424–429 (2013)

27. Jia, X., Yang, H., Lin, A., Chan, K.P., Patras, I.: Structured semi-supervised forest for facial landmarks localization with face mask reasoning (2014)

28. Canny, J.: A computational approach to edge detection. IEEE Trans. Pattern Anal. Mach. Intell. **8**, 679–698 (1986)

29. Dalal, N., Triggs, B.: Histograms of oriented gradients for human detection. In: 2005 IEEE Conference on CVPR, vol. 1, pp. 886–893. IEEE (2005)

30. Andrews, S., Tsochantaridis, I., Hofmann, T.: Support vector machines for multiple-instance learning. In: NIPS, pp. 561–568 (2002)

31. Lazebnik, S., Schmid, C., Ponce, J.: Beyond bags of features: spatial pyramid matching for recognizing natural scene categories. In: 2006 IEEE Conference on CVPR, vol. 2, pp. 2169–2178. IEEE (2006)

32. Quattoni, A., Torralba, A.: Recognizing indoor scenes. In: 2009 IEEE Conference on CVPR. IEEE (2009)

33. Zhu, J., Li, L.J., Fei-Fei, L., Xing, E.P.: Large margin learning of upstream scene understanding models. In: Advances in Neural Information Processing Systems, pp. 2586–2594 (2010)

34. Pandey, M., Lazebnik, S.: Scene recognition and weakly supervised object localization with deformable part-based models. In: 2011 IEEE International Conference on ECCV, pp. 1307–1314. IEEE (2011)

35. Wu, J., Rehg, J.M.: Centrist: a visual descriptor for scene categorization. IEEE Trans. Pattern Anal. Mach. Intell. (PAMI) **33**, 1489–1501 (2011)

36. Li, L.J., Su, H., Fei-Fei, L., Xing, E.P.: Object bank: a high-level image representation for scene classification and semantic feature sparsification. In: Advances in Neural Information Processing Systems, pp. 1378–1386 (2010)

37. Parizi, S.N., Oberlin, J.G., Felzenszwalb, P.F.: Reconfigurable models for scene recognition. In: 2012 IEEE Conference on CVPR, pp. 2775–2782. IEEE (2012)

38. Zheng, Y., Jiang, Y.-G., Xue, X.: Learning hybrid part filters for scene recognition. In: Fitzgibbon, A., Lazebnik, S., Perona, P., Sato, Y., Schmid, C. (eds.) ECCV 2012, Part V. LNCS, vol. 7576, pp. 172–185. Springer, Heidelberg (2012)

39. Sadeghi, F., Tappen, M.F.: Latent pyramidal regions for recognizing scenes. In: Fitzgibbon, A., Lazebnik, S., Perona, P., Sato, Y., Schmid, C. (eds.) ECCV 2012, Part V. LNCS, vol. 7576, pp. 228–241. Springer, Heidelberg (2012)

40. Chatfield, K., Lempitsky, V.S., Vedaldi, A., Zisserman, A.: The devil is in the details: an evaluation of recent feature encoding methods, pp. 1–12 (2011)

Modification of Polyp Size and Shape from Two Endoscope Images Using RBF Neural Network

Yuji Iwahori[1]([✉]), Seiya Tsuda[1], Robert J. Woodham[2],
M.K. Bhuyan[3], and Kunio Kasugai[4]

[1] Faculty of Engineering, Chubu University,
Matsumoto-cho 1200, Kasugai, Aichi 487-8501, Japan
{iwahori,tuda}@cvl.cs.chubu.ac.jp
[2] Department of Computer Science,
University of British Columbia, Vancouver, B.C. V6T 1Z4, Canada
woodham@gmail.com
[3] Department of Electronics and Electrical Engineering,
Indian Institute of Technology Guwahati, Guwahati 781039, India
mkb@iitg.ernet.in
[4] Department of Gastroenterology, Aichi Medical University,
Nagakute-cho, Aichi-gun, Aichi 480-1195, Japan
kuku3487@aichi-med-u.ac.jp
http://www.cvl.cs.chubu.ac.jp/

Abstract. In the medical imaging applications, endoscope image is used to observe the human body. The VBW (Vogel-Breuß-Weickert) model is proposed as a method to recover 3-D shape under point light source illumination and perspective projection. However, the VBW model recovers relative, not absolute, shape. Here, shape modification is introduced to recover the exact shape. Modification is applied to the output of the VBW model. First, a local brightest point is used to estimate the reflectance parameter from two images obtained with movement of the endoscope camera in depth. A Lambertian sphere image is generated using the estimated reflectance parameter and VBW model is applied for a sphere. Then Radial Basis Function Neural Network (RBF-NN) learning is applied. The NN implements the shape modification. NN input is the gradient parameters produced by the VBW model for the generated sphere. NN output is the true gradient parameters for the true values of the generated sphere. Here, regression analysis is introduced in comparison with the performance by NN. Depth can then be recovered using the modified gradient parameters. Performance of shape modification by NN and regression analysis was evaluated via computer simulation and real experiment. The result suggests that NN gives better performance than the regression analysis to improve the absolute size and shape of polyp.

Keywords: Endoscope image · VBW model · RBF-NN · Regression analysis · Shape modification · Reflection factor

© Springer International Publishing Switzerland 2015
A. Fred et al. (Eds.): ICPRAM 2015, LNCS 9493, pp. 229–246, 2015.
DOI: 10.1007/978-3-319-27677-9_15

1   Introduction

Endoscopy allows medical practitioners to observe the interior of hollow organs and other body cavities in a minimally invasive way. Sometimes, diagnosis requires assessment of the 3-D shape of the observed tissue. For example, the pathological condition of a polyp often is related to its geometrical shape. Medicine is an important area of application of computer vision technology. Specialized endoscopes with a laser light beam head [1] or with two cameras mounted in the head [2] have been developed. Many approaches are based on stereo vision [3]. However, the size of the endoscope becomes large and this imposes a burden on the patient. Here, we consider a general purpose endoscope, of the sort still most widely used in medical practice.

Shape recovery from endoscope images is considered. Shape from shading (SFS) [4] and Fast Marching Method (FMM) [5] based SFS approach [6] are proposed. These approaches assume orthographic projection. An extension of FMM to perspective projection is proposed in [7]. Further extension of FMM to both point light source illumination and perspective projection is proposed in [8]. Recent extensions include generating a Lambertian image from the original multiple color images [9,10]. Application of FMM includes solution [11] under oblique illumination using neural network learning [12]. Most of the previous approaches treat the reflectance parameter as a known constant. The problem is that it is impossible to estimate the reflectance parameter from only one image. Further, it is also difficult to apply point light source based photometric stereo [13] in the context of endoscopy.

Iwahori et al. [14] developed Radial Basis Function Neural Network (RBF-NN) photometric stereo, exploiting the fact that an RBF-NN is a powerful technique for multi-dimensional non-parametric functional approximation.

Recently, the Vogel-Breuß-Weickert (VBW) model [15], based on solving the Hamilton-Jacobi equation, has been proposed to recover shape from an image taken under the conditions of point light source illumination and perspective projecction. However, the result recovered by the VBW model is relative. VBW gives smaller values for surface gradient and height distribution compared to the true values. That is, it is not possible to apply the VBW model directly to obtain exact shape and size.

This paper proposes a new approach to improve the accuracy of polyp shape determination as absolute size. The proposed approach estimates the reflectance parameter from two images with small camera movement in the depth direction. A Lambertian sphere model is synthesized using the estimated reflectance parameter. The VBW model is applied to the synthesized sphere and shape then is recovered. An RBF-NN is used to improve the accuracy of the recovered shape, where the input to the NN is the surface gradient parameters obtained with the VBW model and the output is the corresponding, corrected true values. Regression analysis is another candidate for the modification of surface gradients.

Performance of shape modification by NN and regression analysis was evaluated via computer simulation and real experiment. NN gives better performance than the regression analysis to improve the absolute size and shape of polyp.

2   VBW Model

The VBW model [15] is proposed as a method to calculate depth (distance from the viewer) under point light source illumination and perspective projection. The method solves the Hamilton-Jacobi equations [16] associated with the models of Faugeras and Prados [17,18]. Lambertian reflectance is assumed.

The following processing is applied to each point of the image. First, the initial value for the depth $Z_{default}$ is given using Eq. (1) as in [19].

$$Z_{default} = -0.5 \log(I f^2) \qquad (1)$$

where I represents the normalized image intensity and f is the focal length of the lens.

Next, the combination of gradient parameters which gives the minimum gradient is selected from the difference of depths for neighboring points. The depth, Z, is calculated from Eq. (2) and the process is repeated until the Z values converge. Here, (x, y) are the image coordinates, Δt is the change in time, (m, n) is the minimum gradient for (x, y) directions, and $Q = \frac{f}{\sqrt{x^2+y^2+f^2}}$ is the coefficient of the perspective projection.

$$Z(x, y) = Z(x, y) + \Delta t \exp(-2Z(x, y))$$

$$-\Delta t \left(\frac{If^2}{Q} \sqrt{f^2(m(x)^2+n(y)^2)+(xm(x)+yn(y))^2+Q^2} \right) \qquad (2)$$

Here, it is noted that the shape obtained with the VBW model is given in a relative scale, not an absolute one. The obtained result gives smaller values for surface gradients than the actual gradient values.

3   Proposed Approach

3.1   Estimating Reflectance Parameter

When uniform Lambertian reflectance and point light source are assumed, image intensity depends on the dot product of surface normal vector and the light source direction vector subject to the inverse square law for illuminance.

Measured intensity at each surface point is determined by Eq. (3).

$$E = C \frac{(\mathbf{s} \cdot \mathbf{n})}{r^2} \qquad (3)$$

where E is image intensity, $\mathbf{s}$ is a unit vector towards the point light source, $\mathbf{n}$ is a unit surface normal vector, and r is the distance between the light source and surface point.

The proposed approach estimates the value of the reflectance parameter, C, using two images acquired with a small camera movement in the depth direction. It is assumed that C is constant for all points on the Lambertian surface. Regarding geometry, it is assumed that both the point light source and the optical center of lens are co-located at the origin of the (X, Y, Z) world coordinate system. Perspective projection is assumed.

The actual endoscope image has the color textures and specular reflectance. Using the approach proposed by [20] the original input endoscope image is converted into one that satisfies the assumptions of a uniform Lambertian gray scale image.

The procedure to estimate C is as follows.

Step 1. If the value of C is given, depth Z is uniquely calculated and determined at the point with the local maximum intensity [21]. At this point, the surface normal vector and the light source direction vector are aligned and produce the local maximum intensity for that value of C.

Step 2. For camera movement, ΔZ, in the Z direction, two images are used and the difference in Z, Z_{diff}, at the local maximum intensity points in each image is calculated. Here the camera movement, ΔZ, is assumed to be known.

Step 3. Let $f(C)$ be the error between ΔZ and Z_{diff}. $f(C)$ represents an objective function to be minimized to estimate the correct value of C. That is, the value of C is the one that minimizes $f(C)$ given in Eq. (4).

$$f(C) = (\Delta Z - Z_{diff}(x, y))^2 \qquad (4)$$

3.2 NN Learning for Modification of Surface Gradient

The size and shape recovered by the VBW model are relative. VBW gives smaller values for surface gradient and depth compared to the true values. Here, modification of surface gradient and improvement of the recovered shape are considered. First, the surface gradient at each point is modified by a neural network. Then the depth is modified using the estimated reflectance parameter, C, and the modified surface gradient, $(p, q) = (\frac{\partial Z}{\partial X}, \frac{\partial Z}{\partial Y})$. A Radial Basis Function Neural Network (RBF-NN) [12] is used to learn the modification of the surface gradient obtained by the VBW model.

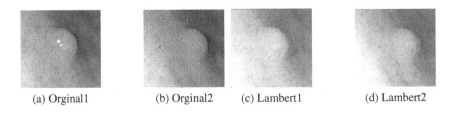

(a) Orginal1   (b) Orginal2   (c) Lambert1   (d) Lambert2

Fig. 1. Endoscope image and Lambertian image.

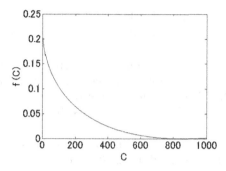

Fig. 2. Objective function $f(C)$.

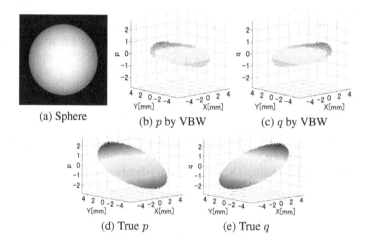

(a) Sphere   (b) p by VBW   (c) q by VBW

(d) True p (e) True q

Fig. 3. Synthesized sphere for NN learning.

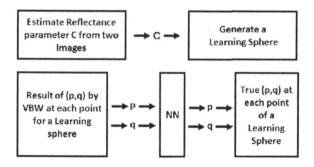

Fig. 4. Learning flow.

Using the estimated C, a sphere image is synthesized with uniform Lambertian reflectance.

The VBW model is applied to this synthesized sphere. Surface gradients, (p, q), are obtained using forward difference of the Z values obtained from the VBW model.

The estimated gradients, (p, q), and the corresponding true gradients for the synthesized sphere, (p, q), are given respectively as input vectors and output vectors to the RBF-NN. NN learning is applied.

After NN learning, the RBF-NN can be used to modify the recovered shape for other images.

Two endoscope images, (a) and (b), and the images assuming Lambertian reflectance, (c) and (d), generated using [20], are shown in Fig. 1.

An example of the objective function, $f(C)$, is shown in Fig. 2.

The synthesized sphere image used in NN learning is shown in Fig. 3(a). Surface gradients obtained by the VBW model are shown in Fig. 3(b) and (c) and the corresponding true gradients for this sphere are shown in Fig. 3(d) and (e). Points are sampled from the sphere as input for NN learning, except for points with large values of (p, q). The procedure for NN learning is shown in Fig. 4.

3.3   NN Generalization and Modification of Z

The trained RBF-NN allows generalization to other test objects. Modification of estimated gradients, (p, q), is applied to the test object and its depth, Z, is calculated and updated using the modified gradients, (p, q).

In the case of endoscope images, preprocessing is used to remove specularities and to generate a uniform Lambertian image based on [20].

Next, the VBW model is applied to the this Lambertian image and the gradients, (p, q), are estimated from the obtained Z distribution.

The estimated gradients, (p, q), are input to the NN and modified estimates of (p, q) are obtained as output from the NN.

Recall that the reflectance parameter, C, is estimated from $f(C)$, based on two images obtained by small movement of endoscope in the Z direction.

The depth, Z, is calculated and updated by Eq. (5) using the modified gradients, (p, q), and the estimated C, where Eq. (5) also is the original equation developed in [8].

$$Z = \sqrt{\frac{CV(-px - qy + f)}{E(p^2 + q^2 + 1)^{\frac{1}{2}}}} \tag{5}$$

Again, $(p, q) = (\frac{\partial Z}{\partial X}, \frac{\partial Z}{\partial Y})$, E represents image intensity, f represents the focal length of the lens and $V = \frac{f^2}{(x^2 + y^2 + f^2)^{\frac{3}{2}}}$.

A flow diagram of the processing described above is shown in Fig. 5.

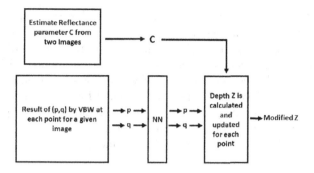

Fig. 5. Flow of NN generalization.

3.4 Regression Analysis for Shape Modification

Here, modification of surface gradient andd improvement of the recovered shape
are considered by another approach for evaluation. Here, the surface gradient
at each point is modified by a regression analysis. The depth is modified using
the estimated reflectance parameter C and modified surface gradient parameters
by regression analysis. The regression analysis estimates the linear coefficients
of regression model and modifies each of the surface gradient parameters for
the recovered result by VBW model. Using the estimated C, a sphere image is
synthesized with uniform Lambertian reflectance. The VBW model is applied
to this synthesized sphere. Ellipsoid is assumed under the consideration to treat
a sphere. The hight z of ellipsoid is differenciated with x and y respectively
and equation of p and q are derived. Using n points on the sphere, p and q is
calculated from the equation. These ideal values are provided as true p and q.

The estimated gradients, (p,q), and the corresponding true gradients for the
synthesized sphere, (p,q), are given respectively as input vectors and output vectors to the regression analysis. Suppose when we assume the 3rd order polynomial
for the relation between the original gradient parameter and the corresponding
true gradient parameter, the regression model becomes Eq. (6) and four coefficients a to d are obtained using linear least squares of Eqs. (7) to (8) with n
samples points on the Lambertian sphere.

$$y = ax^3 + bx^2 + c^x + d \tag{6}$$

$$A = \begin{pmatrix} x_1^3 & x_1^2 & x_1 & 1 \\ x_2^3 & x_2^2 & x_2 & 1 \\ \vdots & \vdots & \vdots & \vdots \\ x_n^3 & x_n^2 & x_n & 1 \end{pmatrix}, \quad x = \begin{pmatrix} a \\ b \\ c \\ d \end{pmatrix}, \quad y = \begin{pmatrix} y_1 \\ y_2 \\ \vdots \\ y_n \end{pmatrix} \tag{7}$$

$$x = (A^T A)^{-1} A^T y \tag{8}$$

Here x is assumed to be each of gradient parameter p or q, respectively and
y is assumed to be each of corresponding modified gradient parameter p or q,

respectively. That is, the mapping of p by VBW to the true p and that of q by VBW to the true q are considered.

Substituting coefficients a to d into Eq. (6) can obtain the modified gradient parameters p or q. Gradient parameters p and q are obtained with numerical difference taken for the recovered result z of VBW. For the real endoscope image, these parameters are used to this regression model for the modification. In the case of endoscope images, preprocessing is used to remove specular points and to generate Lambertian image based on [20]. After modifying p and q respectively. z is updated with Eq. (5) using the modified p and q.

4   Experimental Results

4.1   NN Learning

A sphere was synthesized with radius 5 mm and with center located at $(0, 0, 15)$. The focal length of the lens was 10 mm. The image size was 9 mm $\times$ 9 mm with pixel size 256×256 pixels.

The VBW model was used to recover the shape of this sphere. The resulting gradient estimates, (p, q), are shown in Fig. 3(b) and (c), respectively.

These estimated gradients, (p, q), are used as NN input and the corresponding true gradients, (p, q), output from the NN, are shown in Fig. 3(d) and (e). Learning was done under the conditions: error goal 1.0e-1, spread constant of the radial basis function 0.00001, and maximum number of learning epochs 500. Learning was complete by about 400 epochs with stable status.

The results of learning is shown in Fig. 6.

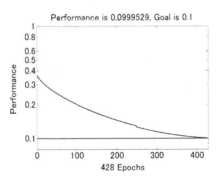

Fig. 6. Learning result.

The reflectance parameter, C, was estimated as 854 from $f(C)$. The difference in depth, Z, was 0.5 [mm] for the known camera movement.

As shown in Fig. 6, NN learning was complete at 428 epochs. The square error goal reached the specified value. Processing time for NN learning was around 30 s.

A sphere has a variety of surface gradients and it is used for the NN learning. After a sphere is used for NN learning, not only a sphere object but also other object with another shape including convex or concave surfaces is also available in the generalization process. This is because surface gradient for each point is modified by NN and this modification does not depend on the shape of target object.

4.2   Order Number in Polynomial with Regression Analysis

Other order number of n in polynomial used in regression analysis is also investigated via simulation. Mean error was evaluated for the height z in each case with substituting the modified (p,q) into Eq. (5). The result by VBW and that after each case of 1st to 5th regression analysis are shown in Table 1, where the number of percentage is the proportion to the radius r = 5 mm of a sphere.

Table 1. Mean error of Z [mm].

VBW	1st-order regression	2nd-order regression	3rd-order regression	4th-order regression	5th-order regression
3.6343	0.3635	0.3635	0.3221	0.3203	0.3203
(72.5 %)	(7.3 %)	(7.3 %)	(6.4 %)	(6.4 %)	(6.4 %)

Table 1 shows every case of regression analysis with different order polynomials gives improvement than the result of VBW. 1st and 2nd order regression gave some difference but 3rd, 4th and 5th order regressions did not give much difference. From these observations, 3rd-order regression analysis is used below for the comparison with NN based modification.

4.3   Computer Simulation

Computer simulation was performed for a second pair of synthesized images to confirm the performance of NN generalization. Synthesized cosine curved surfaces were used, one with center located at coordinates (0, 0, 12) and the other with center at (0, 0, 15). Common to both, the reflectance parameter, C, is 120, the focal length, f, is 10 mm and the waveform cycle is 4 mm and the ± amplitude is 1 mm. Image size is 5 mm × 5 mm and pixel size is 256 × 256 pixels.

The synthesized image whose center is located at (0, 0, 12) is shown in Fig. 7(a) and the one with center located at (0, 0, 15) is shown in Fig. 7(b).

The reflectance parameter, C, was estimated according to the proposed method. Using the learned NN, the gradients, (p, q), obtained from the VBW model were input and generalized. The gradients, (p, q), were modified and the depths, Z, were updated using Eq. (5).

The graph of the objective function, $f(C)$, is shown in Fig. 8 and the true depth is shown in Fig. 9(a). The estimated C was 119 (compared to the true value of 120). The estimagtd Z_{diff} was 2.9953 (compared to the true value of 3).

The result recovered by VBW for Fig. 7(a) is shown in Fig. 9(b). The modified values of depth, using the NN and Eq. (5), are shown in Fig. 9(c) and those by the regression analysis and Eq. (5), are shown in Fig. 9(d).

Table 2 gives the mean errors in surface gradient and depth estimation. The percentages given in the Z column represent the error relative to the amplitude of maximum−minimum depth (=4 mm) of the cosine synthesized function. In Table 2, the original VBW results have a mean error of around 3.8 degrees for the surface gradient while the proposed approach reduced the mean error to about 0.1 degree. Depth estimation also improved to a mean error of 8.3 % from 43.1 %. NN generalization improved estimation of shape for an object with different size and shape. It took 9 s to recover the shape while it took 61 s for NN learning with 428 learning epochs, that is, it took 70 s in total. Althoug regression analysis also improves the depth Z, it is also shown that modified Z by NN gives less mean error in comparison with Z by regression analysis.

Computer simulation was performed for a second pair of synthesized images to confirm the performance of NN generalization. Synthesized cosine curved surfaces were used, one with center located at coordinates (0, 0, 12) and the other with center at (0, 0, 15). Common to both, the reflectance parameter, C, is 120, the focal length, f, is 10 mm and the waveform cycle is 4 mm and the ± amplitude is 1 mm.

Another experiment was performed under the following assumptions. The reflectance factor, C, is 590, the focal length, f, is 10 mm and the object is

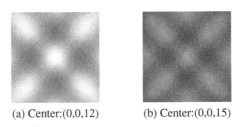

(a) Center:(0,0,12)   (b) Center:(0,0,15)

Fig. 7. Cosine model.

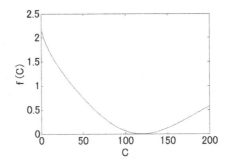

Fig. 8. Objective function $f(C)$.

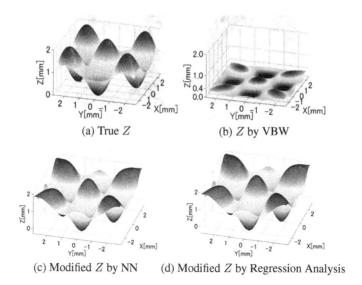

(a) True Z (b) Z by VBW

(c) Modified Z by NN   (d) Modified Z by Regression Analysis

Fig. 9. Results.

Table 2. Mean error.

	p	q	Z[mm]
VBW	23.0406	23.0406	0.8611 (43.1 %)
NN	0.3259	0.3259	0.1695 (8.5 %)
Regression analysis	0.4378	0.4378	0.1823 (9.1 %)

a sphere with radius 5 mm. The centers for two positions of the sphere were set at (0, 0, 15) and (0, 0, 17) respectively, as shown in Fig. 10. The image size was 9 mm × 9 mm with pixel size 360 × 360 pixels. Here, 4 % Gaussian noise (mean 0, variance 0.02, standard deviation 0.14142) is added to each of the two input images. The graph of the objective function, $f(C)$, is shown in Fig. 11 and the true depth is shown in Fig. 12(a). The result recovered by VBW is shown in Fig. 12(b). The improved result is shown in Fig. 12(c). The mean errors in surface gradient and depth are shown in Table 3. Evaluations for 6 % (mean 0, variance 0.03, standard deviation 0.17320) and 10 % (mean 0, variance 0.03, standard deviation 0.17320) Gaussian noise included in Table 4, as well.

Table 3. Mean error.

	p	q	Z[mm]
VBW	17.04	17.45	3.36 (67.2 %)
Proposed	0.45	0.45	0.36 (7.2 %)

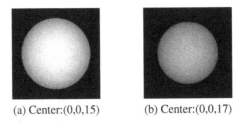

(a) Center:(0,0,15)   (b) Center:(0,0,17)

Fig. 10. Sphere images with Gaussian noise.

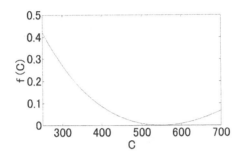

Fig. 11. Objective function $f(C)$.

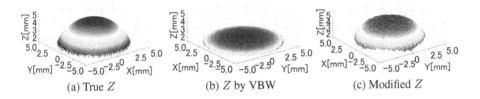

(a) True Z (b) Z by VBW   (c) Modified Z

Fig. 12. Results.

Learning epochs for Gaussian noise 4 %, 6 % and 10 % were 212, 212 and 210. Processing time was around 40 s for every case of different Gaussian noises.

The reflectance parameter, C, estimated from Fig. 11, was 591. Improvement in the estimated results is shown in Fig. 12(a), (b) and (c) and Table 3.

In all three cases, Gaussian noise of 4 %, 6 % and 10 %, the proposed approach reduced the mean error in Z significantly compared to the original VBW model. Table 3 also suggests that modification by NN gives better performancce than that by regression analysis.

This suggests generalization using the RBF-NN is robust to noise and is applicable to real imaging situations, including endoscopy. Result of VBW model gives less errors with noises but this is based on the result that the recovered shape is relative scale and sensitive to the original intensity of each point according to Gaussian noise, while the proposed approach gives much better shape with

Table 4. Mean error of Z for different Gaussian noise.

	4 %	6 %	10 %
VBW	3.2007(64.0 %)	3.2917(65.8 %)	3.3607(67.2 %)
NN	0.4089(8.2 %)	0.4529(9.1 %)	0.5317(10.6 %)
Regression analysis	0.4380(8.8 %)	0.4839(9.7 %)	0.5636(11.3 %)

the absolute size. Although the error increases a little bit according to Gaussian noise, the approach is still robust and stable result is obtained.

4.4   Real Image Experiments

Two endoscope images obtained with camera movement in the Z direction are used in the experiments.

The reflectance parameter, C, was estimated and a RBF-NN was learned using a sphere synthesized with the estimated C. VBW was applied to one of the images which was first converted to a uniform Lambertian image.

Surface gradients, (p, q), were modified with the NN then depth, Z, was calculated and updated at each image point. The focal length, $f = 10$ mm, the image size 5 mm × 5 mm, and camera movement, $\Delta Z = 3$ mm, were assigned to the same known values as those in the computer simulation. The error goal was set to be 0.1.

The two endoscope images are shown in Fig. 13(a) and (b). The generated Lambertian images are shown in Fig. 13(c) and (d), respectively.

The objective function, $f(C)$, is shown in Fig. 14.

The result from the VBW model is shown in Fig. 15(a) and the modified result is shown in Fig. 15(b).

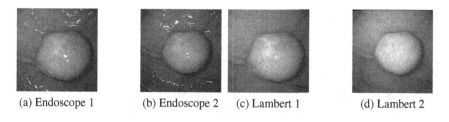

(a) Endoscope 1   (b) Endoscope 2   (c) Lambert 1   (d) Lambert 2

Fig. 13. Endoscope image and generating Lambert image.

The estimated value of the reflectance parameter, C, was 1141. The difference in depth, Z, at the local maximum point was 1 [mm] for the camera movement Z_{diff} between two images. In Fig. 13(c) and (d), specularities were removed compared to Fig. 13(a) and (b). The converted images are gray scale with the appearance of uniform reflectance. Figure 15(b) gives a larger depth range than Fig. 15(a). This suggests depth estimation is improved. The size of the polyp was

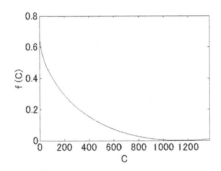

Fig. 14. Objective function $f(C)$.

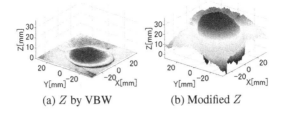

(a) Z by VBW　　　　　(b) Modified Z

Fig. 15. Result for endoscope images.

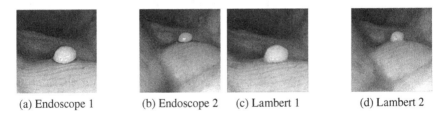

(a) Endoscope 1　　　(b) Endoscope 2　　(c) Lambert 1　　　　(d) Lambert 2

Fig. 16. Endoscope image and generating Lambert image.

1 cm and the processing time for shape modification was 9 s. As it took 117 s for NN learning with 540 epochs, a total processing time was 126 s.

Although quantitative evaluation is difficult, medical doctors with experience in endoscopy qualitatively evaluated the result to confirm its correctness. Different values of the reflectance parameter, C, were estimated in different experimental environments. The absolute size of a polyp is estimated based on the estimated value of C. Accurate values of C lead to accurate estimation of the size of the polyp. The estimated polyp sizes were seen as reasonable by the medical doctor. This qualitatively confirms that the proposed approach is effective in real endoscopy.

Another experiment was done for the endoscope images shown in Fig. 16(a) and (b). The generated gray scale Lambertian images are shown in Fig. 16(c) and (d), respectively. Here the focal length is 10 mm, image size is 5 mm × 5 mm, pixel size is 256×256 and ΔZ was set to be 10 mm.

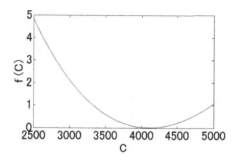

Fig. 17. Objective function $f(C)$.

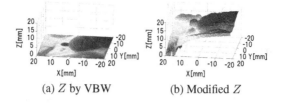

(a) Z by VBW   (b) Modified Z

Fig. 18. Result for endoscope images.

The graph of $f(C)$ is shown in Fig. 17. The VBW result for Fig. 16(d) is shown in Fig. 18(a), while that for the proposed approach is shown in Fig. 18(b).

C was estimated as 4108 from Fig. 17. Figure 18(b) shows greater depth amplitude compared to Fig. 18(a).

The estimated size of the polyp was about 5 mm. This corresponds to the convex and concave shape estimation based on a stain solution. It took 60 s for NN learning with 420 epochs and a total processing time was 70 s.

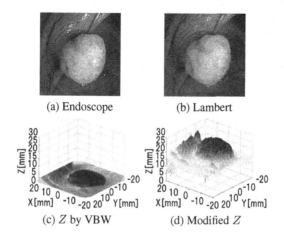

(a) Endoscope   (b) Lambert

(c) Z by VBW   (d) Modified Z

Fig. 19. Result with Gaussian noise.

4 % Gaussian noise was added to a real image and the corresponding results are shown in Fig. 19(a). Shape was estimated from the generated Lambertian image, shown in Fig. 19(b). The corresponding results are shown in Fig. 19(c) and Fig. 19(d). Here the focal length is 10 mm, image size is 5 mm × 5 mm, pixel size is 256 × 256 and ΔZ was 3 mm.

In this paper, it is assumed that movement of the endoscope is constrained to be in the depth, Z, direction only. Here, it is seen that the result is acceptable even when camera movement is in another direction, provided rotation is minimal and the overall camera movement is still small.

The reflectance parameter, C, was estimated as 13244 and Fig. 19(b) shows the final estimated shape. Total of processing time was 90 s including 80 s for NN learning with 480 epochs. The estimated size of this polyp was about 1 cm. Although Gaussian noise was added, shape recovery remained robust.

5   Conclusion

This paper proposed a new approach to improve the accuracy of absolute size and shape determination of polyps observed in endoscope images.

An RBF-NN was used to modify surface gradient estimation based on training with data from a synthesized sphere. The VBW model was used to estimate a baseline shape. However VBW gives the relative shape and cannot recover the shape with absolute size. So this paper proposed modification of gradients with the RBF-NN to recover the size and shape of object. After the gradient parameters are modified, the depth map is updated with the reflectance parameter C. The approach also proposed that this reflectance parameter, C, can be estimated under the assumption that two images are acquired via small camera movement in the depth, Z, direction. The RBF-NN is non-parametric in that no parametric functional form has been assumed for gradient modification. Performance of modification of RBF-NN is evaluated with the regression analysis for the comparison. It is shown that NN gives better modification to the original result of VBW model. The approach was evaluated both in computer simulation and with real endoscope images. Results confirm that the approach improves the accuracy of recovered shape to within error ranges that are practical for polyp analysis in endoscopy.

Acknowledgement. Iwahori's research is supported by Japan Society for the Promotion of Science (JSPS) Grant-in-Aid for Scientific Research (C) (26330210) and Chubu University Grant. Woodham's research is supported by the Natural Sciences and Engineering Research Council (NSERC). The authors would like to thank Kodai Inaba for his experimental help and the related member for useful discussions in this paper.

References

1. Nakatani, H., Abe, K., Miyakawa, A., Terakawa, S.: Three-dimensional measure-men endoscope system with virtual rulers. J. Biomed. Opt. **12**(5), 051803 (2007)
2. Mourgues, F., Devernay, F., Coste-Maniere, E.: 3D reconstruction of the operating field forimage overlay in 3D-endoscopic surgery. In: Proceedings of the IEEE and ACM International Symposium on Augmented Reality (ISAR), pp. 191–192 (2001)
3. Thormaehlen, T., Broszio, H., Meier, P.N.: Three-dimensional endoscopy. In: Falk Symposium, pp. 199–212 (2001)
4. Horn, B.K.P.: Obtaining Shape from Shading Information. MIT Press, Cambridge (1989)
5. Sethian, J.A.: A fast marching level set method for monotonically advancing fronts. Proc. Nat. Acad. Sci. U.S.A. (PNAS U.S.) **93**(4), 1591–1593 (1996)
6. Kimmel, R., Sethian, J.A.: Optimal algorithm for shape from shading and path planning. J. Math. Imaging Vis. (JMIV) **14**(3), 237–244 (2001)
7. Yuen, S.Y., Tsui, Y.Y., Chow, C.K.: A fast marching formulation of perspective shape from shading under frontal illumination. Pattern Recogn. Lett. **28**(7), 806–824 (2007)
8. Iwahori, Y., Iwai, K., Woodham, R.J., Kawanaka, H., Fukui, S., Kasugai, K.: Extending fast marching method un-der point light source illumination and perspective projection. In: ICPR 2010, pp. 1650–1653 (2010)
9. Ding, Y., Iwahori, Y., Nakamura, T., He, L., Woodham, R.J., Itoh, H.: Shape recovery of color textured object using fast marching method via self-caribration. In: EUVIP 2010, pp. 92–96 (2010)
10. Neog, D.R., Iwahori,Y., Bhuyan, M.K., Woodham, R.J., Kasugai, K.: Shape from an endoscope image using extended fast marching method. In: Proceedings of IICAI 2011, pp. 1006–1015 (2011)
11. Iwahori, Y., Shibata, K., Kawanaka, H., Funahashi, K., Woodham, R.J., Adachi, Y.: Shape from SEM image using fast marching method and intensity modification by neural network. In: Tweedale, J.W., Jain, L.C. (eds.) Recent Advances in Knowledge-based Paradigms and Applications. AISC, vol. 234, pp. 73–86. Springer, Heidelberg (2014)
12. Ding, Y., Iwahori, Y., Nakamura, T., Woodham, R.J., He, L., Itoh, H.: Self-calibration and image rendering Using RBF neural network. In: Velásquez, J.D., Ríos, S.A., Howlett, R.J., Jain, L.C. (eds.) KES 2009, Part II. LNCS, vol. 5712, pp. 705–712. Springer, Heidelberg (2009)
13. Iwahori, Y., Sugie, H., Ishii, N.: Reconstructing shape from shading images under point light source illumination. In: ICPR 1990, vol. 1, pp. 83–87 (1990)
14. Iwahori, Y., Woodham, R.J., Ozaki, M., Tanaka, H., Ishii, N.: Neural network based photometric stereo with a nearby rotational moving light source. IEICE Trans. Info. and Syst. **E80–D**(9), 948–957 (1997). ICPR 1990, vol. 1, pp. 83-87 (1990)
15. Vogel, O., Breuß, M., Weickert, J.: A direct numerical approach to perspective shape-from-shading. In: Vision Modeling and Visualization (VMV), pp. 91–100 (2007)
16. Benton, S.H.: The Hamilton-Jacobi Equation: A Global Approach, vol. 131. Academic Press, New York (1977)
17. Prados, E., Faugeras, O.: Unifying approaches and removing unrealistic assumptions in shape from shading: mathematics can help. In: Pajdla, T., Matas, J.G. (eds.) ECCV 2004. LNCS, vol. 3024, pp. 141–154. Springer, Heidelberg (2004)

18. Prados, E., Faugeras, O.: A mathematical and algorithmic study of the Lambertian SFS problem for orthographic and pinhole cameras. Technical report 5005, INRIA 2003 (2003)
19. Prados, E., Faugeras, O.: Shape from shading: a well-posed problem? In: CVPR 2005, pp. 870–877 (2005)
20. Shimasaki, Y., Iwahori, Y., Neog, D.R., Woodham, R.J., Bhuyan, M.K.: Generating lambertian image with uniform reflectance for endoscope image. In: IWAIT 2013, 1C–2 (Computer Vision 1), pp. 60–65 (2013)
21. Tatematsu, K., Iwahori, Y., Nakamura, T., Fukui, S., Woodham, R.J., Kasugai, K.: Shape from endoscope image based on photometric and geometric constraints. KES 2013 Procedia Comput. Sci. **22**, 1285–1293 (2013). Elsevier

Detecting and Dismantling Composite Visualizations in the Scientific Literature

Po-Shen Lee[1]([⊠]) and Bill Howe[2]

[1] Department of Electrical Engineering,
University of Washington, 185 Stevens Way, Seattle, USA
sephon@cs.washington.edu
[2] Department of Computer Science and Engineering,
University of Washington, 185 Stevens Way, Seattle, USA
billhowe@cs.washington.edu
http://www.washington.edu

Abstract. We are analyzing the visualizations in the scientific literature to enhance search services, detect plagiarism, and study bibliometrics. An immediate problem is the ubiquitous use of multi-part figures: single images with multiple embedded sub-visualizations. Such figures account for approximately 35 % of the figures in the scientific literature. Conventional image segmentation techniques and other existing approaches have been shown to be ineffective for parsing visualizations. We propose an algorithm to automatically recognize multi-chart visualizations and segment them into a set of single-chart visualizations, thereby enabling downstream analysis. Our approach first splits an image into fragments based on background color and layout patterns. An SVM-based binary classifier then distinguishes complete charts from auxiliary fragments such as labels, ticks, and legends, achieving an average 98.1 % accuracy. Next, we recursively merge fragments to reconstruct complete visualizations. Finally, a scoring function is used to choose between alternative merge trees. For the multi-chart figure detection, we utilize the output of the splitting algorithm as image features to train a classifier. It can avoid unnecessary time consuming by applying the complete algorithm to determine a multi-chart visualization. To evaluate our approach, we randomly collected 880 single-chart scientific figures and 1067 multi-chart scientific figures from the PubMed database. For the detection, we achieve 90.2 % accuracy via 10-fold cross-validation on the entire corpus. To evaluate the decomposition algorithm, we randomly extracted 261 multi-chart figures as a testing set. Our algorithm achieves 80 % recall and 85 % precision of perfect extractions for the common case of eight or fewer sub-figures per figure. Further, even imperfect extractions are shown to be sufficient for most chart classification and reasoning tasks associated with bibliometrics and academic search applications.

Keywords: Visualization · Multi-chart figure · Chart segmentation · Chart recognition and understanding · Scientific literature retrieval · Content-based image retrieval

© Springer International Publishing Switzerland 2015
A. Fred et al. (Eds.): ICPRAM 2015, LNCS 9493, pp. 247–266, 2015.
DOI: 10.1007/978-3-319-27677-9_16

1   Introduction

The information content of the scientific literature is largely represented visually in the figures — charts, diagrams, tables, photographs, etc. [14]. However, this information remains largely inaccessible and unused for document analysis [1,15] or in academic search portals such as Google Scholar and Microsoft Academic Search. We posit that the structure and content of the visual information in the figures closely relate to its impact, that it can be analyzed to study how different fields of science organize and present data, and, ultimately, that it can be used to improve search and analytics tools [3,16]. All of these applications share requirements around the ability to extract, classify, manage, and reason about the *content* of the figures in the papers rather than just the text alone.

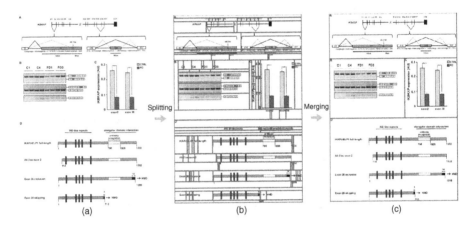

Fig. 1. Dismantling a visualization from the scientific literature. The original source image (a) is first segmented by a splitting method relying on background and layout patterns (b), then the fragments are classified and recursively merged into meaningful visualizations. Image source: (Boone et al., *PLoS ONE, 5.*)

In an initial investigation of these hypotheses, we extracted all figures from a corpus of PubMed papers and developed a classifier to recognize them, building on the work of Savva et al. [11]. We quickly found that about 35 % of all figures were *composite figures* that contained multiple sub-figures, and therefore could not be meaningfully classified directly. Finding it difficult to ignore 35 % of the information before we even began, we decided to tackle the problem of "dismantling" these composite figures automatically; our solution to the dismantling problem is described in this paper.

Figure 1(a) illustrates a simple example of a composite figure. The figure includes a diagram of a molecular sequence (A), a set of photographs of electrophoresis gels (B), an accumulation of a specific type of cells represented as a bar chart (C), and an alternative visualization of molecular sequences (D). Some sub-figures includes additional substructures: Part A breaks the sequences into two zoom-in sections and part D includes four distinct (but related) sub-diagrams. The task to extract the intended sub-figures is hard: The diversity and complexity of the hand-crafted visualizations that appear in the literature resist simple heuristic approaches. (The reader is encouraged to browse some of the real examples in this paper as an illustration of this diversity and complexity). Basic image segmentation techniques are inapplicable; they cannot distinguish between meaningful sub-figures and auxiliary fragments such as labels, annotations, legends, ticks, titles, etc.

Aside from the applications in literature search services, figure classification, and bibliometrics, exposing these multi-part figures for analysis affords new basic research into the role of visualization in science. Consider the following questions:

1. What types of visualization are common in each discipline, and how have these preferences evolved over time?
2. How does the use of visualization in the literature correlate with measures of impact?
3. Do certain disciplines tend to use specialized visualizations more often than others? Does the use of specialized visualizations tend to be rewarded in terms of impact?
4. Can we reverse engineer the data used in the paper automatically by analyzing the text and visualizations of the paper alone?

As a first step toward answering these questions, we present a decomposition algorithm to extract the content of multi-part figures and a detector to determine whether an image is a composite figure or a single figure. The algorithm involves three steps: First, we split a composite figure into small components by reasoning about layout patterns and the empty space between sub-figures (Fig. 1(b)). Second, we merge the split fragments by using an SVM-based classifier to distinguish auxiliary elements such as ticks, labels and legends that can be safely merged into standalone sub-figures that should remain distinct (Fig. 1(c)). Third, we assign a score to alternative initial segmentation strategies and select the higher scoring decomposition as the final output. About the detector, we utilize the output from the splitting step as image features and train an SVM-classifier for the task. To evaluate our approach, we compiled a corpus of 880 multi-chart figures and 1067 single-chart figures chosen randomly from the PubMed databases. For the detector we achieved 90.2 % accuracy via 10-fold cross-validation. For the decomposition algorithm we randomly extracted 261 multi-chart figures as a testing set. We manually decomposed these multi-part figures and found that they were comprised of 1534 individual visualizations. Our algorithm produced 1281 total sub-images, of which 1035 were perfect matches for a manually extracted sub-figure. The remaining 246 incorrect pieces were either multi-chart images that required further subdivision, or meaningless fragments that required further merging. For the 85 % of the images containing eight or fewer sub-figures, we achieved 80.1 % recall and 85.1 % precision of correct sub-images. For the remaining 15 % of densely packed and complex figures, we achieved 42.1 % recall and 68.3 % precision for correct sub-images.

2   Related Work

Content-based image retrieval (CBIR) organizes digital image archives by their visual content [2,9,13], allowing users to retrieve images sharing similar visual elements with query images. This technology has been widely deployed and is available in multiple online applications. However, CBIR has not been used to enhance scientific and technical document retrieval, despite the importance of figures to the information content of a scientific paper. Current academic search systems are based on annotations of titles, authors, abstracts, key words and references, as well as the text content.

Recognition of data visualizations is a different problem than recognition of photographs or drawn images. In early studies, Futrelle et al. presented a diagram- understanding system utilizing graphics constraint grammars to recognize two-dimensional graphs [4]. Later, they proposed a scheme to classify vector graphics in PDF documents via spatial analysis and graphemes [5,12]. N. Yokokura et al. presented a layout-based approach to build a layout network containing possible chart primitives for recognition of bar charts [17]. Y. Zhou et al. used Hough-based techniques [18] and Hidden Markov Models [19] to approach bar chart detection and recognition. W. Huang et al. proposed model-based method to recognize several types of chart images [8]. Later they also introduced optical character recognition and question answering for chart classification [7]. In 2007, V. Prasad et al. applied multiple computer vision techniques including Histogram of Orientation Gradient, Scale Invariant Feature Transform, detection of salient curves etc. as well as Support Vector Machine (SVM) to classified five commonly used charts [10]. In 2011, Savva et al. proposed an interesting application of chart recognition [11]. Their system classifies charts first, extracts data from charts second and then re-designs visualizations to improve graphical perception finally. They achieved above 90 % accuracy in chart classification for ten commonly used charts. These works focused on recognizing and understanding individual chart images. None of these efforts worked with figures in the scientific literature, which are considerably more complex than typical visualizations, and none involved the use of multi-chart images.

3   Decomposition Algorithm

Our algorithm is comprised of three steps: (1) splitting, (2) merging and (3) selecting. In first step, we recursively segment an original images into separate sub-images by analyzing empty space and applying assumptions about layout. In the second step, we use an SVM-based classifier to distinguish complete subfigures from auxiliary fragments (ticks, labels, legends, annotations) or empty regions. In the third step, we compare the results produced by alternative initial segmentation strategies by using a scoring function and select the best choice as the final output.

3.1   Step 1: Splitting

The splitting algorithm recursively decomposes the original figure into sub-images. Authors assemble multiple visualizations together in a single figure to accommodate a limited space budget or to relate multiple visualizations into a coherent argument. We made a few observations about how these figures are assembled that guide the design of our splitting algorithm: First, the layout typically involves a hierarchical rectangular subdivision as opposed to an arbitrarily unstructured collage. Second, authors often include a narrow blank buffer between two sub-figures as a "fire lane" to ensure that the overall layout is readable (Fig. 2(a)). Third, paper-based figures are typically set against a light-colored background. We will discuss figures that violate these assumptions in Sect. 5.

Based on these assumptions, our splitting algorithm recursively locates empty spaces and divide the multi-chart figure into *blocks*. Based on our rectangularity assumption, we locate empty spaces by seeking wholly blank rows or columns as opposed to empty pixels. We first convert the color image into grayscale, and then compute a histogram for rows (and a second histogram for columns) by summing the pixel values of each row (or column). Figure 2(b) gives an example of a figure with its corresponding histogram for the columns. Candidate fire lanes appear as peaks or plateaus in the histogram with a value near the maximum, so we normalize the histogram to its maximum value and apply a high-pass empty threshold θ_e to obtain a candidate set of "blank" rows (or columns). The maximum value does not necessarily indicate a blank row or column, because there may be no entirely blank rows or columns. For example, the green vertical line in Fig. 3(a) is the maximum pixel value sum, but is not blank and is not a good choice as a fire lane. To address this issue, we apply a low-pass variance threshold θ_{var1} to filter such items by their relatively high variances (Fig. 3(b)). We use a second method to detect empty spaces by applying another, stricter

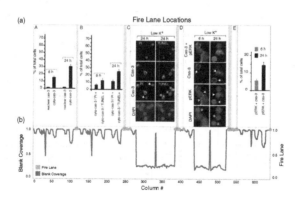

Fig. 2. (a) Fire lanes. We locate the lanes by using the histogram of columns. Orange dots represent qualified columns that pass the thresholds. (b) Histogram of columns. Image source: (Subramaniam et al., *The Journal of cell biology, 165:357-369.*) (Color figure online)

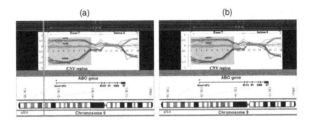

Fig. 3. The identification of fire lanes is non-trivial. (a) Locating fire lanes without applying the variance threshold θ_{var1} leads to an error: since there are no entirely blank columns, the maximum value (highlighted in green) is not a qualified fire lane. (b) The disqualified column is filtered by applying $\theta_{var1} = 100$. Image source: (Hong et al., *BMC Genetics, 13:78*) (Color figure online)

low-pass variance threshold θ_{var2} on rows or columns. The first method provides a wider pass window and the second method is well-suited to handle figures with a dark background.

To set the values of the three thresholds, we collected 90 composite figures (avoiding figures with photographs) and ran the splitting step with different combinations of thresholds against this training set. Since our goal is just to tune these parameters, we make a simplifying assumption that finding the correct number of sub-images implies a perfect split; that is, if the number of divided sub-images equals the correct number of sub-figures determined manually, we assume the division was perfect. The reason for this simplifying assumption is to improve automation for repeated experiments; we did not take the time to manually extract perfect splits for each image with which to compare. Under this analysis, the values for the thresholds that produced the best results were $\theta_e = 0.999$, $\theta_{var1} = 100$, and $\theta_{var2} = 3$.

We group neighboring empty pixel-rows or empty pixel-columns to create empty "fire lanes" as shown in Fig. 2(a). The width of the fire lane is used in the merge step to determine each sub-image's nearest neighbor. Half of each fire lane is assigned to each of the two blocks; each block becomes a new image input to be analyzed recursively. Row-oriented splits and column-oriented splits are alternatively performed, recursively, until no fire lane is found within a block. The recursion occurs at least two times to ensure both orientations are computed at least once.

Different initial splitting orientations can result in different final divisions, so the splitting algorithm is performed twice: once beginning vertically and once beginning horizontally. We individually execute merging for the two results and automatically evaluate the merging results in step 3. The split with higher score is taken as the final decomposition.

3.2   Step 2: Merging

The merging algorithm receives the splitting result as input and then proceeds in two substeps: First, we use an SVM-based classifier to distinguish *standalone*

Table 1. The features used to classify sub-images as either standalone sub-figures or auxiliary fragments. We used $k = 5$ for our experiment; thus the feature vector consists of 15 elements. We achieved classification accuracy of 98.1 %, suggesting that these geometric and whitespace-oriented features well describe the differences between the two categories.

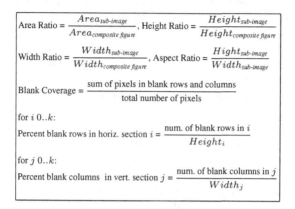

$$\text{Area Ratio} = \frac{Area_{sub\text{-}image}}{Area_{composite\,figure}}, \text{Height Ratio} = \frac{Height_{sub\text{-}image}}{Height_{composite\,figure}}$$

$$\text{Width Ratio} = \frac{Width_{sub\text{-}image}}{Width_{composite\,figure}}, \text{Aspect Ratio} = \frac{Hight_{sub\text{-}image}}{Width_{sub\text{-}image}}$$

$$\text{Blank Coverage} = \frac{\text{sum of pixels in blank rows and columns}}{\text{total number of pixels}}$$

for $i\ 0..k$:

$$\text{Percent blank rows in horiz. section } i = \frac{\text{num. of blank rows in } i}{Height_i}$$

for $j\ 0..k$:

$$\text{Percent blank columns in vert. section } j = \frac{\text{num. of blank columns in } j}{Width_j}$$

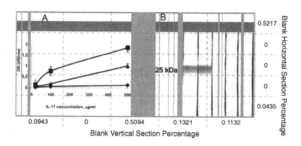

Fig. 4. Blank coverage according to blank rows (red) and blank columns (green). We divided the image into 5 sections horizontally and vertically. In each section, we {computed the percentage of blank-row or blank-column respectively. The 10 vectors form a portion of image feature. Image source: (Kapina et al., *PLoS ONE, 6.*) (Color figure online)

sub-figures representing meaningful visualizations from auxiliary blocks that are only present as annotations for one or more standalone sub-figures. Second, we recursively merge auxiliary blocks, assigning each to its nearest block, until all auxiliary blocks are associated with one (or more) standalone sub-figures. We refer to this process as hierarchical merging. If two neighboring blocks have incongruent edges, a non-convex shape may result. In this case, we perform *T-merging*: we search nearby for sub-figures that can fill the non-convexity in the shape. We will discuss the details of the classifier, Hierarchical Merging, and T-Merging in this section.

Training SVM-Based Binary Classifier. Figure 5 shows an example of an intermediate state while merging, consisting of 18 sub-images from the composite figure. Sub-images labeled (D, F, H, J, N, O, Q, R) are classified as standalone blocks. All others are classified as auxiliary blocks. The goal of the merging algorithm is to remove auxiliary blocks by assigning them to one or more standalone blocks. To recognize auxiliary blocks, we extract a set of features for each block and train an SVM-based classifier. The features selected are based on the assumption that the authors tend to follow implicit rules about balancing image dimensions and distributing empty space within each figure, and that these rules are violated for auxiliary annotations. To describe the dimensions of the block, we compute proportional area, height and width relative to that of the original image, as well as the aspect ratio. To describe the distribution of empty space, we use the same thresholds from the splitting step to locate entirely blank rows or columns and then compute the proportion of the total area covered by the pixels of these blank elements. We do not consider the overall proportion of empty pixels, because many visualizations use an empty background — consider a scatter plot, where the only non-empty pixels are the axes, the labels, the legend, and the glyphs. As a result, blank rows and columns should be penalized, but blank pixels should not necessarily be penalized.

Blank coverage alone does not sufficiently penalize sub-figures that have large blocks of contiguous empty space; a pattern we see frequently in auxiliary sub-images. For example, an auxiliary legend offset in the upper right corner of a figure will have large contiguous blocks of white space below and to the left of it. To describe these cases where empty space tends to be concentrated in particular areas, we divide each sub-image into k equal-size sections via horizontal cuts and another k sections via vertical cuts. We then extract one feature for each horizontal and vertical section; $2k$ features total. Each feature f_i is computed as the proportion of blank rows in section i. To determine a suitable k, we experimented with different values from $k = 0$ to $k = 10$ on the training data and set $k = 5$ based on the results. In this paper, we do not consider further optimizing this parameter. With the combination of the dimensional features and the empty-space features, we obtain a 15-element feature vector for each sub-image. These features are summarized in Table 1.

As an example of how these features manifest in practice, consider Fig. 4. This image has 26.6 % blank coverage; blank columns are colored green and blank rows are colored red. As with most visualizations in the literature, the overall percentage of blank pixels is very high, but the percentage of blank rows and columns is relatively low. We divide the image into horizontal and vertical sections as indicated by the green and red dashed lines. The decimals indicate percentages of blank row or column of their nearby sections. The complete 15-element feature vector of this image is {1, 1, 1, 0.4272, 0.2656, 0.5217, 0, 0, 0, 0.0435, 0.0943, 0, 0.5094, 0.1321, 0.1132}.

To evaluate our classifier, we collected another corpus containing 213 composite figures from the same source for training independence. The splitting algorithm was used to produce 7541 sub-images from the corpus. For evaluation,

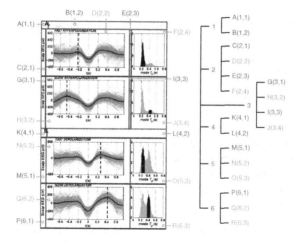

Fig. 5. The tree structure of a decomposition. This multi-chart image was split using column-orientation. The result of the splitting step can form a tree structure. The numbers in parentheses present the sections in each splitting level that the block belongs to. For instance, H(3, 2) refers that block H is in the third section when the original multi-chart figure is split. Then it is the second sub-section when the third section is split again. With the assistance of the classifier, we color standalone blocks by blue and auxiliary blocks by red. Image source: (Botella-Soler et al., *PLoS ONE, 7.*) (Color figure online)

we manually classified a set of 6524 standalone sub-images and 1017 auxiliary sub-images. We used LibSVM [6] and set all parameters as default to train the model. The classification accuracy of 98.1 is achieved.%.

Hierarchical Merging. The result of the splitting step forms a tree structure as shown in Fig. 5. Merging starts from the leaves of the tree. All leaves can only merge with other leaves in the same level. After completing all possible merges among siblings, we transfer the newly merged block to their parents' level as new leaves. Hierarchical merging stops after it finishes merging in the top level.

In each level, we re-run the classifier to determine auxiliary blocks. We then induce a function on the set of blocks, assigning each block to its nearest adjacent neighbor called its *merge target*. A block and its merge target are called a merge pair. Under the assumption that the width of a fire lane indicates the strength of relationship between the two blocks, the merge target for a block is the adjacent neighbor with the narrowest lane between them. Figure 6(a) shows a combination of two blocks. The new block is the smallest rectangle that covers both merging blocks. Only adjacent blocks are allowed to merge together. If there are two or more qualified blocks, we break the tie by the shortest distance between centroids of blocks.

If after merging all auxiliary blocks at one level of the tree we find that the resulting shape is non-rectangular, we attempt to apply T-merging or pass the

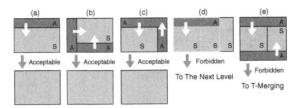

Fig. 6. Examples of hierarchical merging. In all cases, the goal is to merge all auxiliary blocks, (labeled A) into standalone blocks (labeled S). Each merge operation is indicated by a white arrow. (a) An acceptable merge. The new block is the smallest rectangle that covers both merging blocks. (b) Two different merge paths that lead to the same result. (c) Another case of acceptable multi-merging. (d) This merging is forbidden because after merging the auxiliary into the standalone block the resulting shape is non-rectangular. The operation only involves the blocks with yellow outline. Once the local merging in this level is completed, it repeats again in the next level, which will involve the very right standalone block. (e) Another case of forbidden merging because of the same reason. After completing hierarchical merging, the residual auxiliary blocks will be handled by T-Merging (Color figure online).

result on to the next higher level. For example, Fig. 6(b) and (c) can be merged since the result is rectangular. Figure 6(d) and (e) are forbidden. In this case, the merging of the block pair is skipped and the auxiliary block is labeled as standalone and processed using T-Merging, described next. We repeat the local merging until the statuses of all blocks are standalone. We then pass these blocks up to the next level, reclassify them, and repeat the local merging again.

T-Merging. T-Merging handles residual auxiliary blocks ignored in Hierarchical Merging. These are usually shared titles, shared axes or text annotations that apply to multiple sub-figures; e.g., Fig. 6(d) and (e). As shown in Fig. 7, merging the auxiliary block 1 ("the legacy"), to any adjacent standalone block generates a non-rectangular shape. We define block 2 and block 3 as legatees[1] to proportionally share block 1. We find the set of legatees by the following procedure: For each edge e of the legacy, find all blocks that share any part of e and construct a set. If merging this set as a unit produces a rectangle, then it is a qualified set. If multiple edges produce qualified sets, choose the edge with t the narrowest fire lane. In Fig. 7, only the set consisting of block 2 and 3 satisfies the above criteria; blocks 4, 5, 6 are not proper legatees. Figure 8 illustrates the evolution from a source image through Hierarchical Merging to its T-merged output.

3.3   Step 3: Selecting

The splitting and merging steps may produce different results from different initial splitting orientations (Fig. 8). Step 3 scores the two different results and selects the one with higher score as the final output. Under the assumption

---

[1] meaning "those who will receive the legacy".

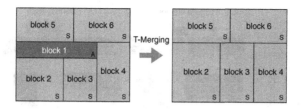

Fig. 7. Examples of T-Merging. The legacy (block 1) is marked by white color in text. According to our algorithm, only block 2 and block 3 are qualified to share block 1.

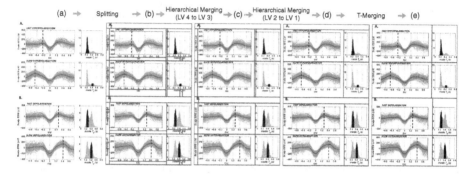

Fig. 8. (a) Composite figure. (b) Splitting result. (c) Intermediate state of hierarchical merging after completing level 3. (d) Hierarchical merging result. The very-top block and the middle block require T-Merging. (e) T-Merging result. Image source: (Botella-Soler et al., *PLoS ONE, 7*.)

that authors tend to follow implicit rules about balancing image dimensions, the decomposition that produces more sub-images with similar dimensions is given a higher score. To capture this intuition, we define the scoring function as

$$S_{decomposition} = 4 \sum_{i \in \text{blocks}} \sqrt{A_i} - 2\alpha \sum_{i,j \in \text{Pairs}} |l_i^{top} - l_j^{top}| + |l_i^{left} - l_j^{left}|,$$

where A_i is the area of the corresponding block, α is a penalty coefficient and l_i^{top} is the length of the top edge of block i (respectively, $left$). Each element of the set $Pairs$ is a pair of blocks i, j, where j is the block that has the most similar dimensions to i for $i \neq j$. The two coefficients normalize the two terms to the full perimeter. The formula enforces a geometric property of composite figures: The first term obtains its maximum value when all blocks are equal in size. The second term subtracts the difference between each block and its most similar neighbor to reward repeating patterns and penalize diversity in the set. The penalty coefficient weights the importance of dimensional difference. We assigned $\alpha = 1$ in our experiment.

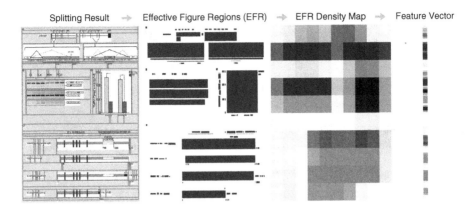

Fig. 9. The pipeline from splitting an image to acquiring it's feature vector. Image source: (Boone et al., *PLoS ONE, 5.*)

4   Composite Figure Detection

To make our algorithm for a broader use that a given image is not pre-labeled as a composite figure, we extended the splitting algorithm to pre-recognize composite figures instead of applying the complete composition algorithm for the task. It can save a large portion of time consuming by the merging algorithm. The output of splitting algorithm can be a feature describing the geometric layout of sub-figures. Figure 9 shows the splitting result where the firelanes obtained from all recursive layers are highlighted by the light red color. It can be regarded as a binary mask. The uncovered area is defined as the effective figure regions (EFR). Next, we subdivide the mask into $N \times N$ blocks and compute the proportion of EFR in each block as shown by the EFR density map. Finally squeeze the values into a 1-D vector with N^2 elements. Furthermore for the features, we average the heights and widths of all training images and calculate the dimension ratio as $(height_n/height_{avg})$ and $(width_n/width_{avg})$ for each image n as simple dimensional features. We concatenate the dimensional features and geometrical features to create the final feature vector. By using the same technique to train the auxiliary classifier, we obtained 90.2 % accuracy from 10-fold cross-validation on the entire corpus comprising 880 composite figures and 1067 single figures.

5   Experimental Evaluation

In this section, we describe experiments designed to answer the following questions: (a) Can our algorithm be used to estimate visualization diversity, a weaker quality metric sufficient for many of our target applications? (Yes; Table 2) (b) Can our algorithm effectively extract correct sub-figures, a stronger quality metric? (Yes; Table 2) (c) Could a simpler method work just as well as our algorithm? (No; Table 2) (d) Is step 3 of the algorithm (selection) necessary and effective? (Yes; Fig. 11).

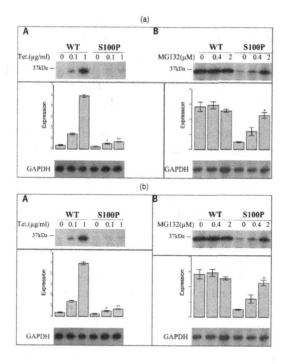

Fig. 10. Results of different initial splitting orientations. (a) Split begins from horizontal (initially row-oriented), and a lower score due to mismatched elements. (b) Split begins from vertical (initially column-oriented), and a higher score. Image source: (Türumen et al. *PLoS Genetics, 5.*)

The corpus we used for our experiments was collected from the PubMed database. We selected a random subset of the PubMed database by collecting all tar.gz files from 188 folders (from //pub/pmc/ee/00 to //pub/pmc/ee/bb); these files contain the pdf files of the papers as well as the source images of the figures, so figure extraction was straightforward. In order to filter non-figure images such as logos, banners, etc., we only used images of size greater than 8 KB. We manually separate the composite figures from the single-chart figures and divided the composite figures into a testing set and a training set. We trained the classifier and performed cross-evaluation with the training set, reserving the test set for a final experimental evaluation. The testing set S for the experiments contains 261 composite figures related to biology, biomedicine, or biochemistry. Each figure contains at least two different types of visualizations; e.g., a line plot and a scatter plot, a photograph and a bar chart, etc. We ignored multi-chart figures comprised of single-type figures in this experiment for the convenience of evaluation, described later in the first question. We evaluated performance in two ways: (1) type-based evaluation, a simpler metric in which we attempt to count the number of distinct types of visualizations within a single figure, and (2) chart-based evaluation, a stronger metric in which we attempt to perfectly recover all sub-figures within a composite figure.

Can Our Algorithm Be Used to Estimate Visualization Diversity? The motivation for type-based evaluation is that some of our target applications in bibliometrics and search services need only know the presence or absence of particular types of visualizations in each figure to afford improved search or to collect aggregate statistics — it is not always required to precisely extract a perfect sub-figure, as long as we can tell what type of figure it is. For example, the presence or absence of an electrophoresis gel image appears to be a strong predictor of whether the paper is in the area of experimental molecular biology; we need not differentiate between a sub-figure with one gel and a sub-figure with several gels. Moreover, it is not always obvious what the correct answer should be when decomposing collections of sub-figures of homogeneous type: Part of Fig. 8(a) contains a number of repeated small multiples of the same type — it is not clear that the correct answer is to subdivide all of these individually. Intuitively, we are assessing the algorithms' ability to eliminate ambiguity about what types of visualizations are being employed by a given figure, since this task is a primitive in many of our target applications.

To perform type-based evaluations we label a test set by manually counting the number of distinct visualization types in each composite figure. For example, Fig. 2 has two types of visualizations, a line chart and a bar chart; Fig. 5 also has two types of visualizations, a line chart and an area chart; Fig. 10(a) also has two types of visualizations, bar charts and electrophoresis gels. We then run the decomposition algorithm and manually distinguish correct extractions from incorrect extractions. Only *homogeneous* sub-images — those containing only one type of visualization — are considered correct. For example, the top block in Fig. 10(a) is considered correct, because both sub-figures are the same type of visualization: an electrophoresis gel image. The bottom two blocks of Fig. 10(a) are considered incorrect, since each contains both a bar chart and a gel.

Using only the homogeneous sub-images (the heterogeneous sub-images are considered incorrect), we manually count the number of distinct visualization types found for each figure. We compare this number with the number of distinct visualization types found by manual inspection of the original figure. For example, in Fig. 10(a), the algorithm produced one homogeneous sub-image (the top portion), so only one visualization type was discovered. However, the original image has two distinct visualization types. So our result for this figure would be 50 %.

To determine the overall accuracy we define a function *diversity* : *Figure* → *Int* as *diversity*(f) = |{*type*(s) | s ∈ *decompose*(f)}|, where *decompose* returns the set of subfigures and *type* classifies each subfigure as a scatterplot, line plot, etc. The final return value is the number of distinct types that appear in the figure. We then sum the diversity scores for all figures in the corpus. We compute this value twice: once using our automatic version of the *decompose* function and once using a manual process. Finally, we divide the total diversity computed automatically by the total diversity computed manually to determine the overall quality metric. The automatic method is not generally capable of finding more types than are present in the figure, so this metric is bounded above by 1. In our experiment, we obtained the diversity score of 591 and 640 respectively from automatic decomposition and manual process. The accuracy by this metric is therefore 92.3 %.

Can Our Algorithm Effectively Extract Correct Sub-figures? For chart-based evaluation, we attempt to perfectly extract the exact subfigures found by manual inspection, and measure precision and recall. For instance, Figs. 2, 5, and 10(b) contain 5, 8, and 6 sub-figures respectively. To obtain ground truth, we manually extracted 1534 visualizations from the entire image set S; about 5.88 visualizations per composite figure on average. In this experiment, a sub-image that includes exactly one visualization is defined as correctly extracted; exceptions are described next. However, a sub-image that crops a portion of a visualization, includes only auxiliary annotations, or includes two or more visualizations is considered incorrect. These criteria are stricter than necessary for many applications of the algorithm; for example, partial visualizations or visualizations with sub-structure will often still be properly recognized by a visualization-type classifier and can therefore be used for analysis. However, this metric provides a reasonable lower bound on quality.

We make an exception to these criteria: We consider an array of photographic images to be one visualization. This exception is to ensure that we do not artificially improve our results: The algorithm is very effective at decomposing arrays of photos, but it is not obvious that these arrays should always be decomposed; the set is often treated as a unit. In this analysis, we also ignore cases where an auxiliary annotation is incorrectly assigned to one owner instead of another. The reason is that we find a number of ambiguous cases where "ownership" of an auxiliary annotation is not well-defined.

Table 2. Chart-based evaluation. Where S_{all} denotes the entire composite figure set, $S_{p\leq 8}$ denotes the subset of composite figures containing eight or fewer sub-figures, and $S_{p>8}$ denotes the subset of composite figures containing nine or more sub-figures. We compared our main approach to a splitting-only method based on our splitting algorithm. The recall and the precision of correct sub-images, as well as the accuracy of decomposition were significantly enhanced. Our technique achieved a better performance for a subset of composite figures that contains eight or fewer sub-figures.

		Recall of correct sub-images	Precision of correct sub-images	Accuracy of perfect decomposition
Splitting-only approach	S_{all}	54.3 % (833/1534)	53.1 % (833/1569)	16.1 % (42/261)
Main approach	S_{all}	67.5 % (1035/1534)	80.8 % (1035/1281)	57.9 % (151/261)
	$S_{p\leq 8}$	80.1 % (811/1002)	85.1 % (811/953)	67.1 % (149/222)
	$S_{p>8}$	42.1 % (224/532)	68.3 % (224/328)	5.1 % (2/39)

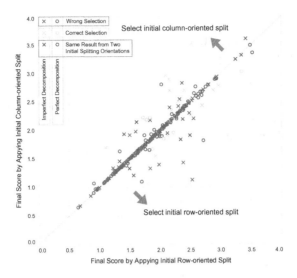

Fig. 11. Step 3 of the algorithm, selection, makes correct decisions. Circles represent perfectly decomposed figures and crosses represent imperfectly decomposed figures. This scatter plot illustrates that figures with perfect decomposition mostly distribute near the line of slope 1, indicating similar solutions were found by our decomposition algorithm regardless of the starting orientation. The selection step deals with the grey plots and the red plot, which are composite figures that have different outputs from the two initial splitting orientations. Only one mistaken selection was made and 95.9 % accuracy was achieved (Color figure online).

We define a notion of recall and precision based on these correctness criteria. To compute recall, the number of correct sub-images returned by the algorithm is divided by the number of correct sub-figures manually counted in the corpus using the same criteria for correctness. To compute precision, we divide the number of correct sub-images returned by the algorithm by the total number of extracted sub-images. Our algorithm achieves recall of 67.5 % and precision of 80.8 %. In addition, the percentage of figures that are perfectly decomposed — the right number of correct images and no incorrect images — is 57.9 %. Table 2 summarizes the chart-based evaluation in more detail. Later in this section we will analyze the mistakes made by the algorithm.

Does a Simpler Method Work Just as Well? For comparison, we measured the performance of our algorithm relative to a simpler split-based algorithm. Here, we modified our splitting step (Sect. 3.1) to make it more viable as a complete algorithm. As presented, our splitting step may produce a large number of auxiliary fragments that need to be merged (e.g., Fig. 8(b)). But a reasonable approach would be to cap the number of recursive steps and see if we could avoid the need to merge altogether. We use two recursive steps — once for vertical and once for horizontal. Also, as a heuristic to try and improve the results, we discarded fire lanes with width less than 4 pixels for the same purpose because most

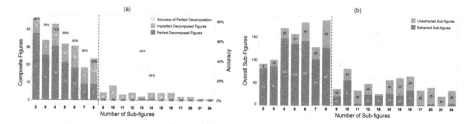

Fig. 12. (a) Histogram of perfectly decomposed and imperfectly decomposed figures. (b) Histogram of extracted sub-figures. Our decomposition algorithm performed better for composite figures with lower number of sub-figures. Entanglement and over-merging are common issues for images of densely packed sub-figures.

lanes between auxiliary fragments or between auxiliary fragments and effective sub-figures are relatively narrow.

Our results show that this splitting-only algorithm extracted 833 correct sub-images and achieved 54.3 % recall and 53.1 % precision. Only 16.1 % of the original composite figures were decomposed perfectly into exact sub-figures without any errors. By both measures, this simpler method performs significantly worse despite optimizations (Table 2).

Is Step 3 of the Algorithm (Selection) Useful and Effective? To evaluate the utility of our selection step, we manually compared the two outputs of different splitting orientations before our algorithm automatically chose one. There are 237 figures that have the same results from the two initial splitting orientations. For the remaining 24 figures that require selecting algorithm, our selection algorithm correctly chose the better output for 23 figures, 11 from initial column-oriented split and 13 from initial row-oriented split. Figure 11 shows an overview of all selection scores as computed by the formula in Sect. 3.3. Each point denotes a composite figure. Circles are figures decomposed perfectly and crosses are figures decomposed imperfectly. Figures with perfect decomposition mostly appear near the line of slope 1, indicating that our decomposition algorithm often finds similar solutions for regardless of the starting orientation. However, for points where one score is different than the other, we conclude that the selection step plays an important role.

Where Does the Algorithm Make Mistakes? To understand the algorithm's performance more deeply, we considered whether the complexity of the initial figure had any effect on the measured performance. Figure 12(a) shows a histogram of composite figures, where each category is a different number of sub-figures. The dark portion of each bar indicates the proportion of composite figures that were perfectly decomposed. The curve, which shows the accuracy of perfect decomposition, decays significantly as the number of sub-figures increases; the algorithm tends to perform significantly better on figures with eight or fewer sub-figures.

Figure 12(b) is a histogram of the total number of sub-figures extracted from each category, regardless of whether or not the entire figure was perfectly decomposed. The black dotted line divides the categories into two subsets. The right subset, comprising composite figures containing nine or more sub-figures, includes only 15.0 % source figures but contributes 61.7 % of the unextracted sub-figures (i.e., the figures the algorithm failed to properly extract). Thus, a relative low recall of 42.1 % was obtained in this subset (Table 2). In the left subset, comprising 222 composite figures with eight or fewer sub-figures, the recall was greatly increased to 80.1 % (Table 2). The two bar charts both show a better performance on composite figures with lower sub-figure populations.

6   Limitations and Future Work

The current decomposition algorithm is suitable for grid-aligned multi-chart visualizations, where there exists at least one edge-to-edge fire lane that can bootstrap the process. Figure 13 shows two examples that do not satisfy this criterion, and for which our algorithm does not produce a result. Our algorithm is also ill-suited for arrays of similar sub-figures for which it is ambiguous and subjective whether or not they should be considered as one coherent unit. We chose to maximally penalize our algorithm by assuming that every individual element should be considered a separate sub-figure.

We are also working to make the classifier used in the merging phase more robust. Although our current binary classifier has achieved 90 % recall to recognize standalone sub-figures, we still receive an exponential decay to perfectly divide figures. For our corpus, 5.88 charts per image on average gave only $1 - 0.9^{5.88} = 46.2\%$ accuracy of perfect decomposition in expectation. The misclassification is mostly on relatively small sub-figures. To improve the binary classifier, we need to consider more features derived from the color information, and from text location (via the use of character recognition algorithms).

For the future work, we are working to use the decomposition algorithm with visualization classification techniques to analyze the use of visualization in the literature by domain, by impact, by year, and by demography. We believe this effort represents a first step toward a new field of viziometrics — the analysis of how visualization techniques are used to convey information in practice.

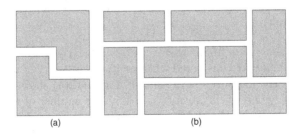

(a)   (b)

Fig. 13. Cases for which the splitting algorithm is not appropriate. (a) Irregular outer bound of sub-figures may form a zigzag fire lane. (b) There is no end-to-end fire lane.

7   Conclusions

We have presented an algorithm that automatically detects composite figures and dismantles them into sub-figures. The decomposition algorithm first splits an image into several blocks via spatial analysis. An SVM-based image classifier with an accuracy of 98.1 % is used to recognize effective visualizations and auxiliary fragments. Next recursively merge fragments to reconstruct complete visualizations. For the detection, we extended the splitting algorithm to obtain geometric features from visualizations and again used the SVM as a classifier. An accuracy of 90.2 % was achieved. For type-based analysis, a weaker metric suitable for understanding the diversity of visualization used in the literature, 78.1 % of the composite figures in our corpus were completely divided into homogeneous, recognizable visualizations. For chart-based analysis, a stronger metric suitable for extracting the original visualizations exactly, we successfully extracted 67.5 % of the sub-images from 261 composite figures. For the 85 % of images in the corpus containing eight or fewer sub-figures, a better performance of 80.1 % recall was achieved. With this technique, we are now poised to unlock the content of multi-part composite figures in the literature and make it available for use in advanced applications such as bibliometrics and academic search.

Acknowledgements. The authors wish to thank the authors of the papers from which we drew the examples in this paper. This work is sponsored in part by the National Science Foundation through S2I2 award 1216879 and IIS award III-1064505, the University of Washington eScience Institute, and an award from the Gordon and Betty Moore Foundation and the Alred P. Sloan Foundation.

References

1. Bergstrom, C.T., West, J.D., Wiseman, M.A.: The Eigenfactor metrics. J. Neurosci. Official J. Soc. Neurosci. **28**, 11433–11434 (2008)
2. Datta, R., Joshi, D., Li, J., Wang, J.Z.: Image retrieval: ideas, influences, and trends of the new age. ACM Comput. Surv. 1–35 (2006)
3. Dean, J., Ghemawat, S.: Mapreduce: simplified data processing on large clusters. Commun. ACM **51**(1), 107–113 (2008). http://doi.acm.org/10.1145/1327452.1327492
4. Futrelle, R., Kakadiaris, I., Alexander, J., Carriero, C., Nikolakis, N., Futrelle, J.: Understanding diagrams in technical documents. Computer **25**, 75–78 (1992)
5. Futrelle, R., Shao, M., Cieslik, C., Grimes, A.: Extraction, layout analysis and classification of diagrams in pdf documents. In: Proceedings of the Seventh International Conference on Document Analysis and Recognition, pp. 1007–1013, August 2003
6. Hsu, C.W., Chang, C.C., Lin, C.J.: A practical guide to support vector classification. Bioinformatics **1**, 1–16 (2010). http://citeseerx.ist.psu.edu/viewdoc/download?doi=10.1.1.6.3096&rep=rep1&type=pdf
7. Huang, W., Tan, C.L.: A system for understanding imaged infographics and its applications. In: DOCENG 2007: Proceedings of the 2007 ACM Symposium on Document Engineering, pp. 9–18 (2007)

8. Huang, W., Tan, C.-L., Leow, W.-K.: Model-Based Chart Image Recognition. In: Lladós, J., Kwon, Y.-B. (eds.) GREC 2003. LNCS, vol. 3088, pp. 87–99. Springer, Heidelberg (2004). http://dx.doi.org/10.1007/978-3-540-25977-0_8

9. Lew, M.S.: Content-based multimedia information retrieval: state of the art and challenges. ACM Trans. Multimedia Comput. Commun. Appl. **2**, 1–19 (2006)

10. Prasad, V., Siddiquie, B., Golbeck, J., Davis, L.: Classifying computer generated charts. In: 2007 International Workshop on Content-Based Multimedia Indexing (2007)

11. Savva, M., Kong, N., Chhajta, A., Fei-Fei, L., Agrawala, M., Heer, J.: ReVision: automated classification, analysis and redesign of chart images. In: UIST 2011, pp. 393–402 (2011)

12. Shao, M., Futrelle, R.P.: Recognition and classification of figures in PDF documents. In: Liu, W., Llads, J. (eds.) GREC 2005. LNCS, vol. 3926, pp. 231–242. Springer, Heidelberg (2006). http://dx.doi.org/10.1007/11767978_21

13. Smeulders, A.W.M., Worring, M., Santini, S., Gupta, A., Jain, R.: Content-based image retrieval at the end of the early years. IEEE Trans. Pattern Anal. Mach. Intell. **22**, 1349–1380 (2000)

14. Tufle, E.: The visual display of quantitative information. CT Graphics, Cheshire (1983). http://www.colorado.edu/UCB/AcademicAffairs/ArtsSciences/geography/oote/maps/assign/reading/TufteCoversheet.pdf

15. West, J.D., Bergstrom, T.C., Bergstrom, C.T.: The eigenfactor metrics: a network approach to assessing scholarly journals. Coll. Res. Libr. **71**, 236–244 (2006)

16. White, T.: Hadoop: The Definitive Guide. O'Reilly Media, Sebastopol (2009)

17. Yokokura, N., Watanabe, T.: Layout-based approach for extracting constructive elements of bar-charts. In: Tombre, K., Chhabra, A. (eds.) Graphics Recognition Algorithms and Systems. LNCS, vol. 1389, pp. 163–174. Springer, Heidelberg (1998). http://dx.doi.org/10.1007/3-540-64381-8_47

18. Zhou, Y.P.Z.Y.P., Tan, C.L.T.C.L.: Hough technique for bar charts detection and recognition in document images. In: Proceedings 2000 International Conference on Image Processing (Cat. No. 00CH37101), vol. 2 (2000)

19. Zhou, Y., Tan, C.L.: Learning-based scientific chart recognition. In: 4th IAPR International Workshop on Graphics Recognition, GREC 2001, pp. 482–492 (2001)

Tensor Deep Stacking Networks and Kernel Deep Convex Networks for Annotating Natural Scene Images

Niharjyoti Sarangi[✉] and C. Chandra Sekhar

Department of Computer Science and Engineering,
Indian Institute of Technology Madras, Chennai, India
niharsarangi@gmail.com, chandra@cse.iitm.ac.in
http://www.cse.iitm.ac.in

Abstract. Image annotation is defined as the task of assigning semantically relevant tags to an image. Features such as color, texture, and shape are used by many machine learning algorithms for the image annotation task. Success of these algorithms is dependent on carefully handcrafted features. Deep learning models use multiple layers of processing to learn abstract, high level representations from raw data. Deep belief networks are the most commonly used deep learning models formed by pre-training the individual Restricted Boltzmann Machines in a layer-wise fashion and then stacking together and training them using error back-propagation. However, the time taken to train a deep learning model is extensive. To reduce the time taken for training, models that try to eliminate back-propagation by using convex optimization and kernel trick to get a closed-form solution for the weights of the connections have been proposed. In this paper we explore two such models, Tensor Deep Stacking Network and Kernel Deep Convex Network, for the task of automatic image annotation. We use a deep convolutional network to extract high level features from different sub-regions of the images, and then use these features as inputs to these models. Performance of the proposed approach is evaluated on benchmark image datasets.

Keywords: Image annotation · Tensor deep stacking networks · Kernel deep convex networks · Deep convolutional network · Deep learning

1 Introduction

Developing techniques for efficient extraction of usable and meaningful information has become increasingly important with the explosive growth of digital technologies. Low level features like color, texture and shape can be used to classify images into different categories. However, in many cases it is not suitable to use a single class label because of the presence of more than one semantic concept in an image. One way to handle this is by assigning multiple relevant keywords to a given image, reflecting its semantic content. This is often referred to as image annotation.

© Springer International Publishing Switzerland 2015
A. Fred et al. (Eds.): ICPRAM 2015, LNCS 9493, pp. 267–281, 2015.
DOI: 10.1007/978-3-319-27677-9_17

Learning techniques such as Binary Relevance [2] and Classifier Chains [21], transform an annotation task into a task of binary classification. Another approach to tackle the problem of annotation is by adapting popular learning techniques to deal with multiple labels directly [23,27]. Multi Label k-Nearest Neighbors (ML-kNN) [26], Multi Label Decision Tree (ML-DT) [24] and Rank-SVM [6] are some of the commonly used methods in this category. Rank-SVM is a ranking based approach coupled with a set size predictor which uses Support Vector Machines to minimize the ranking loss while having a large margin. Among other models, semantic space auto-annotation model [7] constructs a special form of a vector space, called a semantic space, from the labels associated with the images. Images are projected into this space in order to be retrieved or annotated. Latent semantic analysis [11] is used to build this space. The success of these techniques is largely dependent on the effectiveness of the features used.

Learning representations of the data that makes it easier to extract useful information is highly desirable [1] for developing a good classification or annotation framework. Deep learning models are the commonly used techniques for learning representation from raw data. These models aim at learning feature hierarchies with features from higher levels of the hierarchy formed by the composition of lower level features, as illustrated in Fig. 1.

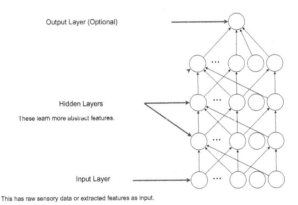

Output Layer (Optional)

Hidden Layers
These learn more abstract features.

Input Layer

This has raw sensory data or extracted features as input.

Fig. 1. Scheme of learning representations in a multilayered network. Raw pixel values or extracted features are given as input. The input layer is followed by multiple hidden layers that learn increasingly abstract representation.

Deep learning models such as Deep Belief Networks (DBN) [10] and Deep Boltzmann Machines (DBM) [19,20,22] have performed well in classification and recognition tasks [15]. These models are formed by pre-training individual layers [8] and then stacking together and training them using error backpropagation. Each layer of a DBN consists of an energy-based model known as Restricted Boltzmann Machine (RBM). An RBM is trained using contrastive divergence to obtain a good reconstruction of the input data [9]. Contrastive divergence and error back-propagation are computationally complex methods.

In Deep Convolutional Networks [14, 16, 17], convolution operation is used to extract features from different sub-regions of an image to learn a better representation. Although Deep Convolutional Networks are trained completely using error back-propagation, they use sub-sampling layers to reduce the number of inputs to each layer. To solve the issue of complexity, a model known as Deep Stacking Network (DSN) [5] that consists of many stacking modules was recently proposed. Each module is a specialized neural network consisting of a single non-linear hidden layer and linear input and output layers. Since convex optimization is used to speedup the learning in each module, this model is also called as Deep Convex Network (DCN) [4]. Tensor Deep Stacking Networks (T-DSN) [13], introduced as an extension of the DSN architecture, captures better representations by using two sets of nonlinear nodes in the hidden layer. The T-DSN model has been shown to perform better than the DSN model for image classification and phone recognition tasks. Kernel Deep Convex Network(K-DCN) [3] on the other hand uses kernel trick so that the number of hidden nodes in each module is unbounded.

In this paper, we propose a framework that uses convex deep learning models (T-DSN and K-DCN) for the task of image annotation. We also propose using the features extracted from a Deep Convolutional Network as input to the convex models. The remainder of this paper is organized as follows: Sect. 2 gives a brief discussion on T-DSN and K-DCN. In Sect. 3 we describe the details of our experiments and compare the results with the existing methods.

2   Convex Deep Learning Models

2.1   Tensor Deep Stacking Networks

A tensor deep stacking network is a generalized form of a deep stacking network. The input data is provided to the nodes in the input layer of the first module. The input to the higher modules is obtained by appending output from the module just below it to the original input data. Unlike DSN, each module of TDSN has two sets of hidden layer nodes and thus, two sets of connections between the input layer and the hidden layer as shown in Fig. 2. The output layer nodes are bilinearly dependent on the hidden layer nodes.

Let the target vectors $\mathbf{t}$ be arranged to form the columns of matrix T, the input data vectors $\mathbf{v}$ be arranged to form the columns of matrix V, and H_1 and H_2 denote the set of matrices of the outputs of the hidden units. There are two sets of lower weight parameters (W_1 and W_2). They are associated with connections from the input layer to the two hidden layers containing L_1 and L_2 sigmoidal nodes respectively.

Since the hidden layers contain sigmoidal nodes, the output of a hidden layer can be expressed as:

$$H_1 = logistic(W_1^T V)$$
$$H_2 = logistic(W_2^T V)$$

$$(1)$$

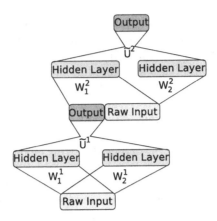

Fig. 2. Architecture of tensor deep stacking network.

Let $\mathbf{h}_1$ be the vector of outputs from the first set of hidden nodes and $\mathbf{h}_2$ be the vector of outputs from the second set of hidden nodes. Let h_{1i} be the ith entry in $\mathbf{h}_1$ and h_{2j} be the jth entry in $\mathbf{h}_2$.

If C is the number of nodes in the output layer, weights of connections from hidden layers to the output layer are represented as a tensor $\mathbf{U} \in \mathbf{R}^{L_1} \times \mathbf{R}^{L_2} \times \mathbf{R}^C$. The tensor $\mathbf{U}$ can be considered as a 3-dimensional matrix.

Let y_k denote the output of kth node in output layer. The output vector can be obtained by computing $(\mathbf{U} \times_1 \mathbf{h}_1) \times_2 \mathbf{h}_2$ where $\times_i$ stands for multiplication along the ith dimension. In a simplified notation

$$y_k = \sum_{i=1}^{L_1} \sum_{j=1}^{L_2} U_{ijk} h_{1i} h_{2j} \tag{2}$$

Let

$$\tilde{\mathbf{h}} = \mathbf{h}_1 \otimes \mathbf{h}_2$$

where $\otimes$ is the Kronecker product. Let $\tilde{\mathbf{u}}_k$ be the vectorized version of matrix U_k in which all columns are appended to form a single vector. The matrix U_k is obtained by setting the third dimension of tensor $\mathbf{U}$ equal to k. Hence, length of $\tilde{\mathbf{u}}_k$ is $L_1 L_2$. Now, we can rewrite Eq. (2) as,

$$y_k = \tilde{\mathbf{u}}_k^T \tilde{\mathbf{h}} \tag{3}$$

Arranging all $\tilde{\mathbf{u}}_k$'s for $k = 1, 2, \ldots, C$, into a matrix $\tilde{U} = [\tilde{\mathbf{u}}_1 \ \tilde{\mathbf{u}}_2 \ \ldots \ \tilde{\mathbf{u}}_C]$, the overall prediction becomes

$$\mathbf{y} = \tilde{U}^T \tilde{\mathbf{h}} \tag{4}$$

where $\mathbf{y}$ is the estimate of target vector $\mathbf{t}$.

Thus, bilinear mapping from two hidden layers can be seen as a linear mapping from an implicit hidden representation $\tilde{\mathbf{h}}$. Aggregating the implicit hidden layer representations for each of the N instances into the columns of an $L_1 L_2 \times N$ matrix $\tilde{H}$, we obtain

$$Y = \tilde{U}^T \tilde{H} \tag{5}$$

where $\tilde{H}$ contains $\mathbf{h}_k$ in k^{th} column.

The convex formulation for $\tilde{U}$ in this case is,

$$min_{\tilde{U}^T}\|\tilde{U}^T H - T\|^2 \qquad (6)$$

where $\|.\|^2$ represents the squared norm operation.

Solving the optimization (6) we get:

$$\tilde{U}^T = T\tilde{H}^T(\tilde{H}\tilde{H}^T)^{-1} \qquad (7)$$

We see that the output of each hidden node in first layer appears L_2 number of times in $\tilde{\mathbf{h}}$. So, we have to add errors due to all those terms in order to get the error caused by this particular node. Hence, the equation for weight update needs to be modified to account for this and the modified equations are:

$$\Delta W_1 = \eta V[H_1^T \circ (\Gamma - H_1^T) \circ \Psi_1] \qquad (8)$$

$$\Delta W_2 = \eta V[H_2^T \circ (\Gamma - H_2^T) \circ \Psi_2] \qquad (9)$$

Here $\circ$ is the element-wise multiplication of two matrices, Γ is a matrix of all ones, η is the learning rate and

$$
\begin{aligned}
\Psi_{1nk} &= \sum_{k=1}^{L_2} H_{2nk}\tilde{\Theta}_{((i-1)L_2+k),n} \\
\Psi_{2nk} &= \sum_{k=1}^{L_1} H_{1nk}\tilde{\Theta}_{((i-1)L_1+k),n}
\end{aligned}
\qquad (10)
$$

$$\tilde{\Theta} = 2\tilde{H}^+(\tilde{H}T^T)(T\tilde{H}^+) - 2T^T(T\tilde{H}^+) \qquad (11)$$

Here H_1 is the matrix of outputs of nodes in the first hidden layer, H_2 is the matrix of outputs of nodes in the second hidden layer. The dimensions of matrices Ψ_1 and Ψ_2 are $N \times L_1$ and $N \times L_2$ respectively. Each of these two matrices Ψ_1 and Ψ_2 acts as a bridge between high dimensional implicit representation $\tilde{\mathbf{h}}$ and low dimensional representations $\mathbf{u}$ and $\mathbf{v}$.

Since T-DSN uses convex optimization techniques to directly determine the upper-layer weights, the training time is greatly reduced. However, computing the lower-layer weights is still an iterative process.

2.2   Kernel Deep Convex Networks

A kernel deep convex network (K-DCN), like a T-DSN, is composed by stacking of shallow neural network modules. This model completely eliminates the non-convex learning for the lower-layer weights using the kernel trick. In case of K-DCN, a regularization term C is included in the expression for computing the upper-layer weights U. This modification helps bound the values of elements of U and prevents the model from over-fitting on the training data.

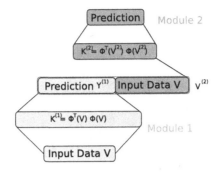

Fig. 3. Architecture of kernel deep convex network with two modules.

The formulation for U takes the form of,

$$min_U[\frac{1}{2} * Tr\{(Y - T)^T(Y - T)\} + \frac{C}{2}U^TU]$$

where Y is the predicted output for the output nodes and T is the target output. The closed form expression for U is obtained by solving this minimization as follows :

$$U = (CI + HH^T)^{-1}HT^T \tag{12}$$

The output of the given module of KDCN is given by,

$$\mathbf{y}_k = TH^T(CI + HH^T)^{-1}\mathbf{h}_i \tag{13}$$

The sigmoidal function of hidden units is replaced with a generic nonlinear mapping function $\Phi(\mathbf{v})$ from the raw input features $\mathbf{v}$. The mapping $\Phi(\mathbf{v})$ will have high-dimensionality (possibly infinite) which is determined implicitly by a chosen kernel function. The unconstrained optimization problem can be reformulated as follows:

$$min_U[\frac{1}{2} * Tr\{(Y - T)^T(Y - T)\} + \frac{C}{2}U^TU]$$

subject to

$$T - U^TG(V) = Y - T$$

where columns of $G(V)$ are formed by applying the transformation $\Phi(.)$ on each input $\mathbf{v}$. Solving this problem gives

$$U = G(V)(CI + K)^{-1}T^T \tag{14}$$

where $K = G^T(V)G(V)$ is the kernel gram matrix of V.

Finally, for each new input vector $\mathbf{v}$ in the test set, the prediction of KDCN module is given by

$$\mathbf{y}(\mathbf{v}) = U^T\Phi(\mathbf{v}) = T(CI + K)^{-1}\mathbf{k}^T(\mathbf{v}) \tag{15}$$

Here $\mathbf{k}(\mathbf{v})$ is the kernel vector such that $k_n(\mathbf{v}) = k(\mathbf{v}_n, \mathbf{v})$ and $\mathbf{v}_n$ is a vector from training set.

For the subsequent modules, the output of nodes in the output layer is appended with the raw input. For l^{th} module $(l > 2)$ Eq. (14) is valid with a slight modification in the kernel function to account for this extra input as follows:

$$K = G^T(Z)G(Z) \tag{16}$$

where $Z = V|Y^{(l-1)}|Y^{(l-2)}|....|Y^1$, Y^m is the prediction of module m, and $U|V$ represents the concatenation of U and V.

Using the Eqs. (15) and (16) we eliminate the need of back-propagation and get a convex expression for training the model. The KDCN model combines the power of deep learning and kernel learning in a principled way. It is fast because there is no back-propagation.

2.3   Framework for Image Annotation

If a concept is present in an image, the corresponding bit in a binary target output vector $\mathbf{t}$ is turned on. Each module of a T-DSN is trained to predict $\mathbf{t}$. Once the module is trained and the weights W_1, W_2, and U are learned, Eq. (4) is used to compute the estimated output. For the higher modules, the input data is concatenated with the output of the module below it (or with the output of n modules below it) and used as an augmented input. This process is repeated for all the modules and the output obtained at the last module is retained. Similarly, in case of a K-DCN, Eq. (15) is used to find predictions for each module.

One of the following methods to obtain the annotation labels from the outputs of a model is used.

1. A threshold value is decided empirically using a held-out validation set. In the estimated output vectors, if the posterior probability value for a particular concept exceeds the threshold, it is considered as an annotation label for the image.
2. Based on the average number of labels present in the images, a value k is selected. An image is annotated with those concepts that correspond to the top k values in the estimated output vector.

3   Experiments and Results

In this section, we present the details of image annotation datasets used and the experimental results for T-DSN and K-DCN. We compare the performance of these models with the state-of-the-art performance.

3.1   Experimental Setup

We used MATLAB on an Intel i7 8-core CPU with 16 GB of RAM for running the Rank-SVM. For T-DSN and K-DCN, we used NVIDIA Tesla K20C GPU with CUDA.

In order to reduce the number of multiplications in the computation of $\tilde{\Theta}$, Eq. (11) is re-written as:

$$\tilde{\Theta} = 2(\tilde{H}^+ \tilde{H} T^T - T^T)(T\tilde{H}^+)$$
$$= 2(\tilde{H}^+ \tilde{H} T^T - T^T)\tilde{U}^+ \qquad (17)$$

In order to reduce the memory requirements for the computation of $\tilde{\Theta}$, Eq. (17) is parenthesized as follows:

$$\tilde{\Theta} = 2(\tilde{H}^+(\tilde{H} T^T) - T^T)\tilde{U}^+ \qquad (18)$$

In this order of multiplication, we avoid computing $\tilde{H}^+ \tilde{H}$, which is a $N \times N$ matrix. In general, the value of N is large (20,000–50,000). Accommodating such a large matrix in the GPU memory is problematic. Many matrices are reused in the process of training. Matrices are allocated memory only when required and freed immediately after their use in order to make the best use of memory available.

For K-DCN, we used three different types of kernel functions, namely, Gaussian kernel, Polynomial kernel and Histogram Intersection Kernel (HIK). The kernel parameters and regularization parameter were tuned to obtain a range of values for the first module. For the later modules, the tuning is done with respect to the range of parameters obtained for the previous module, and a set of globally optimum parameters was obtained.

3.2   Feature Extraction

We used a deep convolutional network to obtain a useful representation from an image. A deep convolutional network consists of several layers. A convolutional layer consists of a rectangular grid of neurons. Each neuron takes inputs from a rectangular section of the previous layer. The weights for this rectangular section are constrained to be the same for each neuron in the convolutional layer. Constraining the weights makes it work like many different copies of the same feature detector applied to different positions. This constraint also helps in restricting the number of parameters. The output of a neuron in the convolutional layer, l for a filter of size $(m * n)$ is given by

$$s^l_{ij} = f\left(\sum_{x=0}^{m}\sum_{y=0}^{n} w_{xy} s^{(l-1)}_{(x+i)(y+j)}\right) \qquad (19)$$

where $f(x) = \log(1 + e^x)$. This nonlinearity was approximated using a simpler function, $f(x) = \max(0, x)$, which is known as the rectifier function. The nodes that use the rectifier function are referred to as Rectified Linear Units (ReLU). Use of ReLU reduced the time taken significantly.

The pooling layer takes outputs of small rectangular blocks in the convolutional layer and subsamples it to produce a single output from that block. The pooling layer can take the average, or maximum, or learn a linear combination of outputs of the neurons in the block. In all our experiments, we used max-pooling. Pooling helps the network achieve small amount of translational invariance at each level. Also, it reduces the number of inputs to the next layer. Finally, after two convolutional and max-pooling layers, we added two fully connected layers. The activity of the nodes in the last fully connected layer was used as input to the T-DSN and K-DCN models.

Apart from this, we also used the SIFT features [18] as input to the deep learning models.

3.3 Datasets Used

We test our models with two real-world datasets that contain color images with their annotations: University of Washington annotation benchmark dataset [25] and the MIRFLICKR-25000 collection [12].

Table 1. List of 45 concepts selected for our study on University of Washington annotation benchmark dataset.

Trees	Bushes	Grass	Sidewalk	Ground
Rock	Flowers	Camp	Sky	Trees
Trunk	People	Water	Dog	Woman
Street	Cars	Pole	House	Beach
Ocean	Clouds	Mountain	River	Building
Lantern	Window	Bridge	Band	Man
Stone	Snow	Boats	Sun	Huskies
Football	Stadium	Stand	Field	Hiker
Mosque	Frozen	Players	Temple	Smoke

Fig. 4. Illustration of images with their annotation labels from the University of Washington annotation benchmark dataset.

The Washington dataset had 1109 color images corresponding to 22 different categories with an average annotation length of 6. Out of all the concepts available, we selected only 45 concepts that had more than 25 images associated with each of them. The list of these 45 concepts is given in Table 1.

Some of the images from this dataset with their annotation labels are shown in Fig. 4. Because of the small number of images, we do not use convolutional features for this dataset.

MIRFLICKR-25000 is a database of 25,000 color images belonging to various categories. The average number of tags per image is 9. Some of the images from this dataset with their annotation labels are shown in Fig. 5. For our studies, we consider the 30 most frequently occurring tags. These tags have at least 150 images associated with each of them. We randomly selected 30 % of the images for testing, and repeated our studies over 5 folds.

3.4   Results

A T-DSN consisting of 3 modules with 100 nodes in each of the hidden layers was used on the University of Washington dataset. In our experiments we observed that having the same number of nodes in both the sets of hidden nodes generally give a better performance.

The precision, recall and F-measure for different thresholds in the threshold based decision logic are reported in Table 2.

We repeated the previous experiment with different values of k in the top-k based decision logic, and the precision, recall, and F-measure values are reported in Table 3.

We repeated these experiments with K-DCN. Best performance was observed for a Gaussian Kernel. The results of these experiments are reported in Tables 4 and 5. It is observed that the F-measure values for K-DCN are slightly lower when compared with that for T-DSN. One of the possible reasons for this could

Fig. 5. Illustration of images with their annotation labels from the MIRFLICKR dataset.

Table 2. Precision, recall, and F-measure for different thresholds in the threshold based decision logic for annotation of images in the University of Washington data with T-DSN.

Threshold	Precision	Recall	F-measure
0.20	0.50	0.87	0.64
0.25	0.58	0.81	0.67
0.30	0.63	0.75	0.68
0.35	0.68	0.70	**0.68**
0.40	0.72	0.64	0.67
0.45	0.73	0.58	0.64
0.50	0.75	0.47	0.58

Table 3. Precision, recall, and F-measure for different values of k in the top-k based decision logic for annotation of images in the University of Washington data with T-DSN.

k	Precision	Recall	F-measure
2	0.42	0.27	0.33
3	0.51	0.46	0.48
4	0.52	0.60	0.56
5	0.50	0.70	0.58
6	0.49	0.79	0.60
7	0.44	0.84	0.58
8	0.41	0.88	0.56

Table 4. Precision, recall, and F-measure for different thresholds in the threshold based decision logic for annotation of images in the University of Washington data with K-DCN.

Threshold	Precision	Recall	F-measure
0.20	0.39	0.90	0.54
0.25	0.50	0.84	0.63
0.30	0.52	0.81	0.63
0.35	0.59	0.76	0.66
0.40	0.64	0.68	0.66
0.45	0.73	0.59	0.65
0.50	0.77	0.49	0.59

be that the kernel parameters used might not be the best. The state-of-the-art methods for image annotation, namely, Rank-SVM and semantic space model give F-measure values of 0.61 and 0.63 respectively. Figure 6 compares the actual

Table 5. Precision, recall, and F-measure for different values of k in the top-k based decision logic for annotation of images in the University of Washington data with K-DCN.

k	Precision	Recall	F-measure
2	0.31	0.21	0.25
3	0.43	0.49	0.46
4	0.48	0.53	0.50
5	0.48	0.61	0.54
6	0.46	0.68	0.55
7	0.41	0.77	0.53
8	0.37	0.85	0.52

annotation labels for some randomly selected images in University of Washington dataset with the annotations generated by the T-DSN model.

It is observed that the number of annotation labels generated by the models were slightly higher than that of the ground truth. In many cases, the extra labels are somehow related to the image.

For the MIRFLICKR dataset, the study is carried out using the SIFT features and convolutional features. Figure 7 shows the precision-recall curves for different models. The best F-measure values for different models are presented in Table 6.

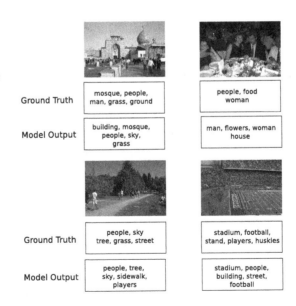

Fig. 6. Illustration of images with actual annotation labels and predicted annotation labels in the University of Washington dataset with T-DSN.

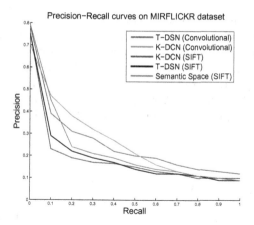

Fig. 7. Precision-recall curves for different models on MIRFLICKR dataset.

Table 6. Performance comparison of models for image annotation task on MIRFLICKR dataset.

Model	Input features	F-Measure
Semantic Space	SIFT	0.26
Rank-SVM	SIFT	0.25
TDSN	SIFT	0.24
TDSN	Convolutional	0.29
KDCN	SIFT	0.26
KDCN	Convolutional	**0.34**

It is observed that K-DCN and T-DSN perform better with convolutional features. It is also noted that convex deep learning methods perform better than the semantic space annotation method.

4   Summary and Conclusions

In this paper, we used the convex deep learning models, such as T-DSN and K-DCN for image annotation tasks. We also used features extracted from a deep convolutional network for this task. Through the experimental studies, it is observed that the T-DSN and K-DCN models with convolutional features as input give an improved performance. Once the convolutional network is trained on a large set of images, it is easy to extract features. The convex networks take less time to train, making them useful for image annotation tasks in practice.

For the K-DCN model, we have used only a single kernel function for a module. We can extend this by using multiple types of kernel functions. Finding a set of globally optimal parameters for K-DCN is difficult. Similarly, for T-DSN we observed that having different number of nodes in each hidden layer is not beneficial. However, we did not find any criterion for selecting the suitable number of hidden layer nodes. A recipe for selecting the number of nodes in T-DSN and globally optimum parameters for K-DCN will be useful.

References

1. Bengio, Y., Courville, A., Vincent, P.: Representation learning: a review and new perspectives. IEEE Trans. Pattern Anal. Mach. Intell. **35**(8), 1798–1828 (2013)
2. Boutell, M.R., Luo, J., Shen, X., Brown, C.M.: Learning multi-label scene classification. Pattern Recogn. **37**(9), 1757–1771 (2004)
3. Deng, L., Tür, G., He, X., Hakkani-Tür, D.Z.: Use of kernel deep convex networks and end-to-end learning for spoken language understanding. In: IEEE Workshop on Spoken Language Technologies, pp. 210–215, December 2012
4. Deng, L., Yu, D.: Deep convex network: a scalable architecture for speech pattern classification. In: Interspeech, August 2011
5. Deng, L., Yu, D., Platt, J.: Scalable stacking and learning for building deep architectures. In: Proceedings of the International Conference on Acoustics, Speech, and Signal Processing, March 2012
6. Elisseeff, A., Weston, J.: A kernel method for multi-labelled classification. Adv. Neural Inf. Process. Syst. **14**, 681–687 (2001)
7. Hare, J., Samangooei, S., Lewis, P., Nixon, M.: Semantic spaces revisited: investigating the performance of auto-annotation and semantic retrieval using semantic spaces. In: Proceedings of the International Conference on Content-based Image and Video Retrieval, pp. 359–368, July 2008
8. Hinton, G.E., Osindero, S., Teh, Y.-W.: A fast learning algorithm for deep belief nets. Neural Comput. **18**(7), 1527–1554 (2006)
9. Hinton, G.E., Osindero, S., Welling, M., Teh, Y.W.: Unsupervised discovery of nonlinear structure using contrastive backpropagation. Cogn. Sci. **30**(4), 725–731 (2006)
10. Hinton, G.E., Salakhutdinov, R.R.: Reducing the dimensionality of data with neural networks. Science **313**(5786), 504–507 (2006)
11. Hofmann, T.: Probabilistic latent semantic analysis. In: Proceedings of the Uncertainty in Artificial Intelligence, pp. 289–296 (1999)
12. Huiskes, M.J., Lew, M.S.: The mir flickr retrieval evaluation. In: Proceedings of the 2008 ACM International Conference on Multimedia Information Retrieval (2008)
13. Hutchinson, B., Deng, L., Yu, D.: Tensor deep stacking networks. IEEE Trans. Pattern Anal. Mach. Intell. **35**(8), 1944–1957 (2013)
14. Krizhevsky, A., Sutskever, I., Hinton, G.E.: Imagenet classification with deep convolutional neural networks. Proc. Neural Inf. Process. Syst. **22**, 1106–1114 (2012)
15. Le Roux, N., Bengio, Y.: Representational power of restricted Boltzmann machines and deep belief networks. Neural Comput. **20**(6), 1631–1649 (2008)
16. LeCun, Y., Kavukcuoglu, K., Farabet, C.: Convolutional networks and applications in vision. In: Proceedings of International Symposium on Circuits and Systems, pp. 253–256 (2010)

17. Lee, H., Grosse, R., Ranganath, R., Ng, A.Y.: Convolutional deep belief networks for scalable unsupervised learning of hierarchical representations. In: Proceedings of the 26th Annual International Conference on Machine Learning, pp. 609–616 (2009)
18. Lowe, D.G.: Distinctive image features from scale-invariant keypoints. Int. J. Comput. Vis. **60**(2), 91–110 (2004)
19. Montavon, G., Braun, M.L., Mller, K.-R.: Deep Boltzmann machines as feedforward hierarchies. Proc. Int. Conf. Artif. Intell. Stat. **22**, 798–804 (2012)
20. Ranzato, M., Krizhevsky, A., Hinton, G.E.: Factored 3-way restricted Boltzmann machines for modeling natural images. J. Mach. Learn. Res. Proc. Track **9**, 621–628 (2010)
21. Read, J., Pfahringer, B., Holmes, G., Frank, E.: Classifier chains for multi-label classification. Mach. Learn. **85**(3), 333–359 (2011)
22. Salakhutdinov, R., Hinton, G.: Deep Boltzmann machines. Proc. Int. Conf. Artif. Intell. Stat. **5**, 448–455 (2009)
23. Tsoumakas, G., Katakis, I.: Multi-label classification: an overview. Int. J. Data Warehouse. Min. **3**(3), 1–13 (2007)
24. Vens, C., Struyf, J., Schietgat, L., Džeroski, S., Blockeel, H.: Decision trees for hierarchical multi-label classification. Mach. Learn. **73**(2), 185–214 (2008)
25. Washington, U.: Washington ground truth database. http://www.cs.washington.edu/research/imagedatabase (2004)
26. Zhang, M.-L., Zhou, Z.-H.: Ml-knn: a lazy learning approach to multi-label learning. Pattern Recogn. **40**(7), 2038–2048 (2007)
27. Zhang, M.-L., Zhou, Z.-H.: A review on multi-label learning algorithms. IEEE Trans. Knowl. Data Eng. **26**(8), 1819–1837 (2014)

MOSAIC: Multi-object Segmentation for Assisted Image ReConstruction

Sonia Caggiano[1], Maria De Marsico[2]([envelope]), Riccardo Distasi[3], and Daniel Riccio[4]

[1] Master of Architecture and PhD in Digital Painting Restoration, Salerno, Italy
soniacaggiano@yahoo.it
[2] Sapienza University of Rome, Via Salaria 113, 00198 Rome, Italy
demarsico@di.uniroma1.it
[3] University of Salerno, 84084 Fisciano, SA, Italy
ricdis@unisa.it
[4] University of Naples "Federico II", 80121 Campi Flegrei, NA, Italy
daniel.riccio@unina.it

Abstract. This paper presents a tool targeted at archaeologists and cultural heritage operators. The tool assists the process of reconstructing broken pictorial artifacts from their physical fragments. The fragments are organized into a database indexed on features such as color distribution, shape and texture. The system can be queried using any fragment as the key, and the results are displayed from the most similar to the most dissimilar. The system provides the operator with complete workflow from photoacquisition onwards. The performance has been assessed with computer simulations and a real use case. Two of the simulations are discussed, as well as the real use case, based on an actual XV century fresco that needed reconstruction.

Keywords: Image processing · Feature extraction · Feature-based indexing · Jigsaw puzzle · Cultural heritage

1 Introduction

Imagine being an archaeologist at work on site. The remnants of a church are being dug out, and it is known that the wall that used to be right in front of your present standing position was painted with a fresco. The scene depicted on the fresco is not known, however. All you can hope to have access to—given enough patience and careful, slow work—will be a collection of painted fragments that crumble to dust if not handled with the utmost gentleness. Reconstructing at least part of the original design involves much manipulation: the pieces have to be repeatedly rotated, tentatively aligned, moved in batches and so on. Going through all the available fragments to locate a possible candidate piece for a particular spot of your puzzle is no trivial task, either: each time a piece is touched or moved it might break, or at least its edge might be ground away, depending on the materials used to build and coat the old wall. It would be nice to have some digital tool to ease the whole process.

© Springer International Publishing Switzerland 2015
A. Fred et al. (Eds.): ICPRAM 2015, LNCS 9493, pp. 282–299, 2015.
DOI: 10.1007/978-3-319-27677-9_18

Enter MOSAIC—an automated system for computer-aided reconstruction of jigsaw puzzles. Such puzzles are grouped into two broad categories, requiring different approaches: apictorial puzzles, where fragment shape is the only kind of available information, and pictorial puzzles, where texture and color information is available and can be meaningfully used—which, unlike most commercial puzzles, does not imply that the solution image is known a priori. Pictorial or apictorial, it has been shown that providing an exact algorithmic solution is an NP-complete problem: the computing time grows super-polinomially with problem size [8]. However, if a non-exact solution is acceptable, there are techniques that provide approximate solutions in a shorter time [4]. The available literature offers several options for tackling both types of jigsaw puzzles—and several applications as well, mostly in the fields of cultural heritage and ancient document reconstruction.

In recent times, advances in hardware and software design have provided improved performance. For example, the Roman site in Tongeren, Belgium contained a number of artifacts that have been at least partially reconstructed [2]. The fragments have been acquired using an ad hoc 3D scanner, and the extracted shapes have been matched by a custom piece of software. The total cost, however, remained quite high—around $25,000 for the scanner only. Such a massive employ of work and money did improve the number of true matches nearly six-fold, but this result is still far from satisfactory, considering the actual numbers: 3 true matches had been found manually, and the system proposed 6103 tentative matches that became 17 true matches after human screening.

Another semi-automatic system was made by Brown et al. [3]. It was used for recontruction frescoes at the Akrotiri site in Thera, Santorini. Greece. Although the sistem does use 3D data, most of the results are obtained via 2D image feature extraction.

Freeman and Gardner's classic paper was one of the earliest to tackle the problem of apictorial jigsaw puzzles [8]. Five fundamental puzzle properties were pinpointed: orientation (unknown a priori), connectivity (presence or absence of internal "holes"), perimeter shape (known/unknown a priori), uniqueness (does the problem admit only one solution?), radiality (topology of fragment juncture). The contours of the fragments are represented as chain codes, and code length is used as a heuristic for search space dimensionality reduction.

Papaodysseus et al. face this problem in the context of reconstructing wall paintings [12]. This context makes their paper particularly relevant for the present discussion. The focus is on specific real-world issues that arise when dealing with a fresco: lack of information about the original content of the painting, non-uniqueness, and especially non-connectedness arising from the presence of very small fragments that are not available to the problem solver. The technique for correspondence deals with missing information using local curve matching.

Better solutions can be obtained by fully exploiting all available information. Most techniques actually used in the cultural heritage field regard the problem not as an apictorial, but as a pictorial puzzle. For instance, Chung et al. use color [5], while Sagiroglu and Ercil use texture [13]. However, actual testing has been limited to problems involving a relatively small number of fragments.

On the other hand, for an example that disregards fragment shape altogether, consider Nielsen et al. [11]. Their method relies on features of the whole represented pictorial scene without using any information pertaining to single pieces. The reported results for this technique show low error margins: the solution to a 320-fragment problem only had 23 pieces out of place—an error margin of 7.2 % obtained by using only color and texture information.

As we have seen, the virtual reconstruction of pictorial fragments is an intrinsically hard problem, and approximate solutions are often all we can get. For this reason, several sophisticated image processing techniques are being incorporated in most recent systems. The most promising are based on local texture analysis, chrominance analysis and contour analysis on single fragments. Methods based on properties of the whole scene are quite powerful, when the original appearance is known or can be at least partially inferred, and can provide further features to consider. Such techniques can produce multimodal representations that allow users to refine the solution progressively, adding detail and information to the features of the solution search space.

The present paper proposes a system for the segmentation and indexing of pictorial fragments: Multi-Object Segmentation for Assisted Image reConstruction (MOSAIC). MOSAIC supports the rebuilding of a fresco from fragments by a human operator. No information about the original appearance of the whole artwork is assumed to be available. The system has been tested with several computer simulations and on a real case study: the reconstruction of a fresco from fragments found in the St. Trophimena church in Salerno (Italy).

2   MOSAIC Procedures

Mosaic can be classified among techniques for Jigsaw pictorial puzzle solving where texture and color information is available. This information can be meaningfully used together with shape information. As mentioned above, it has been shown that an exact algorithmic solution to puzzle solving is NP-complete [7]. Therefore, a number of models supporting alternative approaches have been investigated—e.g., [4]. Our proposal relies on a human-in-the-loop approach.

In particular, MOSAIC was expressly designed to support archaeologists and restorers facing fresco recomposition from fragments. Its software modules implement a procedure for image acquisition and processing. The aim of this procedure is not to perform a completely automatic reconstruction, but rather to relieve the expert from most of the burden implied by reordering fragments and grouping them in similar clusters. The result is a catalog of the single fragments, which are grouped according to their texture/color and shape characteristics. This grouping allows answering user queries, so that reconstruction is made easier, quicker and more effective. The application provides a virtual workspace; here, among the other actions, the user can perform the actions that would have been performed in a real reconstruction attempt, i.e., virtually rotate, translate and search for similar fragments. Figure 1 shows a schema of the sequence of steps executed by the system. After image acquisition, segmentation and feature

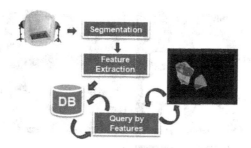

Fig. 1. Steps executed by the MOSAIC system.

extraction modules build the fragment archive. A query engine allows searching the archive for relevant fragments, and a manipulation interface allows the user to manipulate them virtually to attempt recomposing the broken picture.

In the starting phase—image acquisition—the physical fragments are laid in a white tray, whose bottom is covered by a dark foam to reduce reflexions. The tray containing the fragments is placed inside a box for photographic acquisition, which is made by a white curtain and two lateral spotlights. Close to the tray, a colorimeter is used to check for the need for automatic color corrections. The tray is then photographed by a suitable device (in this work, we used an 8-Mpixel Canon camera), orthogonally pointing it from a height of 90 cm.

2.1 Fragment Catalogue

Segmentation. Segmentation accuracy significantly affects the rest of the procedure. The purpose of this operation is to correctly separate each fragment, so that individual features can be extracted from each of them. Segmentation is carried out in two steps. In the first one, the image is binarized and turned into B/W with no shades of gray. Nowadays, binarization of a color photo might appear as a trivial task. However, using naive thresholding on a set of different raw images does not work. In particular, in our case no single threshold value is effective across all trays, unless some pre-processing occurs to enhance the image color separability. Values that are too low fail in separating one piece from another, while values that are too high ones produce "holes" inside pieces. In some simple cases, a morphological fill operation can fill such holes, but in other cases the piece may even come out as two separate fragments—an error that is quite hard to correct later. An example of problematic binarization is shown in Fig. 2.

The process of binarization that we carry out first needs to pre-process the raw image in order to amplify the difference between the brighter pixels (fragments) and the darker ones (the background foam is a dark shade of gray—almost black, but not quite). This pre-processing is described below.

The original image is represented in RGB space, whose single channels are stored in three separate matrices r, g, and b. Two new matrices are then created: M and m. The element $M(i,j)$ of the new matrix M is equal to the maximum

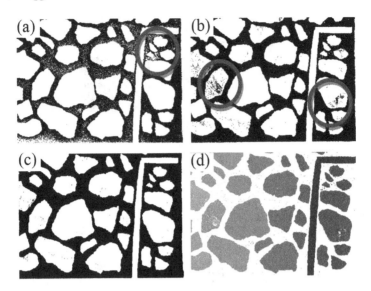

Fig. 2. Effect of threshold parameter t_B choice on the segmentation of a tray image: (a) too low; (b) too high; (c) optimal value $t_B = 0.1$; (d) connected components detected after binarization.

over the three channels in (i, j) position, i.e. the maximum among $r(i, j)$, $g(i, j)$ and $b(i, j)$. In practice, matrix M represents a new image where each pixel contains the largest (brightest) component of the original image. The new matrix m is built so that the image that it represents is a greyscale version of the original image: its element $m(i, j)$ is equal to the mean of $r(i, j)$, $g(i, j)$ and $b(i, j)$. Matrices M and m are used to derive the enhanced image I as follows. First, the transformation described in (1) is applied:

$$I(i, j) \leftarrow M(i, j) \cdot |m(i, j) - \delta| . \tag{1}$$

The value δ in Eq. (1) is a constant offset which is experimentally determined and which is constant over all images. In our specific case, $\delta = 50$. In practice, this value is chosen to be close to the mean luminance value of all pixels in the tray (both fragments and background). Then, the pixel values in I are scaled by dividing them by their mean value $\bar{I}$ as in (2):

$$I(i, j) \leftarrow I(i, j) / \bar{I} . \tag{2}$$

After these operations that enhance color separability, we can apply a single threshold value for all trays. A value of $t_B = 0.1$ has been found to be suitable for all tray images in our pool. The new I is then turned into a 0–1 binary image by thresholding it according to the t_B binarization threshold value:

$$I(i, j) \leftarrow \begin{cases} 0, & \text{if } I(i, j) < t_B \\ 1, & \text{if } I(i, j) \geq t_B \end{cases} . \tag{3}$$

We now explain the rationale for the above operations. The grayscale image m and the maximum component image M are pointwise multiplied in order to enhance the pixels where both m and M have larger values. The resulting image is then divided by its mean value to perform a kind of normalization, so that the value of the threshold t_B used for binarization does not depend on the particular image anymore. The final result of binarization can be seen in Fig. 2(c).

An algorithm for detecting connected components takes the binary image just obtained as input. In an ideal situation, each fragment should correspond to a single connected component and vice versa, as in Fig. 2(d). To obtain this, a morphologic fill operator is applied to each fragment shape extracted from the binary image, in order to fill possible existing gaps or holes. Afterwards, specific information about the newfound fragment is computed, namely its area, its perimeter, and its orientation. We finally obtain a binary connected component, that will be used as a mask to retrieve the fragment from the original image by a pixel-wise logical AND operation.

Feature Extraction. The module for feature extraction takes the fragments resulting from segmentation, so that these can be indexed to allow a convenient successive retrieval. The features chosen for indexing/retrieval are the (basic) shape(s) depicted on the fragment and a spatiogram, which describes the spatial distribution of colors on the fragment surface. The user can search the fragment database according to similar color, similar shape, or similar spatial color distribution, in order to retrieve fragments similar to a given "key" fragment.

Color. Information about color is represented here by a spatial histogram, or spatiogram, computed over the fragment [1]. A histogram can be considered as a zeroth-order spatiogram, while second-order spatiograms also contain spatial means and covariances for each histogram bin, i.e., information about the spatial domain. In other words, a spatiogram is a histogram where the count of occurrences of each color or group of colors (bin) is augmented with spatial information related to it, namely the mean vector and covariance matrices, respectively, of the coordinates of the pixels containing that color (bin). Given an image I and a quantization of its colors (or gray levels) in B bins, we denote the histogram of I as h_I, and the number of pixels in bin b as $h_I(b)$. The second order spatiogram is then

$$h_I^{(2)}(b) = \left\langle n_b, \mu_b, \sum_b \right\rangle, \tag{4}$$

where $b = 1, \ldots, B$, n_b is the number of pixels that fall in the b-th bin (as in the histogram), and μ_b and $\sum_b$ are the mean vector and covariance matrices, respectively, of the coordinates of those pixels. The similarity between two spatiograms can be computed as the weighted sum of the similarity between the two corresponding histograms, where the weight depends on the order of the spatiogram, and the histogram similarity can be computed by a number of techniques, as for example histogram intersection or the Bhattacharyya coefficient, as shown in (5) and (6), where n'_b and n''_b are the values of the two histograms for bin b:

$$int(n'_b, n''_b) = \frac{min(n'_b, n''_b)}{\sum_{j=1}^{B} n'_j} \tag{5}$$

$$bhatt(n'_b, n''_b) = \frac{\sqrt{n'_b \cdot n''_b}}{\sqrt{(\sum_{j=1}^{B} n'_j)(\sum_{j=1}^{B} n''_j)}} . \tag{6}$$

For more details, see [1]. Similarly to histograms, spatiograms allow simple manipulations and are especially useful for comparisons between image areas, because they do not need a geometric mapping between the areas involved. In addition to histograms, the further spatial information about color distribution provides increased matching accuracy.

Shape. The information extracted describes the predominant shapes of the pictorial elements depicted in a fragment. The pixels first undergo a clustering procedure based on color. Clusters resulting in each fragment are then analyzed to extract shape information.

Color clustering is performed by a mean-shift based method [6]. Such methods have been chosen because they are non-parametric and fairly insensitive to noise or similar low-level distortions. Moreover, mean-shift based clustering does not require to predetermine the number of clusters. However, in most cases the result obtained is over-segmented for our purposes. This is exemplified in Fig. 3.

Over-segmentation is corrected by determining a thresholding "color radius" t_C: the distinct RGB colors that lie at a distance less than t_C are coalesced into a single color by re-labeling them. In our specific case, the value for this radius threshold has been determined experimentally and is set as $t_C = 32$. This correction significantly improves clustering accuracy, as demonstrated by comparing Fig. 3(c) and (d).

Each color cluster is considered independently in the context of a single fragment. The pixels belonging to one cluster (and then appearing to the system as being of the same color) are fed to an algorithm for the detection of connected components. This process is depicted in Fig. 3(e). Each detected connected component is in turn processed independently, in order to determine the shapes represented. The smaller components—those whose area is less than 4 % of the total fragment area—are discarded as not significant (small color holes, noise, as well as processing defects occurred during either binarization, segmentation, color clustering, or detection of connected components). At the end of this process, each fragment F_h is characterized by a variable number s_h of shapes S_{h_i}, $i = 1 \ldots s_h$, whose surface area equals at least 4 % of the total fragment surface. These components undergo contour detection so their shapes can be geometrically described. Since this processing is performed on a per-fragment basis, and since the following analysis is performed on single shapes in the fragment, in order to simplify the notation we will drop from now on the subscript identifying fragment and shape.

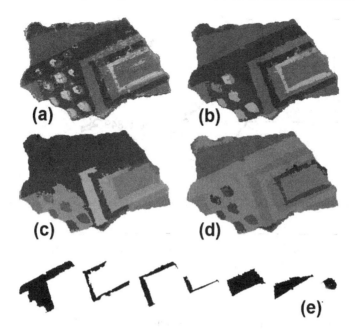

Fig. 3. Shape extraction: (a) the original fragment; (b) the fragment after color clustering, using original colors; (c) false colors reveal oversegmentation; (d) thresholding and re-labeling; (e) extracted shapes.

Shape S is analyzed through its contour C. A further consideration is needed before describing such analysis. Not all components correspond to relevant shapes, even if their area is over the 4 % threshold: some of them are just stains and contribute nothing but noise to the system. Therefore, we investigated possible relevance criteria to assign a higher weight to relevant shapes than to stains. In this context, contour smoothness is a useful relevance criterion. The underlying assumption, which is supported by experts collaborating to MOSAIC design, is that the contour of a pictorial element is usually smoother than the contour of a stain or blemish, which on the contrary is more jagged. Figure 4 shows an example. Since they still contribute to the spatiogram, their possible color information content is not lost in any case. According to this, the shape processing continues as follows.

The contour C of a shape is represented as an ordered sequence of n_C points:

$$C = \{P_1, P_2, \ldots, P_{n_C}\}, \tag{7}$$

where the contour step count n_C differs from shape to shape (and may also be a first cue of its length/regularity). Given a point P_k in C, let us consider another contour point P_{k+l} located l steps further along the same path. In other words, the path from P_k to P_{k+l} has step count l, which also corresponds to the lowest distance between them. Let $d(\cdot, \cdot)$ be the Euclidean distance between any two

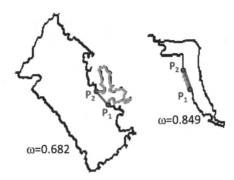

Fig. 4. Smoothness ranks shape contours along the jagged/smooth axis.

points. The smoothness for the subpath starting at P_k and spanning l points $C_{k,l} = \{P_k, P_{k+1}, \ldots, P_{k+l}\}$ is computed as

$$\omega(k, l) = \mathrm{d}(P_k, P_{k+l})/l. \tag{8}$$

The actual value used for smoothness calculation depends on n_C and is $l = \lfloor 4 \log_2 n_C \rfloor$. The smoothness for the whole contour C is given by

$$\omega(C) = \sum_{k=1}^{n_C} \omega(k, l). \tag{9}$$

It assumes values in $[0, 1]$ and is used as a weight in matching operations, as will be shown shortly.

2.2   Fragment Search

After the process of feature extraction, color information is represented by the fragment spatiogram, and comparison between a query and the fragments in the catalogue is performed with related techniques. It is more interesting here to consider search by shape.

After feature extraction carried out on single shapes, each shape is represented as a triple

$$S = \langle \mathbf{v}, \omega, c \rangle. \tag{10}$$

In this characterization, $\mathbf{v} = (v_1, \ldots, v_7)$ is the vector of the first 7 central moments of the shape (a thorough discussion of central moments and some of their applications to pattern recognition can be found in [9,10]). The other two elements in the triple, ω and c, are the shape smoothness and mean color value, respectively. A fragment F_h containing s_h shapes is therefore characterized by s_h such triples.

We compare two shapes $S_1 = \langle \mathbf{v_1}, \omega_1, c_1 \rangle$ and $S_2 = \langle \mathbf{v_2}, \omega_2, c_2 \rangle$, by computing their similarity as the normalized dot product of their moment vectors (i.e., the cosine of the angle between them), weighted by the product of their smoothness values:

$$\text{sim}(S_1, S_2) = \omega_1 \omega_2 \frac{\mathbf{v_1} \cdot \mathbf{v_2}^T}{|\mathbf{v_1}| |\mathbf{v_2}|} \,. \tag{11}$$

Since each fragment contains more shapes, the similarity between two fragments F_1 and F_2 is given by the maximum shape-to-shape similarity:

$$\text{sim}(F_1, F_2) = \max_{\substack{S \in F_1 \\ T \in F_2}} \text{sim}(S, T) \,.$$

In the most common type of query, the key is represented by a single shape S and the search goes through each fragment indexed in the database, looking for shapes with high values of similarity to S. The similarity score assigned to a fragment is the maximum similarity score achieved by a shape it contains. Smaller shapes are discarded as not relevant. When a query is based on shape, the fragments are returned in decreasing order of similarity (of the included shapes) to S.

3   MOSAIC Interface

MOSAIC offers a graphical user interface (GUI) that allows the operator to create a work session to manipulate the virtual fragments. These can be handled and reconstruction can be performed by retrieving relevant fragments through queries based on shape, color or a combination thereof. The interface is shown in Fig. 5. A work session created from scratch appears as a blank workspace where fragments can be brought in. Here is an outline of the main functionalities offered by the interface.

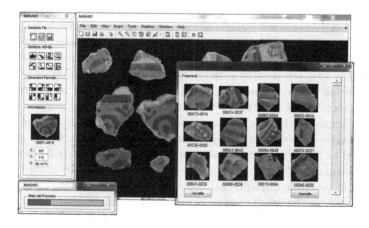

Fig. 5. MOSAIC: the graphical user interface and the workspace.

- **Workspace Management.** Fragments can be picked from the database, manually or via queries, and brought into the workspace. Useless fragments can be removed from the workspace.
- **Manipulation.** The operator, possibly an archaeologist or other cultural heritage specialist, can translate or rotate fragments, in a way resembling the real work with physical pieces.
- **Search.** It is possible to query the system by selecting a fragment from the workspace. The query can address a number of features, among which a specific pictorial shape among those represented in the fragment, the whole set of shapes represented, color distribution, texture (both shape and color are relevant), and more.

4 Experimental Results

Given the nature of the MOSAIC system, it is significantly easier to assess its performance qualitatively than quantitatively; therefore, custom experiments have been designed. The main objective was the evaluation of precision/recall achieved by MOSAIC in the retrieval of fragments close to a given one. This can give a measure of the usefulness of the system, since it is related to the intended use of the system by a human operator, who is trying to reconstruct shattered pictorial artworks.

It is reasonable to assume that adjacent pieces often tend to have similar textural features: searching for features similar ti those characterizing the query key fragment should return fragments lying in the same picture area. In a symmetric way, given a key fragment, the user wants to retrieve the fragments that most probably should be put again close to it, as in the original picture: one can reasonably assume that they present a similar texture. Following these considerations, jigsaw puzzle solution can be simplified, by suggesting to the operator the fragments to pick from a smaller returned subset rather than from the full set. In order to verify the effectiveness of the indexing process, each fragment was characterized with the canvas coordinates of its barycenter, computed assuming uniform mass distribution. Obviously, this is not possible in real situations. The main problem to solve is often exactly related to the lack of knowledge of the original position of the fragments. If this is the case, during reconstruction, we might only recover the position of the already identified fragments, but only if we have an image of the original artwork. However this spatial annotation was useful to analyze some issues, and as a kind of ground truth. Given a key fragment A used as a query, let its barycenter be P_A; a fragment B retrieved by the system is deemed to be "close enough"—that is, significant—if its barycenter P_B falls within a circle of radius t_M centered in P_A.

We report here the results of two selected experiments, that were ran on two pictures that were artificially "broken" into fragments. These two experiments have been chosen among the others because the two pictures present different features. The first chosen picture is the *Madonna della Misericordia* or *Madonna delle pietre*, by an unknown author of the late XV century. The original figure is shown in Fig. 6 while its virtually fragmented version is shown in Fig. 7. The original size of the painting is $140 \times 90\,\mathrm{cm}$. The acquisition was performed at 300 dpi, 24 bit RGB, yielding 1185×1566 pixels. This picture presents a single human figure in the centre, but in some regions of the surface very small details are depicted.

Fig. 6. Madonna della Misericordia: the first image chosen for showing the quantitative performance assessment.

The virtual fragmentation consisted of irregularly shaped small pieces; the smallest ones were discarded to simulate more realistically the true fragmentation and loss that may result from a traumatic event. The 2900 resulting pieces had sizes ranging from roughly $2 \times 2\,\mathrm{cm}$ to $8 \times 8\,\mathrm{cm}$. Such pieces were entered into MOSAIC as they were acquired by a camera and indexed.

The choice of a suitable value for the threshold t_M is strongly dependent on image scaling/resolution. To take this into account, the value of t_M is expressed as the product of a proportionality factor δ in the range $[0, 1]$ times a representative length quantifying the image resolution, namely the diagonal length L_d, in

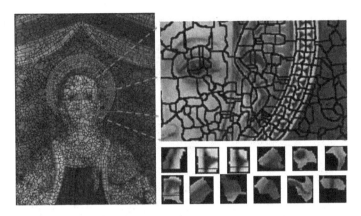

Fig. 7. Madonna della Misericordia: the fragmented version with a detail of some of the fragments.

our case 1964 pixels. Experiments have been performed with $\delta = 0.02$, $\delta = 0.05$ and $\delta = 0.1$, with resulting values of $t_M = 39$, $t_M = 98$ and $t_M = 196$ pixels respectively. Each fragment was used in turn as the key to a MOSAIC query. The average precision and recall curves over all fragments are plotted in Fig. 8.

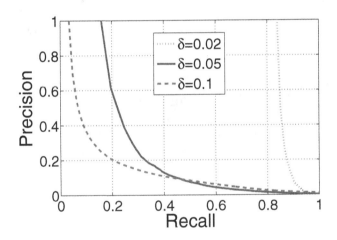

Fig. 8. Madonna della Misericordia: average precision/recall curves for MOSAIC with $\delta = 0.02$, $\delta = 0.05$ and $\delta = 0.1$.

The diagram shows that smaller values of δ generally yield better retrieval: for any given key fragment, the results of interest are usually contained in a relatively small space neighborhood of the key rather than spread around widely. However, looking at Fig. 7, we can notice that there is an intrinsic issue that harms precision and recall: it is represented by the fairly extended regions with

Fig. 9. Giotto's Assunzione di S. Giovanni Evangelista: the second image chosen for quantitative performance assessment.

homogenous features. For example, the whole halo area has nearly homogeneous textural and chromatic features. As a consequence, any two fragments belonging to that area will be "close" in the feature space, while actually lying at a potentially large distance in the reconstructed picture.

The second chosen experiment was carried out on a picture of a fresco by Giotto: "Assunzione di S. Giovanni Evangelista" from Cappella Peruzzi. The processed image is 72 dpi, 24 bit RGB, with size 2560 × 1622 pixels. The same experimental parameters were used, e.g., binarization threshold, color radius, etc., and similar processing considerations were adopted. The original picture is shown in Fig. 9. The virtually fragmented picture is shown in Fig. 10. We can notice the different spatial and chromatic composition with respect to the Madonna: many figures, many architectonic elements, but even wider homogeneous regions. Notwithstanding this, we decided to stress MOSAIC by using the same parameters as above. The results are shown in Fig. 11.

It is interesting to notice that, notwithstanding the higher complexity of the second depicted scene, the general trend of results (i.e., the shape of the curves and their relative positions) remains stable even with the same parameters used for the first picture. This testifies the robustness of the image processing results.

4.1   An Experiment on a Real Life Case

In simulated experiments, it is possible to perform both a qualitative and a quantitative assessment of the retrieval results, since a ground truth is available. This is not always possible in real situations, i.e., in field operations that a system of this kind aims at supporting. One of the main difficulties in many of these cases is just the lack of knowledge of the preexisting appearance of the artwork at hand, i.e., of a reference "model" that can guide the reconstruction. Actually, the

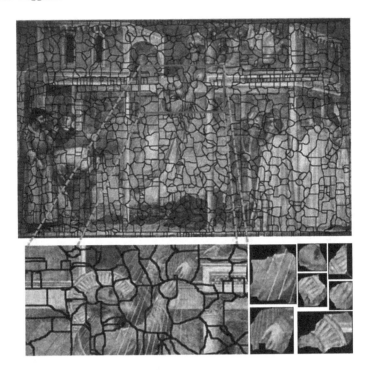

Fig. 10. Giotto's Assunzione di S. Giovanni Evangelista: the fragmented version, with a detail of some of the fragments.

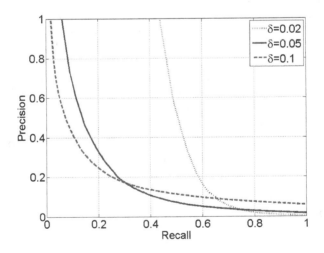

Fig. 11. Assunzione di S. Giovanni Evangelista: average precision/recall curves for MOSAIC with $\delta = 0.02$, $\delta = 0.05$ and $\delta = 0.1$.

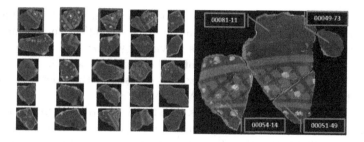

Fig. 12. A query, its result, and a partial reconstruction made from some of the fragments returned.

MOSAIC system has also been tested in such a kind of real case study, when only a qualitative evaluation is possible, since the original appearance of the frescos was unknown. The experiment involved the reconstruction of a set of 6419 fresco fragments that were found during a restoration work at the St. Trophimena Church site in Salerno. Since no information was available about the original appearance of the frescoes, virtual reconstruction was the only option to recover at least parts of the original work without further damaging to the fragments. Examples from a work session are shown in Fig. 12, depicting a reconstruction in progress, and Fig. 13, illustrating a shape-based query.

Fig. 13. Query by shape: strips.

As anticipated, the lack of information about the "real" solution to this jigsaw puzzle makes it impossible to obtain an objective/quantitative measure of the quality of the retrieved set of fragments provided by MOSAIC, but a qualitative assessment of its effectiveness is possible. The end users were involved both in the interface design phase and in the experiments, and their detailed feedback confirmed the practical value of MOSAIC system.

In Fig. 12, each label has two values. The first is the tray where the fragment lies, while the second is the serial number inside that tray. As the figure shows, it is quite usual that pieces close to each other in the original picture can end up in quite distant trays when they are picked up from the original site or during cataloging.

5   Conclusions

The reconstruction of fragmented pictorial artworks is a task always very demanding and often unfeasible, depending on the conditions of the fragments and on the experience of the operator. MOSAIC (Multi-Object Segmentation for Assisted Image reConstruction) is a system aiming at supporting this delicate work by allowing the computer aided reconstruction of virtual fragments. The extraction of relevant features related to color and shape allows cataloging and indexing of the fragments. Queries can be formulated through a simple GUI. Different kinds of selection of the query key are available: it is possible to select a whole fragment as example of the searched companions, or even a single shape represented on a fragment. The results of the comparison with the stored virtual fragments are sorted by similarity to the query key, and provide candidates for puzzle solving in the area of the relevant fragment. This can speed up the reconstruction process significantly, and improve the quality of the result. The system has been tested first via computer simulations. We report here two examples of such simulations. In both cases the solutions were known a priori, so to have a ground truth to assess the retrieval results. Later on the system has also been tested in a real world situation, where the solution was unknown and therefore results can be appreciated only qualitatively. It is interesting to notice the consistent results obtained on the two simulations, notwithstanding the different complexity of the pictures and the use of exactly the same parameters. This assesses the robustness and flexibility of MOSAIC procedures. Domain experts participated in both the design and testing phase of MOSAIC, and have provided precious feedback for tuning the system parameters. Future work will entail more detailed interaction with restorers, archaeologists, and other cultural heritage operators to better understand their needs, the problems related to specific situations, and offer them and even better support.

References

1. Birchfield, S.T., Rangarajan, S.: Spatiograms versus histograms for region-based tracking. In: Proceedings of the IEEE Conference on Computer Vision and Pattern Recognition (CVPR). pp. 1158–1163, Jun 2005

2. Brown, B., Laken, L., Dutré, P., Gool, L.V., Rusinkiewicz, S., Weyrich, T.: Tools for virtual reassembly of fresco fragments. In: Proceedings of the 7th International Conference on Science and Technology in Archaeology and Conservations. pp. 1–10. SCITEPRESS (2010)

3. Brown, B., Toler-Franklin, C., Nehab, D., Burns, M., Dobkin, D., Vlachopoulos, A., Doumas, C., Rusinkiewicz, S., Weyrich, T.: A system for high-volume acquisition and matching of fresco fragments: reassembling theran wall paintings. ACM Trans. Graph. (Proc. SIGGRAPH) 27(3), 1–10 (2008)

4. Cho, T.S., Avidan, S., Freeman, W. T.: A probabilistic image jigsaw puzzle solver. In: CVPR, pp. 183–190. IEEE (2010). http://dblp.uni-trier.de/db/conf/cvpr/cvpr2010.html#ChoAF10

5. Chung, M.G., Fleck, M., Forsyth, D.: Jigsaw puzzle solver using shape and color. In: Proceedings of the 4th International Conference on Signal Processing (ICSP 1998). vol. 2, pp. 877–880 (1998)

6. Comaniciu, D., Meyer, P.: Mean shift: a robust approach toward feature space analysis. IEEE Trans. Pattern Anal. Mach. Intell. (PAMI) 24(5), 603–619 (2002)

7. Demaine, E.D., Demaine, M.L.: Jigsaw puzzles, edge matching, and polyomino packing: connections and complexity. Graphs Combinatorics 23(Supplement), 195–208 (2007). (special issue on Computational Geometry and Graph Theory: The Akiyama-Chvatal Festschrift)

8. Freeman, H., Garder, L.: Apictorial jigsaw puzzles: the computer solution of a problem in pattern recognition. IEEE Trans. Electron. Comput. 2(EC–13), 118–127 (1964)

9. Hu, M.: Visual pattern recognition by moment invariants. IRE Trans. Inf. Theor. IT–8, 179–187 (1962)

10. Mercimek, M., Mumcu, K.G.T.V.: Real object recognition using moment invariants. Sadhana Acad. Proc. Eng. Sci. 30(6), 765–775 (2005)

11. Nielsen, T.R., Drewsen, P., Hansen, K.: Solving jigsaw puzzles using image features. Pattern Recogn. Lett. 14(29), 1924–1933 (2008)

12. Papaodysseus, C., Panagopoulos, T., Exarhos, M.: Contour-shape based reconstruction of fragmented, 1600 bc wall paintings. IEEE Trans. Signal Process. 6(50), 1277–1288 (2002)

13. Sagiroglu, M., Ercil, A.: A texture based matching approach for automated assembly of puzzles. In: Proceedings of the 18th International Conference on Pattern Recognition (ICPR 2006). pp. 1036–1041 (2006)

Author Index

Printed in the United States
By Bookmasters